Practical Computer Analysis of Switch Mode Power Supplies

Practical Computer Analysis of Switch Mode Power Supplies

JOHNNY C. BENNETT

Boca Raton London New York Singapore

A CRC title, part of the Taylor & Francis imprint, a member of the Taylor & Francis Group, the academic division of T&F Informa plc.

Published in 2006 by
CRC Press
Taylor & Francis Group
6000 Broken Sound Parkway NW, Suite 300
Boca Raton, FL 33487-2742

Printed in the United States of America on acid-free paper
10 9 8 7 6 5 4 3 2 1

International Standard Book Number-10: 0-8247-5387-9 (Hardcover)
International Standard Book Number-13: 978-0-8247-5387-0 (Hardcover)

Library of Congress Cataloging-in-Publication Data

Catalog record is available from the Library of Congress

Taylor & Francis Group
is the Academic Division of T&F Informa plc.

Visit the Taylor & Francis Web site at
http://www.taylorandfrancis.com

and the CRC Press Web site at
http://www.crcpress.com

Preface

For many years prior to the 1970s, engineers designed and built switch mode power supplies (SMPSs) using methods based largely on intuitive and experimentally derived techniques. In general, these power supplies were able to achieve their primary goal of high-efficiency power conversion; unfortunately, due to the lack of adequate theoretical analysis techniques, many of these power supplies only marginally met their desired performance requirements. In many cases, they were considered to be unreliable. Although they appeared to be very simple in concept, these switching regulators exhibited phenomena that were not understood and certainly could not be analyzed.

Things began to improve, however, in the early 1970s, when Dr. R.D. Middlebrook and his group of students at the California Institute of Technology developed the powerful circuit-averaging techniques, thus opening the door for the application of conventional linear circuit analysis methods. With these tools brought to bear, the many subtle complexities of the conceptually "simple" switching regulator were soon understood, allowing engineers to design SMPSs with improved performance and higher reliability. At this point, the power electronics field began to expand rapidly as better components were developed, power conversion technology advancements were made, and sophisticated computer-aided design and analysis methods were utilized.

Having been in the power electronics field for many years, I have had the good fortune to be involved with the design and analysis of many different types of power supplies. Here in this book, one of my goals is to provide the reader with a good understanding of the essential requirements for analyzing the switching regulated power supply performance characteristics. Another goal is to further demonstrate the power of the circuit-averaging technique by using computer circuit simulation programs to provide the desired performance analyses. At this point, I would like to reference the very important work of Dr. Vincent Bello, who, in his seminal paper,[9] pointed the way to using the SPICE-based computer circuit simulator to perform linear small signal analysis and nonlinear large signal transient performance analysis as well. The simulation techniques presented in this book are based almost entirely on Dr. Bello's approach. There have been several theoretical and practical contributors to the advancement of the circuit-averaging techniques over the years and I hope that the information presented here can help to further these advancements.

- Chapter 1 is a refresher of the basics of SMPS fundamentals and circuit-averaging modeling. This may also be a primer for the newcomer, but it is recommended that the beginner read the referenced works to obtain a more complete understanding.

- Chapter 2 provides information on the general analysis requirements of a power supply. This is deemed necessary because it is equally important to know what questions to ask as it is to provide the answers.
- Chapter 3 gives information on how to develop the general types of SMPS models and demonstrates the analysis approach using a SPICE-based circuit simulator.
- Chapter 4 looks, in a practical way, at most of the basic first-order types of analysis generally associated with SMPS performance.
- Chapter 5 provides more practical and detailed information on developing an SMPS and SMPS component models.
- In Chapter 6, three power supplies are analyzed in practical detail. In these examples, emphasis is placed on using the circuit-averaging macromodel of the integrated circuit PWM controller. This is felt to simplify and expedite the analysis of a particular design that uses these commercially available controllers. As circuits and systems become larger and more complex, the macromodel approach will continue to increase in importance in almost all areas of electronic circuit analysis. The PWM macromodeling effort presented here will hopefully lead to the future development of many more such macromodels for commercially available PWM controllers, as has been the case with macromodels for transistors, op amps, etc.
- Appendix A deals with the optimal design of SMPS input filters. This is included here simply because of the fundamental importance of this subject to any power supply.
- Appendix B provides the first-order approach used in developing the macromodel for two commercially available PWM controllers. As was stated earlier, this is only a first step and hopefully will lead to the advancement and further development of these macros.

Although they are very important aspects of any switch mode power supply, the analyses of actual switching circuits per se are not specifically addressed in this book. For our purposes, these switching circuits are considered to be in the realm of conventional electronic circuit transient analysis and not implicitly related to the performance of an SMPS. There may be exceptions to this, of course.

I would like to express my thanks and appreciation to all the wonderful people with whom I have worked over the years who have shared their invaluable knowledge and experiences. I would also like to expressly thank Col. William T. McLyman of the Jet Propulsion Laboratory for his encouragement and for being instrumental in producing this book.

Johnny C. Bennett

Author

Johnny C. Bennett is a native of the state of Louisiana and is a graduate of Louisiana Tech University. He received his BSEE in 1964 and has worked as an electronics engineer over the past 40 years specializing in analog and power electronics circuit design. He has worked for many of the top technological companies both as an employee and as a consultant. Johnny's hobbies are reading, music, genealogy, and traveling. He currently resides in the city of South Lake Tahoe, California, and is the father of two adult sons.

Contents

1

Review of Switch Mode Power Supply Fundamentals

A switch mode power supply (SMPS) may in general be defined as any type of electronic circuit that converts and/or regulates voltage or current by utilizing switching circuits and energy storage elements (capacitors and inductors). These circuits are ideally lossless with 100% energy transfer. This beginning chapter will provide a review of the most basic power converter topologies in their simplest form and an explanation of all their various modes of operation and control.

1.1 Basic Topologies

There are basically three fundamentally defined switch mode power supply converter topologies: the buck, the boost, and the combined form generally referred to as the buck–boost converter. Each of these converters has its unique properties and, in general, is applied in a complementary manner to each of the others. Also, each may have the capability of operating in one of two fundamental modes: the continuous mode or the discontinuous mode. These will be discussed in detail in the following sections.

1.1.1 Buck Converter — Continuous Mode

As an immediate initiation to the fundamentals, consider an illustration using the simplest example of all: the elementary buck converter. (Presumably, the defining word "buck" is deduced from the fact that the input voltage is bucked, or attenuated, in amplitude and a lower amplitude voltage appears at the output.) Figure 1.1a shows the circuit topology and Figure 1.1b shows the defining current and voltage waveforms. The switch positions d and d' represent the fraction of time that the periodically toggling switch remains in each position. (The period dT_P is generally referred to as the converter ON time and the period $d'T_P$ is called the converter OFF time.) By assigning a period duration of unity, it can then be seen that $d + d' = 1$. The switching period, dT_P, occurs at a frequency that is much greater than the cut-off

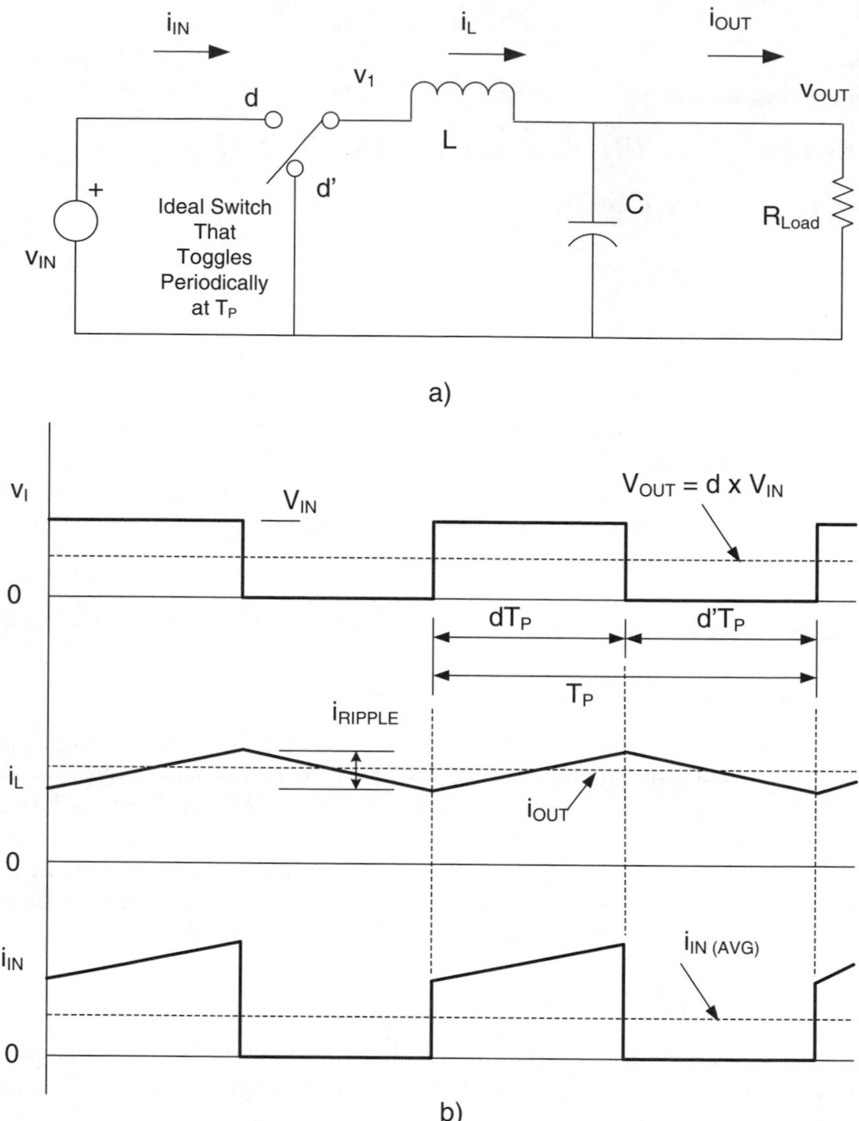

FIGURE 1.1
Basic buck converter: a) topology; b) continuous mode waveforms.

frequency of the LC low-pass filter; thus, it provides the average or DC component of the switched or "chopped" input voltage, v_{IN}, to the load with an attenuated and desired very low AC ripple component. The large signal nonlinear transfer function of this converter is indicated in Equation 1.1.

$$v_{OUT} = dv_{IN} \tag{1.1}$$

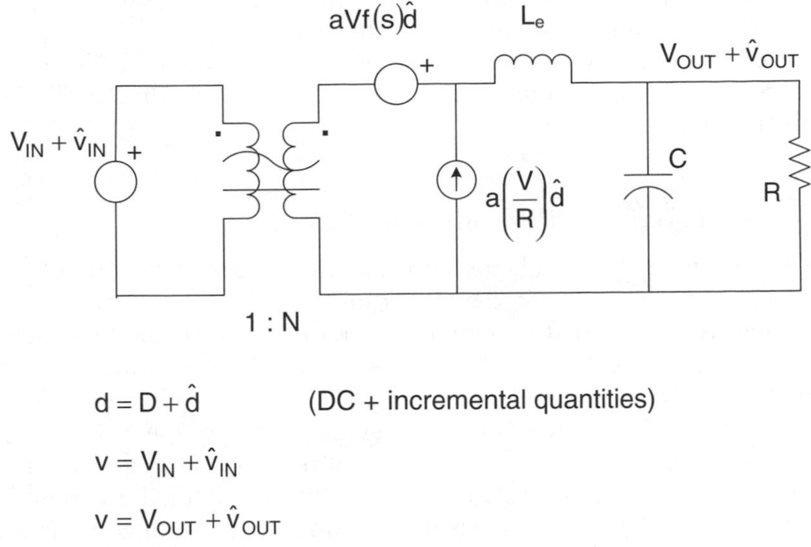

	N	a	f(s)	L_e
Buck	D	$\frac{1}{D}$	1	L
Boost	$\frac{1}{1-D}$	$\frac{1}{1-D}$	$1-s\left(\frac{L_e}{R}\right)$	$\frac{L}{(1-D)^2}$
Buck Boost	$\frac{D}{1-D}$	$\frac{D}{D(1-D)}$	$1-s\left(\frac{DL_e}{R}\right)$	$\frac{L}{(1-D)^2}$

FIGURE 1.2
Basic SMPS continuous conduction mode canonical model.

A small signal linear model developed by Middlebrook and Cuk[1,2] that has achieved wide acceptance is shown in Figure 1.2. This canonical model is applicable to all three basic topologies operating in the continuous mode; the different circuit component parameters are noted in the table in this figure. The different facets of each topology will be discussed in the following corresponding sections. This linearized model is used to allow all of the linear circuit analysis techniques developed over the years to be applied when analyzing a power supply at a particular DC operating point. Note the dependent voltage and current generators. Some important observations are immediately noted.

First, when a feedback control from the output is used to control $\hat{d}$, it is immediately obvious that the two-pole LC filter presents a "sticky" AC stability concern that must be dealt with. Next, with an ideal input voltage source — that is, with zero source impedance — the dependent current generator is essentially shorted and has no effect on the output control; the dependent voltage generator provides output voltage control through the two-pole LC filter. It is then recognized that when real-world power sources with finite source impedances are used, the control, $\hat{d}$, to output, v, will be affected by this dependent current generator.

1.1.2 Buck Converter — Discontinuous Mode

Figure 1.3a shows a more realistic representation of the buck converter in which the ideal switch is replaced by the transistor switch and a catch diode combination. When the load current, i_{OUT}, is reduced to a value that causes the average inductor current to be less than one-half the inductor ripple current, Δi_L, the inductor current wants to flow negatively through the inductor. However, with the transistor switch off and the catch diode reversed biased, the negative inductor current has no path along which to flow. At this point, no current will flow in the inductor for the remainder of the converter OFF time. With this discontinuity in the inductor current, this mode of operation is generally referred to as the discontinuous mode. (Sometimes this zero inductor current condition is referred as "the inductor running dry.")

Figure 1.3b shows the defining current and voltage waveforms with obvious differences noted between those of the continuous mode in Figure 1.1b. The converter OFF time for this mode of operation is generally designated in a different way. The portion of the OFF time during which inductor current is still flowing is designated as $d_2 t_p$, and it is now recognized that $d + d_2 < 1$. The convention for this was established in Cuk and Middlebrook[3] in the course of their pioneering work in the development of analytical power converter models. The transfer function for the discontinuous mode is noted in Equation 1.2 and is not as simple a relationship as that of the continuous mode. It is now a function of d, output load current, i_{OUT}, and the ratio of L/t_p. Middlebrook[3] defines a "conduction parameter," $k = 2L/Rt_P$, that denotes the boundary between continuous and discontinuous modes. For the buck converter, $k = D'$ at this boundary. This relationship can be derived very easily from the waveforms shown in Figure 1.1b and Figure 1.3b.

$$v_{OUT} = v_{IN}\left(\frac{2}{1+\sqrt{1+\frac{4k}{d^2}}}\right) \tag{1.2}$$

where

$$k = \frac{2L}{RT_P} \tag{1.2a}$$

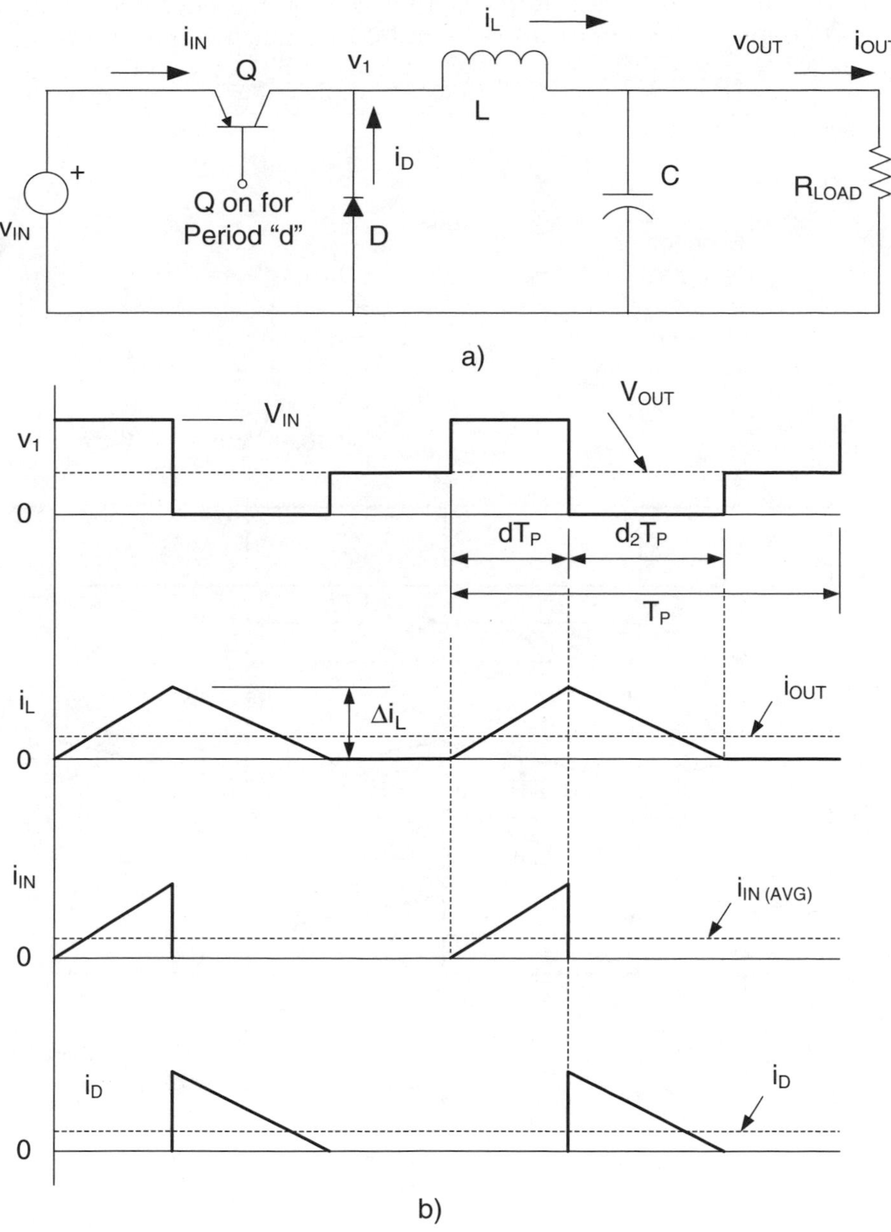

FIGURE 1.3
Basic buck converter: a) topology; b) discontinuous mode waveforms.

and

$$R \equiv \frac{V_{OUT}}{I_{OUT}} \tag{1.2.b}$$

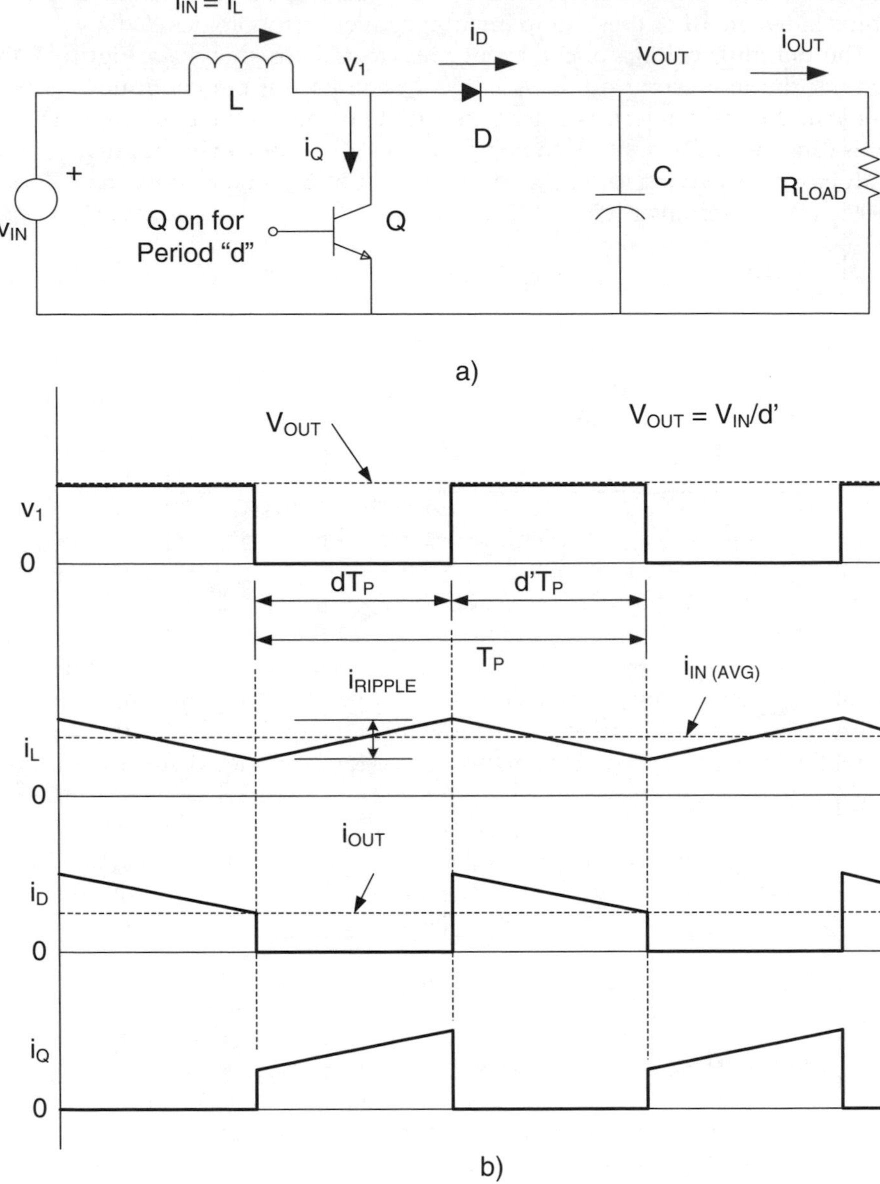

FIGURE 1.4
Basic boost converter: a) topology; b) continuous mode waveforms.

1.1.3 Boost Converter — Continuous Mode

The simple boost converter is shown in Figure 1.4a; as its name implies, it steps up or "boosts" the input voltage to a level higher than that of the input voltage. This topology is considered the "dual" or complement of the buck converter. See Cuk and Middlebrook[4] for more detail on the unique relationships between all of these fundamental converter topologies.

The defining voltage and current waveforms are shown in Figure 1.4b. One of the interesting dualities noted in comparing the continuous mode boost and buck converters is that the input current to the buck converter is pulsating or "chopped," as noted in Figure 1.1b; however, because of the inductor on the input side, the input current to the boost converter is nonpulsating or relatively smooth, as noted in Figure 1.4b. Conversely, the output current of the boost converter, i_D, which is averaged by the output filter capacitor, is chopped, but the output current of the buck converter, i_L, is smooth as a result of the inductor on the output side. The transfer function of this converter is indicated in Equation 1.3.

$$v_{OUT} = \frac{v_{IN}}{d'} \tag{1.3}$$

Note in Figure 1.2 that the boost converter small signal model is a little more complex than that of the buck converter in that the equivalent output filter inductor, L_e, is not simply L but is actually a function of the operating point duty ratio, D. Also note that the dependent voltage source generator has a right-half plane (RHP) zero associated with it. This makes the AC stability problem even more difficult.

A brief word regarding this RHP zero might be made here. This zero results because, when the duty ratio changes with time, the output voltage actually shifts in the opposite direction during this change because of the lagging inductor current change. When this converter is modeled for a computer small signal circuit simulation, this RHP zero can sometimes present a problem. If so, the way to deal with it is to manipulate the circuit model by sliding the dependent current generator to the right of the inductor; the RHP zero will disappear from the model. This is essentially backing through the derivation of the canonical model. See Middlebrook and Cuk[1] for details.

1.1.4 Boost Converter — Discontinuous Mode

As one might expect, the discontinuous mode boost converter exhibits the same dual relationships with the buck converter in the discontinuous mode as it did with the buck converter in the continuous mode. Figure 1.5a shows the basic topology again, and Figure 1.5b shows the defining current and voltage waveforms. Again, the discontinuous or zero inductor current condition is noted. The transfer function of this converter is indicated in Equation 1.4 and has the same parametric dependencies as those of the buck discontinuous mode converter.

$$v_{OUT} = v_{IN}\left(\frac{1+\sqrt{1+\frac{4d^2}{k}}}{2}\right) \tag{1.4}$$

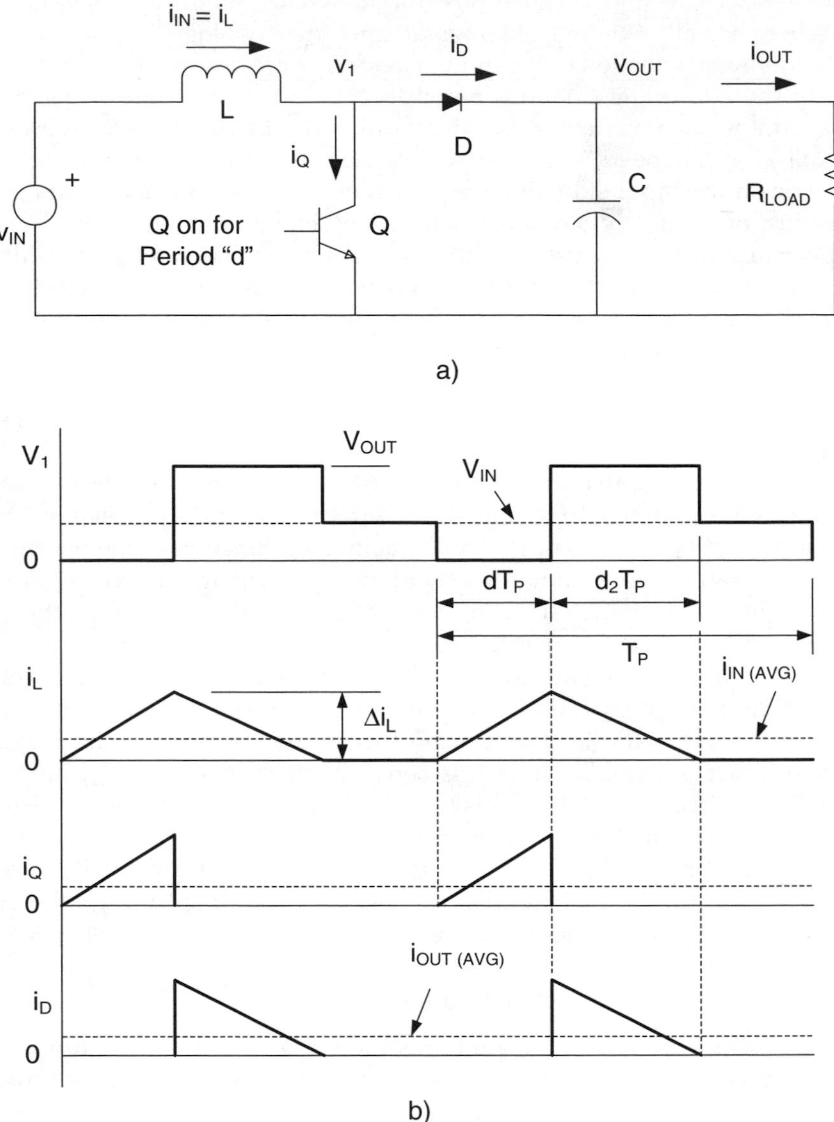

FIGURE 1.5
Basic boost converter: a) topology; b) discontinuous mode waveforms.

1.1.5 Buck–Boost Converter — Continuous Mode

The continuous mode buck–boost converter is in essence a cascaded version of the buck and boost converter with a transfer function of:

$$v_{OUT} = v_{IN}\left(\frac{d}{d'}\right) \tag{1.5}$$

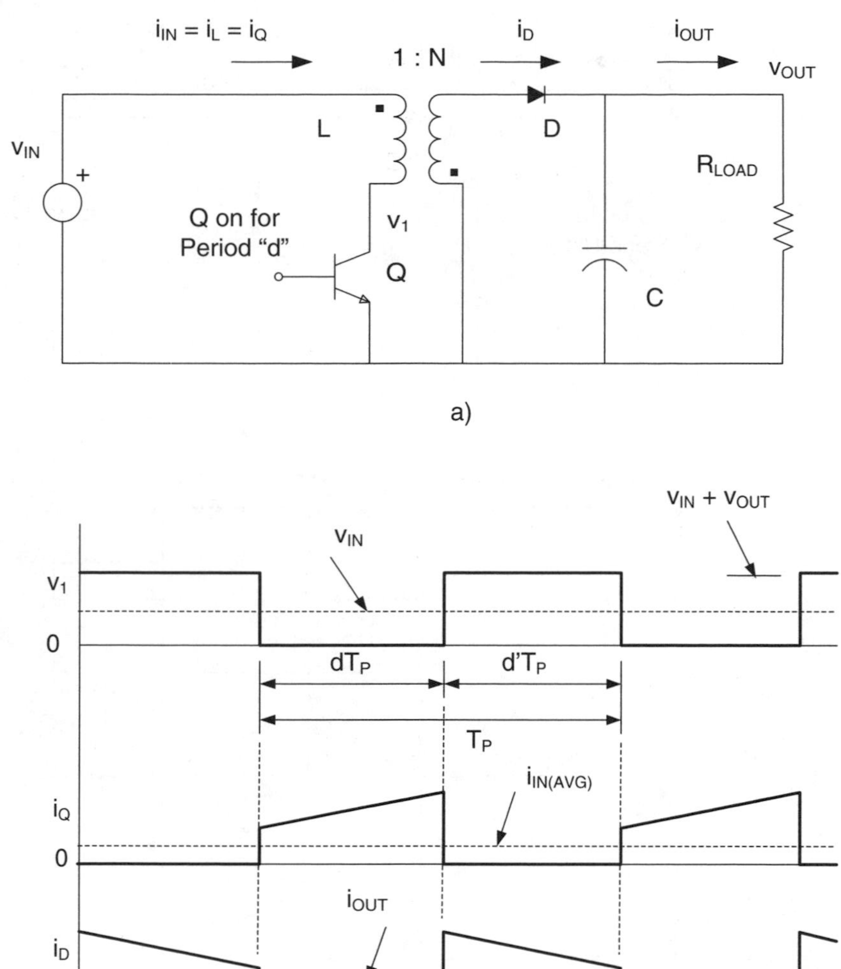

FIGURE 1.6
Basic buck–boost converter: a) topology; b) continuous mode waveforms.

The conventional, simple noninverting version of this topology is shown in Figure 1.6a. A coupled inductor is required for this representation. The accompanying waveforms are shown in Figure 1.6b. Note that the input current, i_Q, is a pulsating current and the current into the output filter capacitor, i_D, is also a pulsating current. These pulsating currents are undesirable because they require more capacitive filtering to keep the ripple voltages to an acceptable level. This particular buck–boost topology, or conventional "flyback" converter as it is generally called, unfortunately incorporates the worst ripple current characteristics of its constituent buck

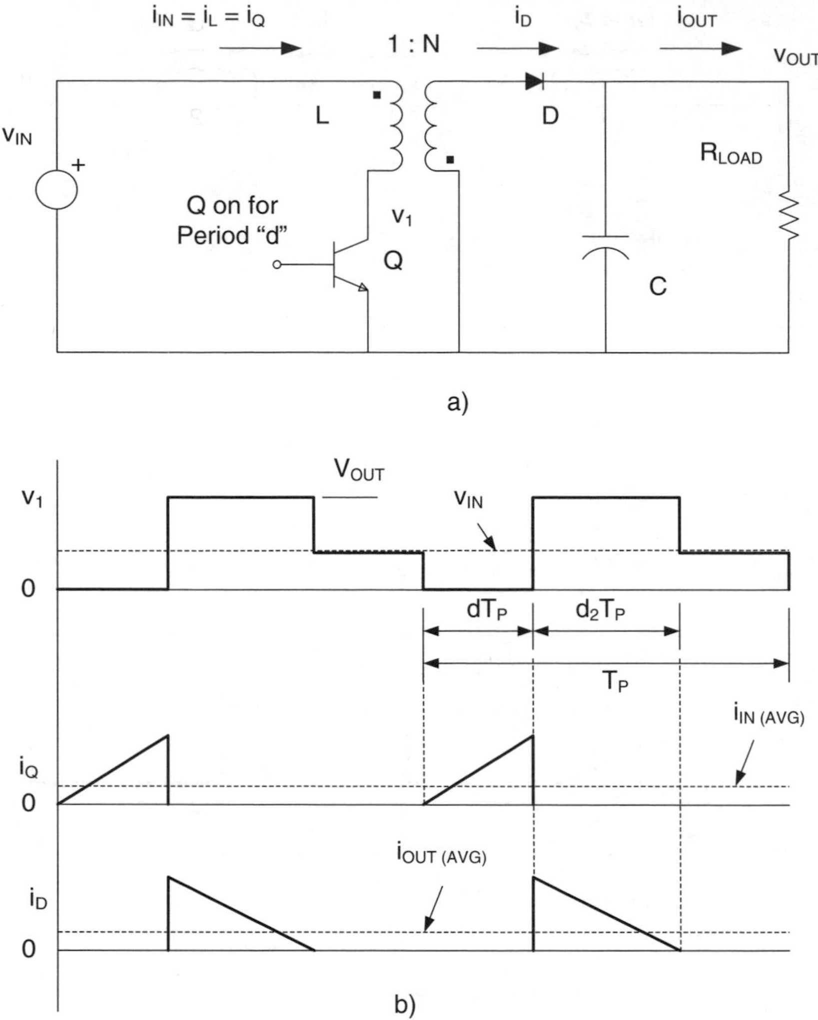

FIGURE 1.7
Basic buck–boost converter: a) topology; b) discontinuous mode waveforms.

and boost topologies. Around 1976, Dr. Slobodan Cuk developed his namesake converter, the Cuk converter,[4] which provides the same buck–boost transfer function as shown in Equation 1.5, but has smooth input and output currents instead of the pulsating or chopped currents in the flyback converter. Figure 1.8a shows the fundamental inverting topology of the Cuk converter and Figure 1.8b shows the accompanying waveforms.

1.1.6 Buck–Boost Converter — Discontinuous Mode

Figure 1.7a shows the basic conventional, noninverting flyback converter with the accompanying discontinuous mode waveforms in Figure 1.7b. The

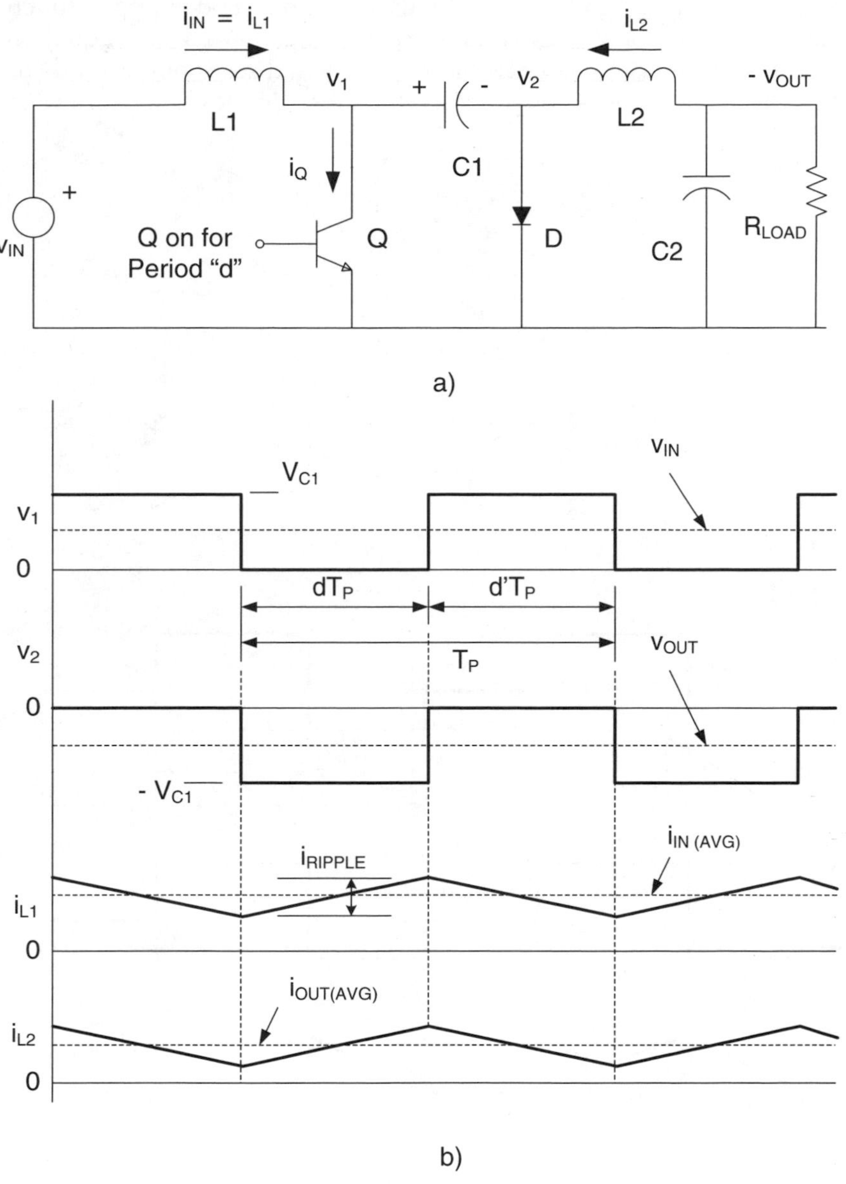

FIGURE 1.8
Cuk converter: a) topology; b) continuous mode waveforms.

transfer function is:

$$v_{OUT} = v_{IN} d \sqrt{\frac{1}{k}} \tag{1.6}$$

A general comparison between the discontinuous mode transfer functions of the three basic topologies of Equation 1.2, Equation 1.4, and Equation 1.6 shows that the flyback converter has the simplest discontinuous mode transfer function and the buck converter has the most complex. This is true despite

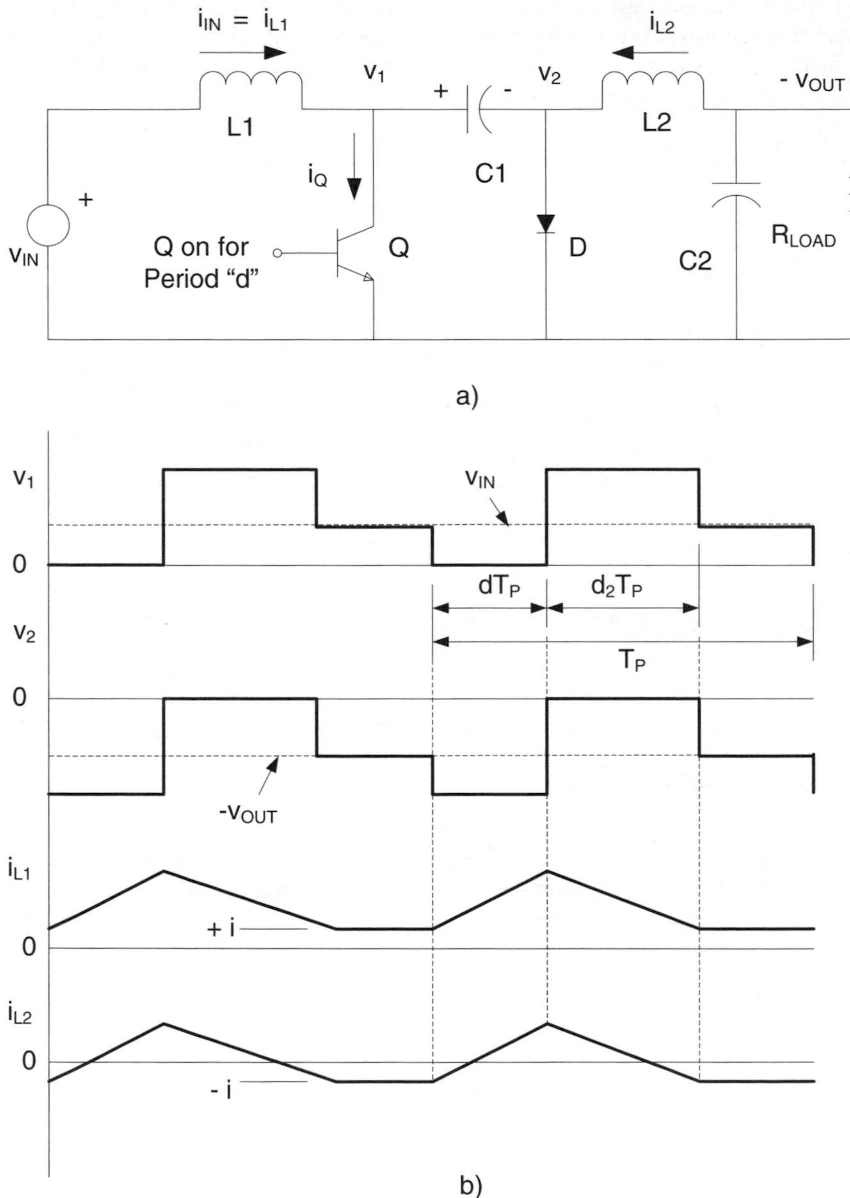

FIGURE 1.9
Cuk converter: a) topology; b) discontinuous mode waveforms.

the fact that the buck converter is the simplest derived converter and the flyback converter is more complex. The Cuk converter has the same discontinuous mode transfer function as the flyback (see Cuk[6] for details). When evaluating the parameter *k*, however, the value of inductance used in the calculation is the parallel combination of *L*1 and *L*2. Figure 1.9a again shows the fundamental inverting topology of the Cuk converter and Figure 1.9b shows the accompanying waveforms. It is interesting to note from Cuk[6] that the inductor currents i_{L1} and i_{L2} are both continuous or both discontinuous and that, in the general case, these currents do not individually become zero for the discontinuous part of the switching cycle; rather, the sum of i_{L1} and i_{L2} becomes zero.

1.2 Basic Control Methods

When controlling the output of an SMPS, the proportion of time during which the power switch is ON to the time during which it is OFF must be set by some controlling mechanism. The most commonly used defining term is the duty ratio, *d*, which is the ratio of the switch ON time to the total switch period. Several methods are used in accomplishing this pulse width modulation (PWM); they may be categorized into one of four categories chosen to be identified by a "frequency of operation factors" definition:

- Constant frequency
- Constant ON time
- Constant OFF time
- Constant hysteresis

Also, these four PWM techniques may generally be applied to either one of two additional categories of control techniques known as voltage or current mode control. Voltage mode control uses the duty ratio control mechanism to control the output voltage directly; current mode control uses the duty ratio controller to regulate the current in the energy storage inductor. An additional outer control loop may then be implemented to regulate the output voltage if desired.

1.2.1 Frequency of Operation Factors (FOFs)

Constant frequency. As the name implies, the switch cycle period is held constant with changes in converter ON time complementary to changes in converter OFF time. This is the most commonly used control because the control circuit may use a fixed frequency clock providing a very straightforward and easy way to implement a control scheme. Also, frequency control may be desirable in efforts to control electromagnetic interference (EMI) phenomena emanating from the power converter. Frequency synchronization with other sources may also be desirable and easily implemented. Most of the available technical literature relating to SMPSs uses these constant frequency schemes.

Constant ON time. This scheme of duty ratio control maintains a constant ON time while varying the OFF time as its method of PWM control. Of course, the frequency varies with changes in duty ratio. This method of control might be desirable in order to optimize the magnetic component design or possibly to meet some particular control law, ripple, or stability criterion.

Constant OFF time. The duty ratio control in this case maintains a constant OFF time while varying the ON time as its method of PWM control. The control scheme here is complementary to that of the constant ON time converter and is generally applied for somewhat the same reasons.

Constant hysteresis. This control technique occurs when the ON time and OFF time are allowed to vary, but not necessarily in any fixed time period. This mode is almost always associated with a class of regulators known as ripple regulators. A hysteresis comparator is used to sense a voltage or current waveform that contains a small ramping up and down ripple component and compares it to some reference. This is probably the most elementary of all regulation schemes; unfortunately, it depends on the presence of an undesirable ripple component in the controlled output. Continuous conduction mode is almost always necessarily used in these designs.

1.2.2 Voltage Mode Control

Voltage mode control for continuous or discontinuous mode is the most original or natural method of controlling an SMPS. The controlling signal is thought of as directly controlling the duty ratio, d, and thus the output voltage. This is glaringly pointed out in Equation 1.1 through Equation 1.6. The several ways of accomplishing this depend on the FOF defined earlier. For the constant frequency control, the analog control signal is simply compared to a triangular or sawtooth waveform to provide the PWM function as shown in Figure 1.10a. The constant ON or OFF time techniques obviously use a fixed time pulse generator along with additional circuitry, which varies the controlled pulse width. This control generally uses ramping techniques to accomplish this as shown in Figure 1.10b and Figure 1.10c. The constant hysteresis is, as explained previously, the most basic and "primitive" method of PWM control. Figure 1.10d shows an example of this scheme, which may be used for an output voltage or an inductor current control converter.

1.2.3 Current Mode Control

Current mode control may be thought of as more or less an extension of voltage mode control in the sense that any duty ratio control directly controls

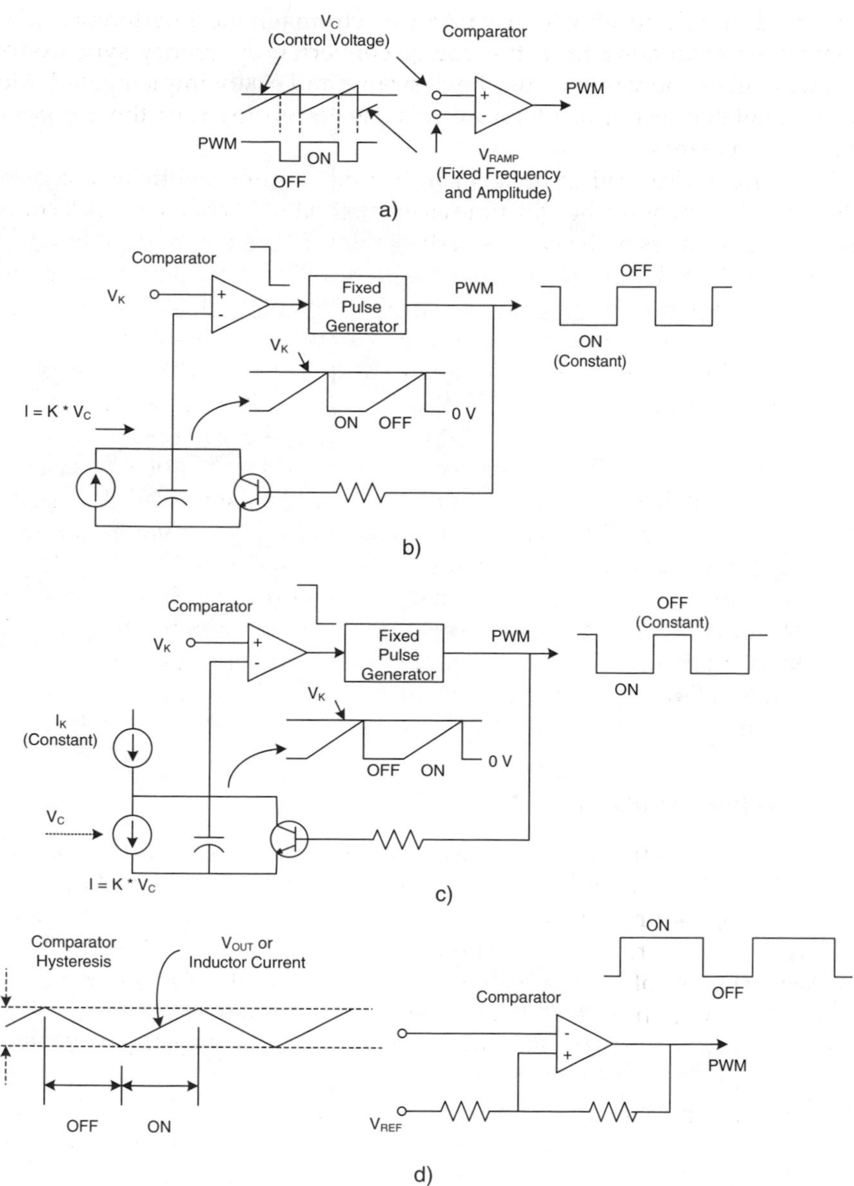

FIGURE 1.10
Voltage mode control schemes: a) constant frequency; b) constant ON time; c) constant OFF time; d) constant hysteresis.

the voltage transfer ratio of the converter and thus the output voltage for a fixed input voltage. The unique feature of current mode control is that the inductor current, with its associated triangular ripple current, is used to provide the triangular (ramping) signal with which to compare the controlling

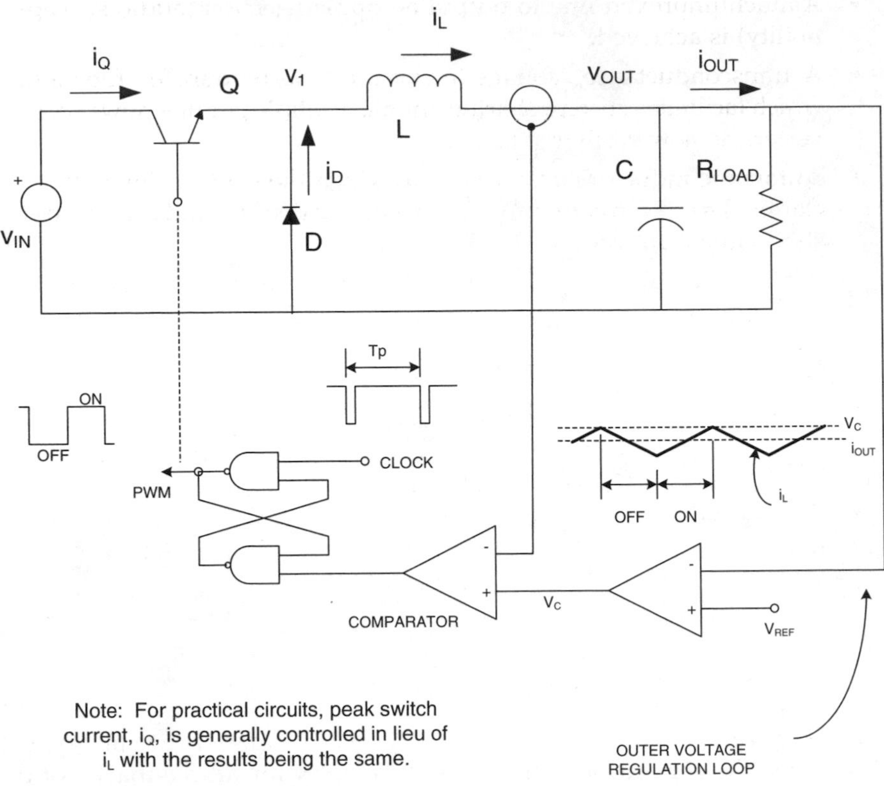

FIGURE 1.11
Conceptual current mode control scheme.

voltage signal, v_C, rather than an extra generated triangular waveform as is the case for voltage mode control.

When the inductor current DC component and ripple are used in this comparison, as shown in Figure 1.11, it is noted now that the control voltage, v_C, regulates the peak inductor current. When the ripple component is small compared to the DC component of the inductor current, the actual DC inductor current is essentially regulated with this "new" control signal, v_C. With this current control, the converters now assume a transconductance type of transfer function. To provide output voltage regulation, an outer voltage control loop, as shown in Figure 1.11, is required. In the continuous mode, some significant advantageous features are now obtained:

- An essentially single-pole control to output voltage transfer function response is obtained as opposed to the two-pole response noted when using voltage mode control. This generally simplifies the control loop design considerably.

- A much improved line to output ac ripple rejection (audio susceptibility) is achieved.
- A transconductance, voltage-to-current type of transfer function, which facilitates current sharing among multiple parallel power converters, is now easily realizable.
- Automatic inductor current limiting when the control signal, v_C, is clamped results in converter output current limiting and thus output short circuit protection.
- Dramatic reduction in transformer flux imbalance tendencies in push–pull converters now makes these topologies function more ideally.

One noted drawback to the current mode topology is that a cyclical instability can occur for converter duty ratios greater than $d = 0.5$ when the peak inductor current feedback control scheme is utilized. An external ramp signal is required to be summed to that of the inductor current ramp component to achieve stable operation over the entire duty ratio range of zero to one if desired. Hsu et al.[5] provide more detail.

1.2.4 Feedback and Feedforward Control Additions

A few general comments about control of these converters are made here. Obviously, the regulation techniques developed for all feedback control systems are applicable to SMPS. Type 1 and type 2 control loops are generally the ones encountered. Multiple feedback loops are sometimes encountered in the form of derivative control such as output capacitor current sensing or simply the derivative of the output voltage. Differentiator circuits, however, are generally undesirable due to noise pick-up limitations.

Several methods of feedforward control can also provide improved line voltage ripple rejection. The most obvious is that of the classical input voltage feedforward compensation added to the linear feedback control signal. In SMPSs, a more desirable feedforward compensation technique is to have the PWM ramp generator slope modified by the instantaneous magnitude of the input voltage. This provides cycle-by-cycle feedforward compensation. With the proper slope compensation, this technique can sometimes provide sufficient line voltage regulation without using any output feedback for some voltage mode converters operating in the continuous mode. Dixon[7] and Arbetter and Maksimovic[28] offer details on this. The excellent line rejection of current mode converters is essentially provided by the implementation of this feedforward property with the instantaneous detection of the change of the inductor current ramp slope with a change in input voltage.

1.3 Conclusion

There are many variations of the basic ideal circuit topologies presented here, but the ones illustrated provide a good review of the fundamental knowledge necessary to understand and analyze switching power supplies. More detail and insight will be provided on the SMPS in the analysis examples used in the following chapters. At first glance, an SMPS often appears deceptively simple; however, when it is analyzed, unexpected subtleties are encountered. This is perhaps the main reason that the power electronics field is so challenging and interesting.

2

SMPS Analysis Requirements

Consider the basic task at hand from the most primitive point of view. All power supplies have inputs and outputs; the outputs are specified to have certain qualities that are obviously different from those of the inputs. The power supply to be analyzed should ensure that, with a given set of inputs, the outputs should be bounded within specified limits. The basic question is then, "What analyses are necessary to ensure that a particular power supply will meet its performance requirements?" First, a listing of as many general requirements as is practical will be presented, along with some typical switch mode power supply (SMPS) circuits; this will help to illustrate the necessity of the various analyses. Because most SMPSs are DC voltage regulators, they will be used as primary examples. However, in principle, the same analytical processes could be applied to DC current regulators, switch mode power amplifiers, or any other circuits that utilize these switching power converter concepts.

2.1 DC Requirements

In this section, all the DC requirements will be listed, as well as what must be done to ensure that they are met. For the purposes of illustration, the elementary push–pull buck voltage regulator of Figure 2.1 will be used. This very basic topology will be expanded to illustrate many of these general analysis considerations.

2.1.1 Output Regulation

Output regulation is perhaps the most important parameter to be verified because almost all other necessary analyses generally delineate from the regulation requirements. The output voltage varies as a function of three stimuli:

- Input or, as it is sometimes called, line voltage perturbations
- Load current variations
- Circuit component variations

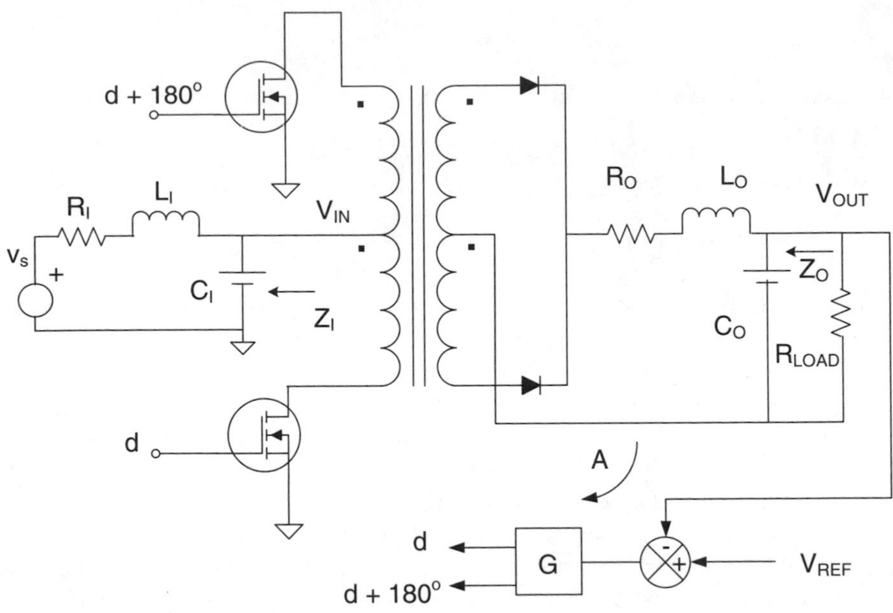

FIGURE 2.1
Elementary push–pull buck voltage regulator.

The line voltage and load current variation effects are caused by external stimuli directly; the effects caused by circuit element variations are "internal" and the designer has more control over them. Deviations from the nominal values of the components result primarily from operating temperature variations, aging, and manufacturers' specified initial tolerances of the devices. There are also numerous other environmental effects but these are the ones that are generally of primary concern and most often necessary to consider when doing an analysis. Now, using the example power supply of Figure 2.1, consider some practical generalizations about the regulation stability.

Probably one of the first things that the analyst would do is to verify the stability of the reference, V_{REF}. This reference may be a highly stable, temperature-compensated zener diode or it may exist in the form of an integrated circuit precision reference, which may exist in many forms. Also, the reference may originate from the output of a digital to analog converter for digitally programmed applications. In any case, this circuit must be analyzed to determine the range of reference voltage that it presents to the voltage regulator because the regulator is only as good as the reference.

The voltage reference may also be considered a voltage regulator and may require some of the same types of regulation analyses required for the main power supply. In other words, it appears that a power supply exists within a power supply; that is exactly the case, although the reference supply is of a much lower power level. If the reference supply is powered from the main supply-regulated output, a certain analytical digressiveness may be noted here; however, generally, the effects diminish to third order and higher rapidly and an iterative analysis is seldom required. In any case, the reference supply should probably be analyzed first using all the required analysis methods indicated in this chapter.

Now that a reference has been defined, it is possible to continue with the required regulation analysis for the main power supply. As was noted in Chapter 1, a switching regulated power supply has a nonlinear large signal control law. The conventional technique for most analysis is to linearize the circuit at a particular operating point; this allows the vast knowledge of linear circuit theory to be brought to bear.[1]

Consider some fundamental background on what is necessary for a regulation analysis. When this analysis is eventually performed with a computer simulation, what is actually occurring will be transparent; however, some basic knowledge is considered essential in understanding the power supply and also to help in troubleshooting a particular model when the need arises.

First, consider the effects of line voltage variations. Assuming that the power supply analysis model has been linearized at a particular operating point of line voltage and load current, it is possible to consider the analysis from a conventional linear circuit point of view. Basic linear circuit theory indicates that the normal forward transfer function of a circuit is effectively reduced by the factor of one plus the loop gain of an applied regulating feedback loop. This is expressed mathematically as:

$$\frac{\Delta V_{OUT}}{\Delta V_{IN}} = \frac{H}{1+A} \tag{2.1}$$

where H is the open loop forward transfer function and A is the regulation loop gain.

This equation is valid for AC as well as DC considerations. This offers a very simple way of making a quick generalization about the power supply if a quick first-order evaluation and possible verification that the computer simulation is correct are desired. For most supplies, the loop gain, A, is very high at DC and the lower frequencies by design, thus making output voltage variations approach zero as a function of line voltage variations.

The effects of load variations have a similar type of relationship on output voltage regulation. The output voltage variations with load current changes are caused by the output impedance, Z_O. With no feedback regulation loop, Z_O is identified as the output impedance of the power supply. When the feedback loop is added, the effective impedance is reduced again by one plus the loop gain and is expressed as Z'_O in the following equation:

$$Z'_O = \frac{Z_O}{1+A} \tag{2.2}$$

Again, at DC and the lower frequencies, the output impedance, Z'_O, approaches zero.

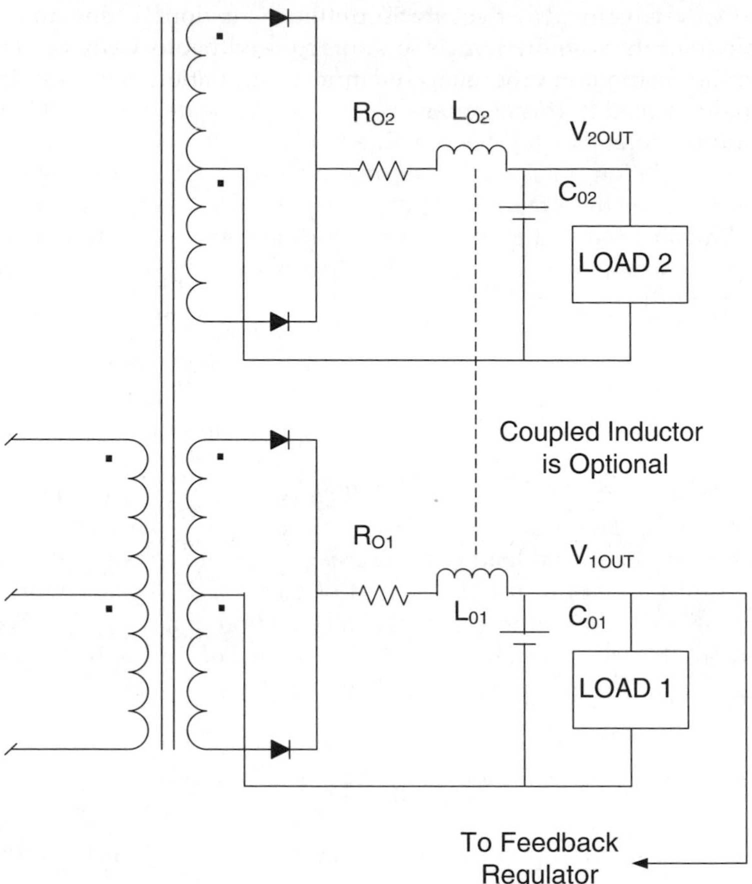

FIGURE 2.2
Push–pull regulator with multiple outputs.

2.1.2 Cross Regulation with Multiple Outputs

Take the push–pull buck regulator of Figure 2.1 and add an additional output to the secondary as shown in Figure 2.2. This second output is not regulated directly, as the one with the feedback is, so it will not be as well regulated. Also, as the component and circuit operating point conditions change for the regulated output, the regulation loop causes different compensating voltages to appear around the loop. The transformer voltage changes and thus causes a change in the unregulated output even though no load or component changes necessarily occurred on this output. This effect is generally referred to as cross regulation.

Sometimes L_{O1} and L_{O2} are coupled on the same core to achieve better dynamic control with line and load changes.[8] A computer analysis with a line voltage and load current variation matrix can be a very useful and labor-saving method of analyzing cross-regulation effects. Obviously, any number of multiple outputs from the same transformer may theoretically be analyzed in a similar fashion. In any case, a computer simulation can provide valuable analysis results when accurate component models are used for this simulation.

2.1.3 Efficiency

When the power-consuming elements are accurately modeled, a computer simulation can sometimes be used to aid in performing an efficiency analysis of an SMPS. Although SMPSs are theoretically 100% efficient, real-world inefficiencies are always present. The internal losses in an SMPS are caused primarily by the following factors:

- *Magnetic component losses.* These are manifested in a number of ways. There are the normal conduction losses in the resistance of the current carrying windings. The resistances of these windings have a low-frequency value, generally noted as the DC resistance. At the higher frequency, the resistance increases with frequency in two phenomena known as "skin effect" and "proximity effect."[16,17] AC magnetic core losses are primarily caused by the cyclical hysteresis losses in the magnetic material; eddy current losses are caused by the induction of circulating currents flowing in the core.
- *Power switch losses.* In a conventional PWM SMPS, these losses are generally thought of as saturation losses and switching losses. The latter occur during the finite transition time between the switch transition from ON to OFF and vice versa. The saturation losses occur during the static part of the switching cycle. Figure 2.3 shows representative periodic waveforms of voltage and currents in these power switches and the portions of the cycle, which attribute to saturation or switching losses. Although rectifier diodes are not generally thought of as switches, they do provide a switching function and their power dissipations must be considered as a power switch loss.
- *Quiescent power losses.* These losses are consumed in the control and power switch drive circuits and other miscellaneous peripheral circuits. These miscellaneous circuits may exist as start-up supplies, power factor correction circuits, inrush current limiters, power monitoring circuits, and even external interface circuitry.

It would be nice if one could develop circuit-averaged models of the power dissipation components of an SMPS that would be realistic enough to allow performing the efficiency analysis entirely on the computer. Unfortunately, a wide diversity of "parasitic" circuit elements that have second- or third-order effects on the overall circuit must sometimes be considered. Also, they

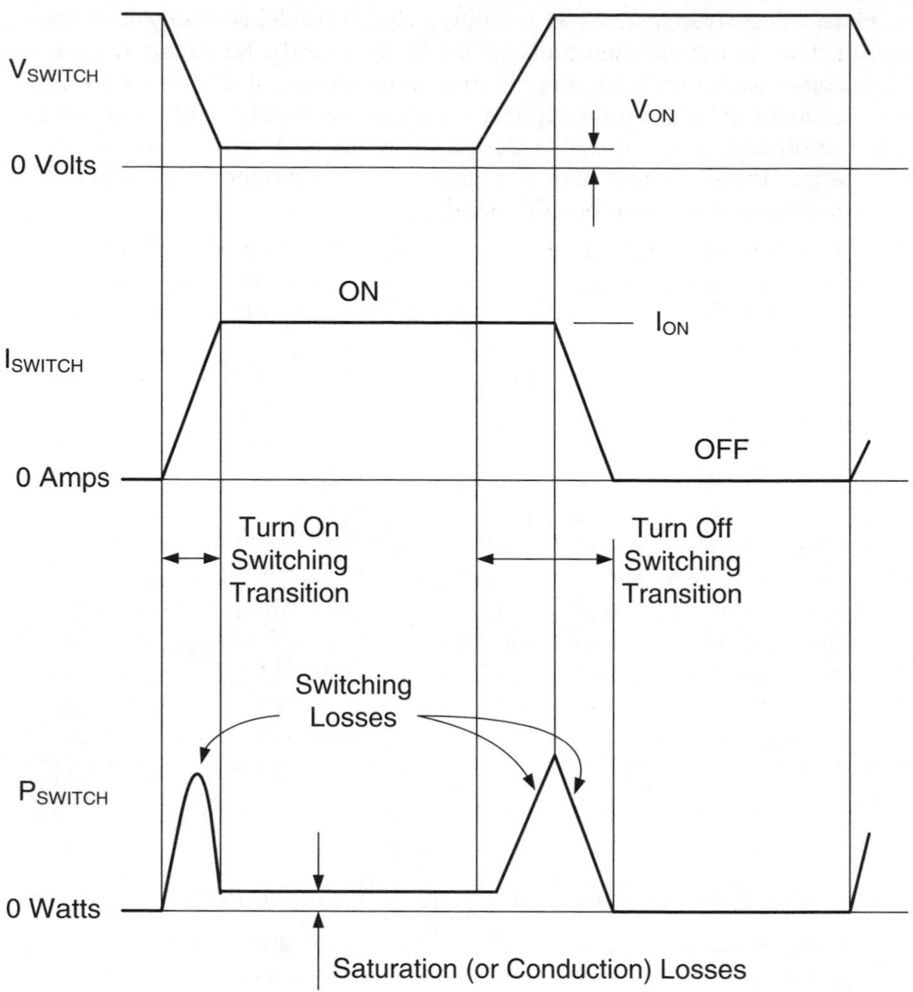

FIGURE 2.3
Illustration showing power switch losses.

are generally difficult to quantify and model to the level of accuracy necessary to provide a reasonable estimation of the power supply efficiency. This makes it difficult to generalize about the best way to conduct an efficiency analysis of an SMPS.

Consider three ways of approaching this problem:

- Do a "hand–calculated" or spreadsheet type of tabulation of the losses in the circuit. These loss effects may be determined from manufacturer's data sheets or from test data. Unfortunately, this approach is tedious and generally requires evaluation at many different operating conditions.

- Develop a replicate exact switching circuit model that simulates the actual circuit and then compute the losses directly. This again requires accurate component models and may require a lot of "number-crunching" computing power if the power supply model is of any significant size. Also, the physical layout and wiring, which can affect the switching action in an uncertain manner that is difficult to assess, must also be addressed.
- Develop an accurate large signal circuit-averaged model. This would be a very desirable method and could possibly delineate from the model that is developed for the analysis of other performance parameters. Unfortunately, accurate large signal circuit-averaged component loss parameters must be obtained and this may be very tedious and difficult, particularly in the case of the switching losses of the power switches. The same physical problems as those in the preceding case also apply for this switching loss case.

As might be surmised, there is no one easy way to calculate the efficiency of an SMPS and obtain a level of accuracy that would be as good as an actual measurement. In the past, the first approach (the spreadsheet method) has been applied and will probably be the one most applied in the future until advanced techniques are developed to allow other approaches such as those proposed in the second and third bullets to be used to any extent. If only a few dominant first-order conduction and switching losses are obvious, the spreadsheet approach will provide the most practical determination of efficiency.

2.1.4 Miscellaneous Range and Threshold Considerations

This area of analysis is sometimes overlooked but may be essential, especially for problems that may exist during power-up and power-down of the SMPS. These problems may be in the form of inrush surge currents causing component stress conditions; power source overloading; output voltage overshoot; or simply anomalous circuit operation, which may lead to component destruction, fuse blowing, or other malfunctioning effects. These threshold analyses may also apply for large load or line transients or even temporary power dropouts that create operating conditions out of the normal range of the SMPS similar to those encountered during initial power-up. A large signal (nonlinear) model containing the appropriate discontinuities must be generated.

The situations encountered are somewhat complex to analyze (and design) because they involve sequential events that are functions of different threshold levels and time delays. Figure 2.4a and Figure 2.4b show a general scenario of the sequence of events that may occur during initial power-up. This example is for a DC power source. An AC power source creates additional concerns because input rectification, filtering, and power factor correction

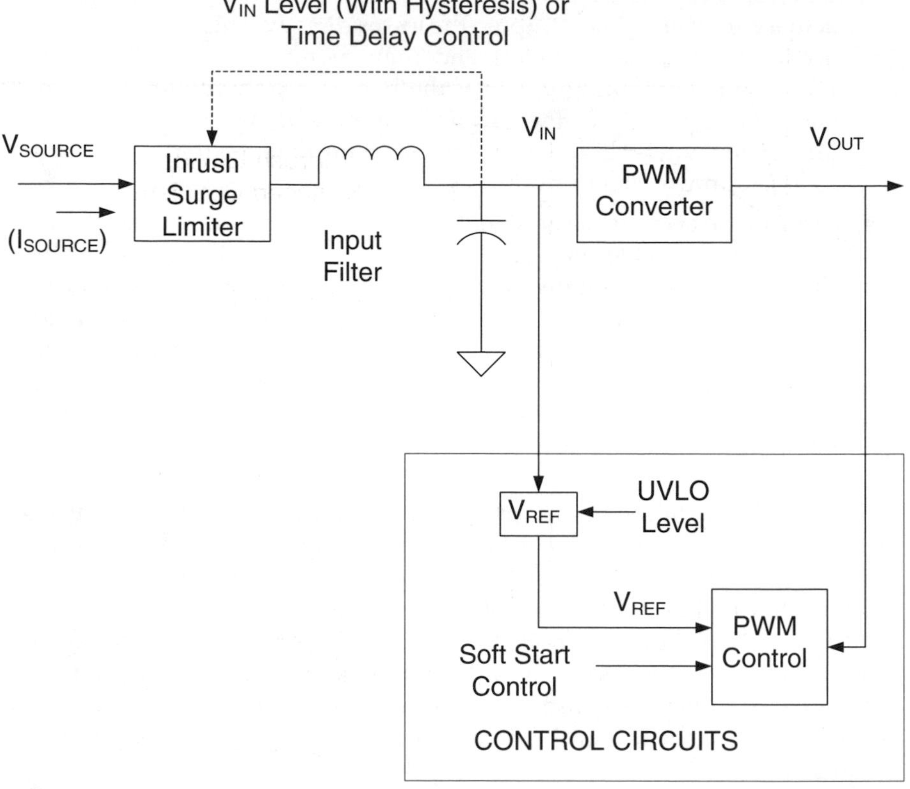

FIGURE 2.4a
An example of an SMPS start-up scenario (circuit).

provide added circuit complexity. In general, however, it is simply an extension of the DC case. For now, consider the DC power source case.

The initial application of input voltage may be specified as a step function or some function of time, such as a ramp, or maybe some nonlinear function. In any case, the design is required to accept this condition and analysis needs to verify this. A step input is probably the most typical and generally the most severe application of power if no severe voltage overshoots or sags result. The example in Figure 2.4a assumes this step input.

2.2 AC Requirements

Although most SMPSs provide a DC power conversion function, there are a number of AC analysis concerns. The bulk of these issues involves control loop stability issues or noise-related issues such as electromagnetic compatibility.

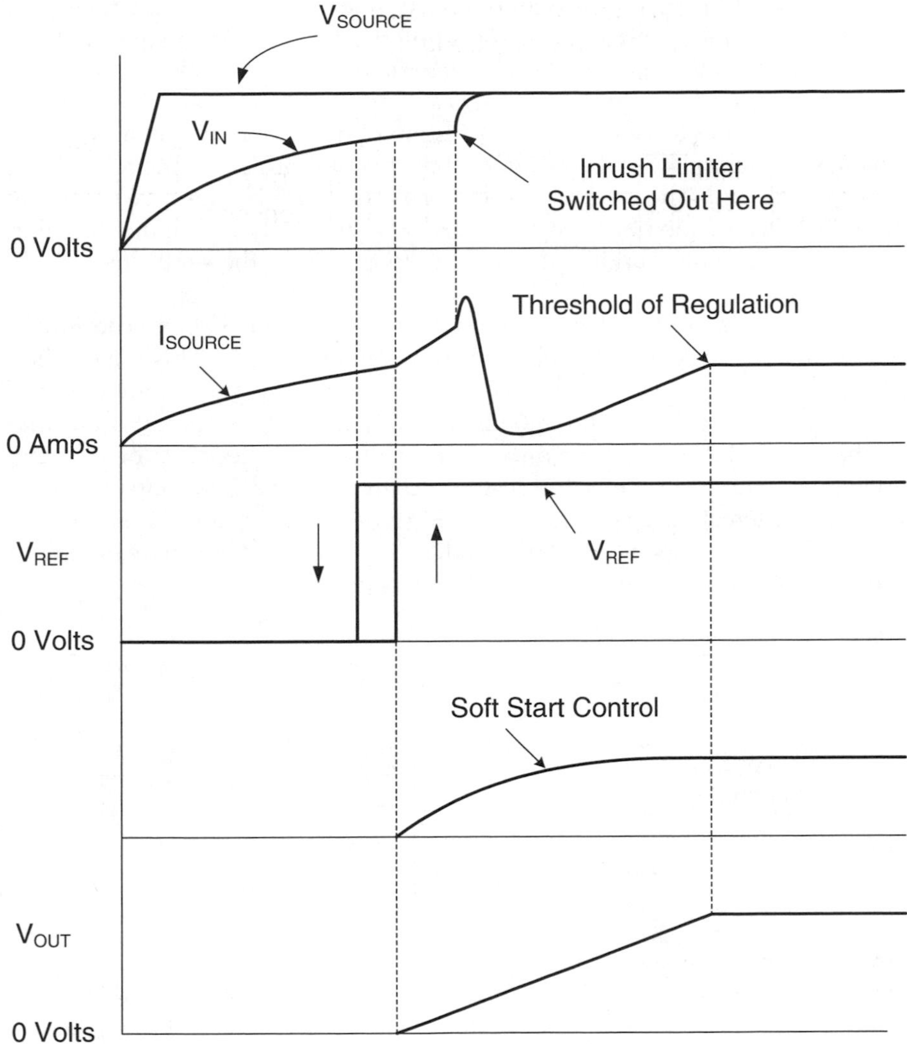

FIGURE 2.4b
An example of an SMPS start-up scenario (waveforms).

2.2.1 AC Control Loop Stability Margins

Of primary concern to the design of any feedback control circuit is the matter of AC control loop stability. In most cases, the verification of the desired stability margin is achieved by showing that the Nyquist stability criterion[10,11] is satisfied. Other stability indicators, such as the Routh–Hurwitz[10,11] criteria or perhaps some time domain criteria, may be applied, but in the vast majority of cases, the linear techniques of the Nyquist criteria are applied for analysis and actual test verification.

The classical Nyquist plot is a polar plot of loop gain vs. loop phase with frequency the independent variable. The criteria's goal is to show that no positive real roots occur in the characteristic equation of the closed loop transfer function.[10,11] Figure 2.5a shows a simple feedback control circuit and Figure 2.5b shows a Nyquist plot representation of the loop gain and phase. The basic criterion for stability states that this plot must not encircle the –1, –180° coordinate. The degree of stability is provided by the phase margin, ϕ_M, which is the loop phase angle occurring at the unity loop gain crossover point, and the gain margin, which is the loop gain at the –180° loop phase point.

The most practical and conventional representation of this criterion is, however, in the form of the Bode plot shown in Figure 2.5c. This is a semilog plot of loop gain, expressed in decibels, vs. frequency and another plot of loop phase vs. frequency. It contains basically the same information as that of the classical Nyquist plot but allows a much simpler way of plotting, analyzing, and interpreting the results. Analyzing feedback loops is one of the more interesting challenges of an SMPS analysis or any linear control system for that matter. See Middlebrook's very good paper[12] on dealing with the concepts of multiple loop regulation systems.

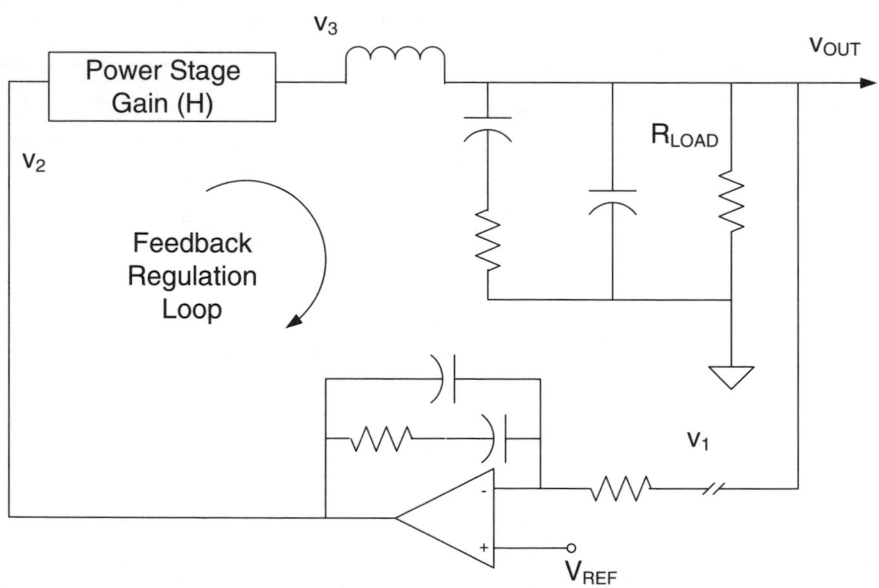

FIGURE 2.5a
Basic SMPS feedback control circuit.

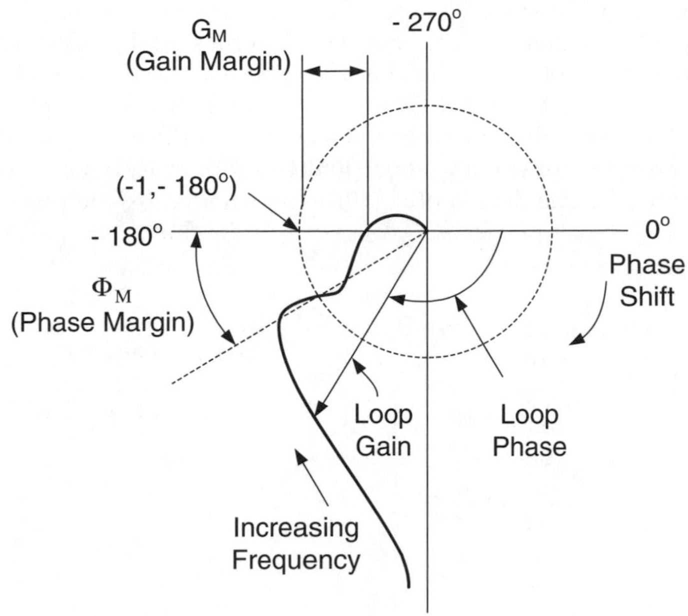

FIGURE 2.5b
Nyquist plot of loop gain.

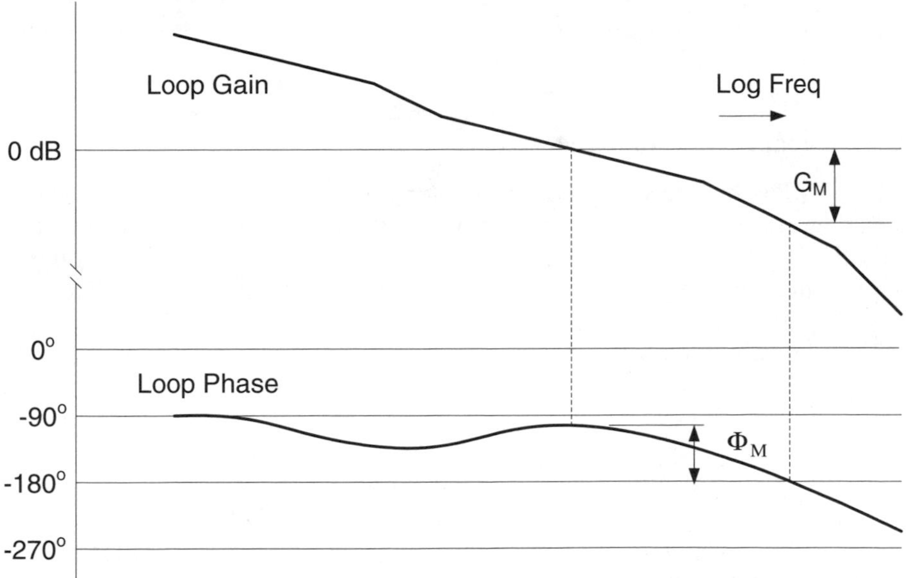

FIGURE 2.5c
Bode plot of loop gain and phase.

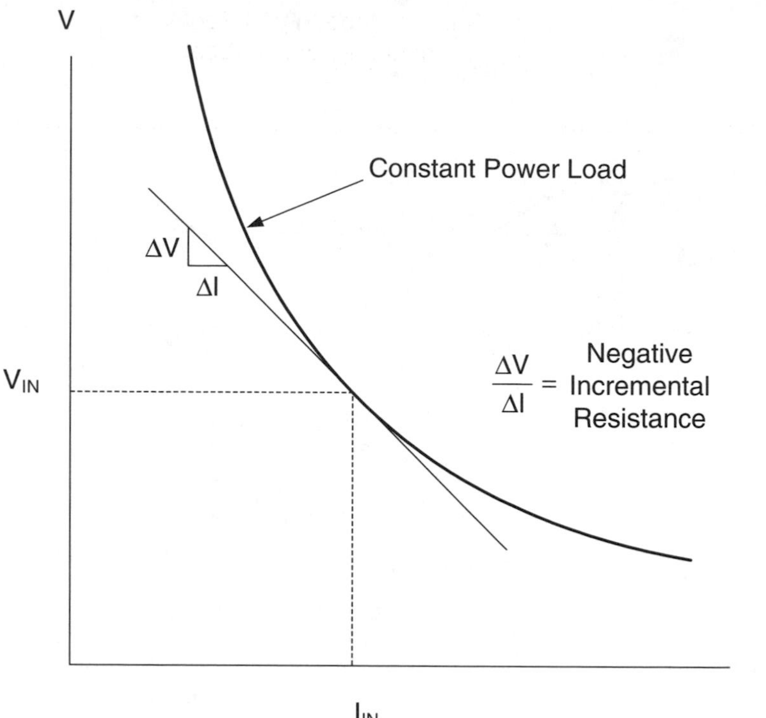

FIGURE 2.6
Negative resistance loading on SMPS input filter.

2.2.2 Input Filter Stability Margins

Most SMPSs have a low-pass input filter; this is required for a number of reasons, but in essence it provides noise isolation between the power source and the SMPS. (See Appendix A for a detailed synopsis of SMPS input filters.) Unfortunately, the insertion of this filter has some undesirable effects on circuit performance. One of the most serious is that this filter has the capability of producing an AC instability on the input voltage line presented to the power converter. The reason for this is that the SMPS presents a constant power load to the source and input filter and, because it is a constant power load, it has a negative incremental resistance characteristic (Figure 2.6). When considered as a negative resistance load connected to the equivalent parallel resonant circuit of the input filter, it can negate any "small" positive resistances in the input circuit and produce a classical electronic oscillator.

Middlebrook and Cuk[2] and Middlebrook[13] offer very good treatments of this subject and provide information on dealing with voltage mode converters. These are generally more complex to optimize than current mode converters because the AC input impedance of the voltage mode converter can have a considerable variation with frequency, particularly at the lower

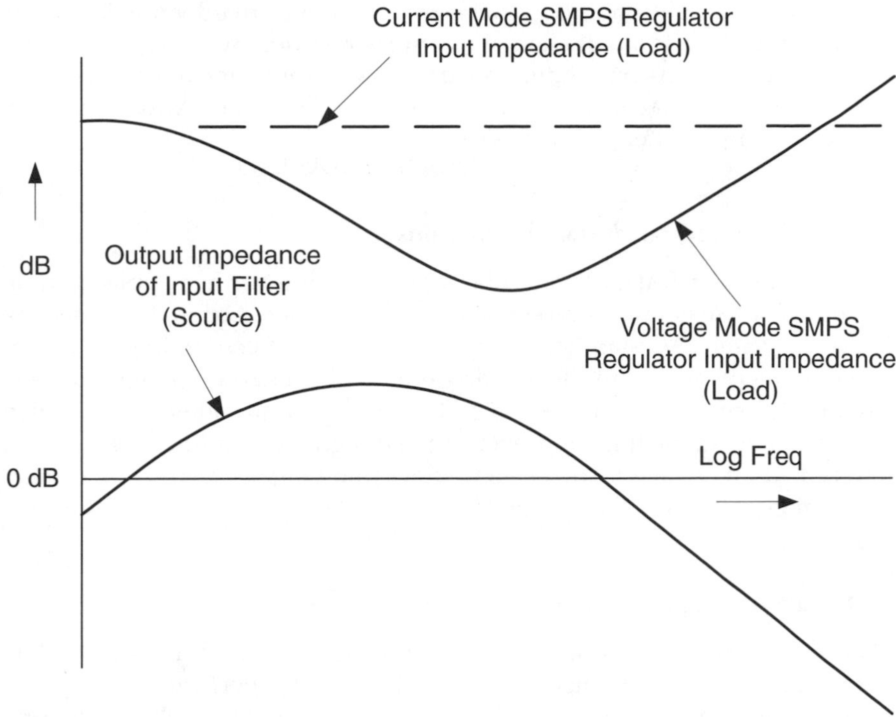

FIGURE 2.7
Typical SMPS source/load impedance comparison.

frequencies near the resonant frequencies of the input and output filters. The current mode converter input impedance, on the other hand, is generally considered constant up to the loop gain crossover frequency of the inner current loop.[12] Figure 2.7 shows representative SMPS regulator input impedance plots illustrating this comparison. With this relatively constant negative input impedance throughout the lower frequencies, the current mode converter input filter problem is generally an easier and more straightforward issue with which to deal.

The solution to this problem is to insert additional damping elements in the input filter; these alter the circuit to the extent that the negative resistance component of the resonant circuit is negated and the net resistance is positive. At the same time, it is vital that the necessary compromises for stability do not reduce the performance of the input filter to the point at which its requirements are not met. Middlebrook and Cuk[2] show a number of different damping circuit approaches that may be taken. The main generalized conclusion is that the parallel resonant peak output impedance of the input filter presented to the SMPS should be at least three times less than the input impedance of the SMPS. This not only ensures AC stability but also negates any potential adverse effects on SMPS performance. These negative effects are manifested in voltage mode converters as an effective loop gain reduction. This could cause a reduction in line noise rejection by the regulator and also a possible increase in output impedance. Multiple stage filters are sometimes necessary to meet the requirements. See Appendix A for a general treatment of input filter circuit design.

2.2.3 Source and Load Stability Margins

The concerns pertaining to this subject are simply a further application of the impedance-matching criteria noted in the previous section. It is essential for power system AC stability to analyze any system containing a number of SMPSs to ensure that these impedance-matching criteria are not violated. Systems that consist of a large central SMPS regulator and several smaller post switching regulators, and even a postmagnetic amplifier, need to be analyzed. It may take a lot of computer modeling to analyze a system containing several SMPSs accurately.

2.2.4 Electromagnetic Compatibility

When an accurate model of the power supply, along with its input filter, is created, the electromagnetic compatibility (EMC) performance aspects are easily analyzed. The general list of items relegated to this category comprises:

- *Conducted susceptibility.* This analysis is to ensure that specified levels of noise interference on the input power lines do not cause any appreciable degradation of SMPS performance. This response requirement is generally specified as an AC sine wave voltage added to the normal input power lines at various voltage amplitudes over various frequency bands. The frequency bands may start as low as 30 Hz and extend into the megahertz region. Also, these input AC noise sources may be differential or common mode. Various transient phenomena such as line spikes with specified pulse shapes and energy content also need to be analyzed. (Ott[15] provides very good general reference source materials for almost any EMC concern associated with electronic equipment.)
- *Conducted emissions.* This analysis is to ensure that an acceptable level of conducted current noise emanating from the SMPS and going back to the power source will not be exceeded. This current may be differential and/or common mode. Load current perturbations reflected back through the SMPS to the source are also to be considered. In many cases, a transient time domain analysis is necessary for these emission analyses.

In all probability, radiated susceptibility and radiated emission requirements will need to be met. The analysis required for this generally exists in the area of electromagnetic fields analysis. This is not specifically considered in this book, but might be considered if conducted effects are determined to result from the presence of these fields.

2.3 Transient Requirements

The transient performance requirements of an SMPS generally exist in the form of specifying a limit for output voltage variations with line voltage and/or load current perturbations. A circuit-averaged model or the actual time domain model may be necessary for these transient analyses. The examples of Chapter 4 and Chapter 6 provide considerable general treatment of these effects.

2.3.1 Load Transients

Load current transients are one of the most commonly specified stimuli for any power supply. Load current perturbations may occur in all sorts of ways, such as single-event load changes or steady-state periodic load changes (e.g., pulse trains and sinusoidal wave shapes, etc.). The single-event load changes are in general the turn-on (and turn-off) load surges of equipment powered by the power supply. The magnitude of some of these changes may be small enough that the SMPS stays within its existing operating mode, or it may be large enough to traverse the continuous–discontinuous mode condition boundary. Still other load changes may be large enough to send the power supply into a current limit mode. Output short-circuit conditions may need to be analyzed if this is a requirement. An accurate large signal model is required to analyze all of these various modes of operation.

2.3.2 Line Transients

Input line voltage transients are also very commonly specified stimuli for a power supply. These transients may be single-event voltage changes and also steady-state periodic pulse trains. The crossing of the SMPS continuous–discontinuous conduction mode boundary may occur for some transients. These transients may in many cases be specified as part of an EMC specification (see Section 2.2.4) or they may be specified in a separate listing of requirements. In any case, SMPS performance must be analyzed for all conditions.

2.3.3 Power-Up and Power-Down Transients

Power-up and power-down transients are considered a special case of input line voltage transients. The input voltage may be applied as a step function resulting from a relay or switch closure, with the possibility of a sag in input voltage caused by large SMPS inrush currents and finite power source impedances. This could cause problems with undervoltage lockout circuits, inrush current-limiting circuits, and even "housekeeping" power supplies. On the other hand, a slow rising ramp or some other voltage profile caused by a slow source power generator start-up could cause anomalous start-up SMPS problems, which may even be destructive. Many times, this type of problem is discovered and dealt with in the lab experimentally; however, an accurate SMPS model along with the computer analysis may help in providing valuable insight and a solution when it is difficult to make an assessment of the actual circuit operation. Sometimes, power-down can cause the same problems, in reverse, as those for power-up; anomalous destructive circuit operation occurs. (See Section 2.1.4 for more general information on a typical power-up sequence of events.)

2.3.4 Energy Storage for Line Dropouts

Line dropouts that require external components to be brought into play to maintain SMPS performance requirements are also considered a special case of transient operation. When these dropouts occur and continued output power to the loads is required, a back-up power source is necessary. This power source may be required to provide power for short transitory periods that are only a little longer than the energy storage capability of the SMPS filter capacitors, or the requirement may be for longer periods of time, which would suggest a near steady or continuous type of operation. The energy storage for shorter periods of time is generally provided by a charged-up capacitor bank, but for longer periods of time, a back-up battery may be necessary. These types of backup are generally referred to as uninterruptable power supplies (UPSs).

In most cases, when a battery back-up is used, a simple diode ORing of the two supplies is all that is required; little or no special analysis is needed for this case. When a charged-up capacitor bank is used for the shorter time period dropouts, the situation generally becomes a little more complex. The capacitor bank is charged to a higher voltage than that of the input voltage for adequate energy storage. When this is done, it becomes necessary to control or regulate the discharging of these capacitors onto the input power lines. To ensure adequate SMPS performance, an analysis of this condition is necessary. This capacitor bank scenario may be implemented on any or all of the output power lines as well.

2.4 Summary

This chapter has attempted to present most of the fundamental analysis tasks that may be required to verify the design integrity of a power supply. The requirements presented are by no means all inclusive; many other specialized requirements may need to be analyzed. In any case, it is the intent of this chapter to emphasize the importance of identifying all of the necessary analysis requirements for a particular power supply. The succeeding chapters will present the methods available to the analyst to facilitate a practical computer analysis of the power supply to verify that the requirements are met.

3

Fundamental Switch Mode Converter Model Development

This chapter develops the fundamental models necessary to analyze the essential requirements of switch mode power supplies (SMPSs) as noted in Chapter 2. The large signal continuous and discontinuous mode models for the buck and the boost converters are developed. Then the continuous and discontinuous mode models are integrated into a single unified model for each of the buck and boost converters. The unified buck–boost (conventional flyback converter) and Cuk converter models are next developed. After the development of these models, an example analysis is provided to illustrate the ease and simplicity that this approach affords in analyzing the performance of SMPSs.

3.1 Buck and Boost Converter Continuous Mode Large Signal Models

When a switch mode power converter is modeled, an electrical circuit equivalent model of the duty ratio controller must be created when the converter operates in the continuous mode. From Middlebrook and Cuk,[1] an ideal (AC and DC) transformer equivalent model is conceived and shown in Figure 3.1. The flyback and Cuk converter continuous mode models are in essence cascaded versions of the buck and boost converters and will be dealt with later.

Converting the ideal transformers to a system of dependent generators makes the models more adaptable to most circuit simulation software. Figure 3.2 shows a SPICE model equivalent. The power converter large signal continuous mode models are thus implemented in Figure 3.3. See Cuk and Middlebrook[4] for more information.

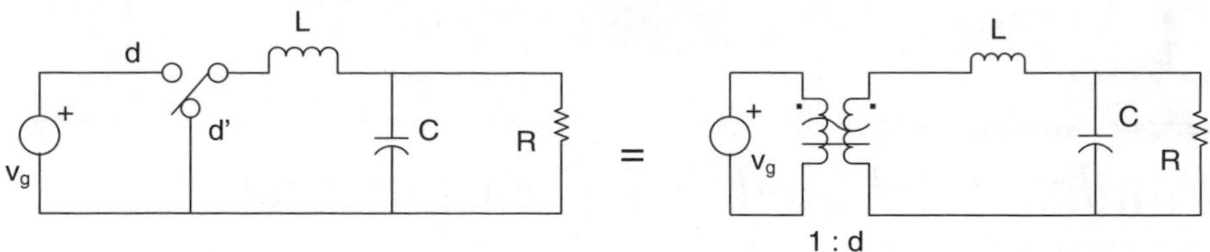

a) Buck Converter

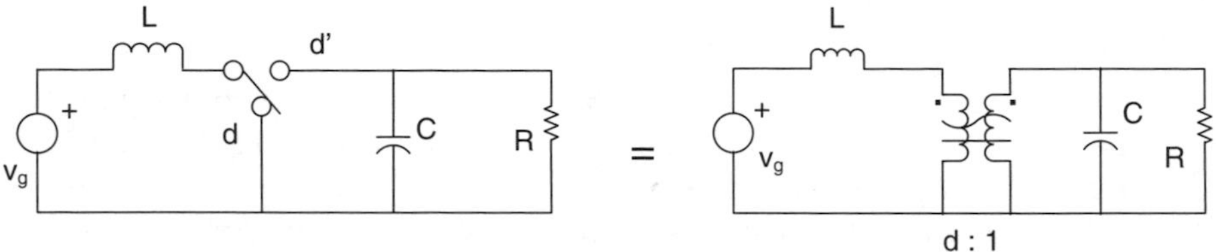

b) Boost Converter

FIGURE 3.1
Duty ratio controller ideal transformer.

3.2 Buck and Boost Converter Discontinuous Mode Large Signal Models

Now consider the discontinuous mode of operation. From Figure 3.4, the average discontinuous mode inductor current, i_{LD}, is expressed by:

$$i_{LD} = \frac{I_P}{2}(d + d_2) \tag{3.1}$$

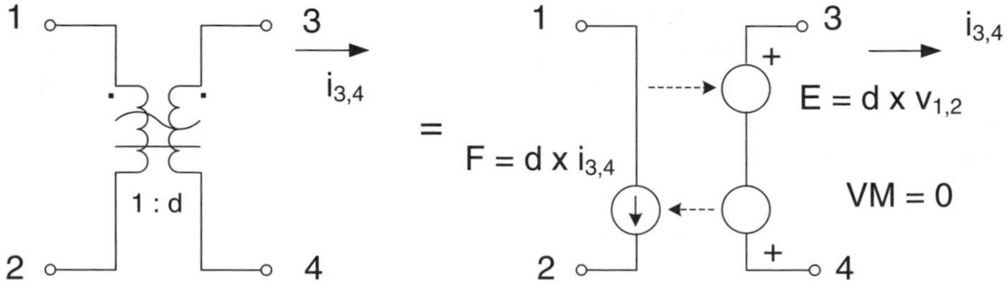

FIGURE 3.2
Ideal transformer equivalent model.

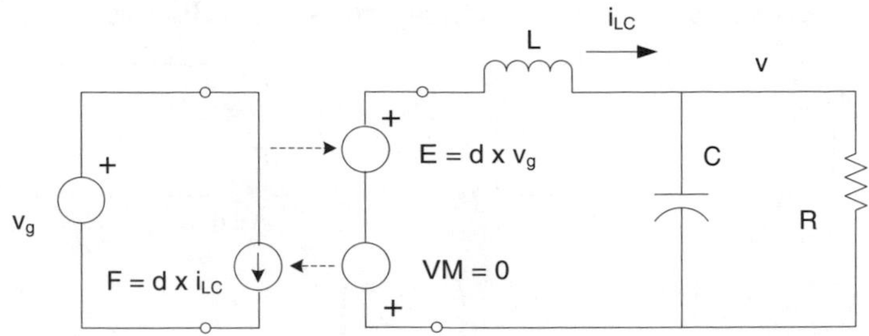

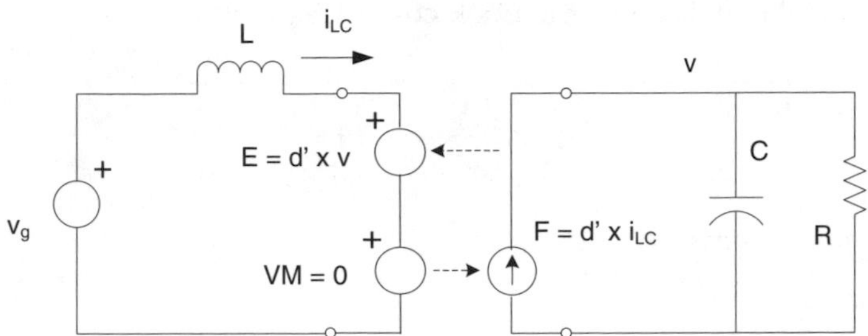

FIGURE 3.3
Power converter continuous mode models.

For the buck converter,

$$I_P = \frac{(v_g - v)dT_S}{L} \tag{3.2}$$

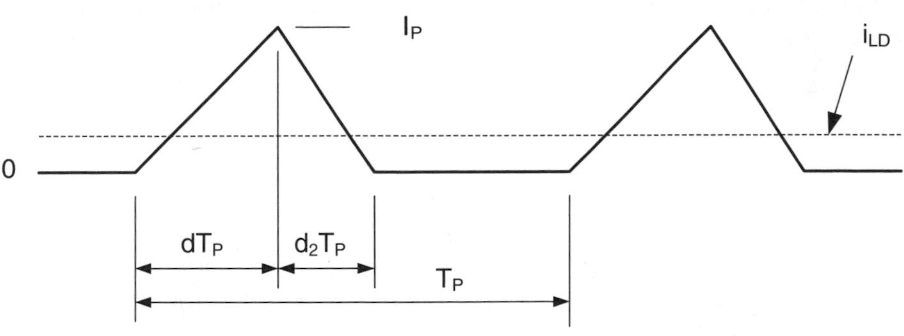

FIGURE 3.4
Discontinuous mode inductor current.

Also, for periodic volt-second balance across the inductor,

$$d(v_g - v) = d_2 v$$

or

$$d_2 = d\left(\frac{v_g - v}{v}\right) \tag{3.3}$$

Substituting Equation 3.2 and Equation 3.3 into Equation 3.1 yields the desired control equation for the buck converter, i_{LD}:

$$i_{LD} = d^2 \frac{v_g}{v}(v_g - v)\frac{T_s}{2L} \tag{3.4}$$

For the boost converter,

$$I_P = \frac{v_g d T_s}{L} \tag{3.5}$$

and for volt-second balance

$$d v_g = d_2(v - v_g)$$

$$d_2 = d\left(\frac{v_g}{v - v_g}\right) \tag{3.6}$$

Substituting Equation 3.5 and Equation 3.6 into Equation 3.1 yields the control equation for the boost converter, i_{LD}:

$$i_{LD} = d^2\left(\frac{v v_g}{v - v_g}\right)\frac{T_S}{2L} \tag{3.7}$$

The next step is to develop a large signal discontinuous mode model from the previous equations. Figure 3.5 shows buck and boost topologies.

The terms i_{LID} and i_{LOD} are as yet undefined. With the inductor current starting and returning to zero during each cycle, there is no cyclical energy storage in the inductor. Its properties as an inductive circuit element are nonexistent at the lower frequencies; therefore it does not appear in the models.

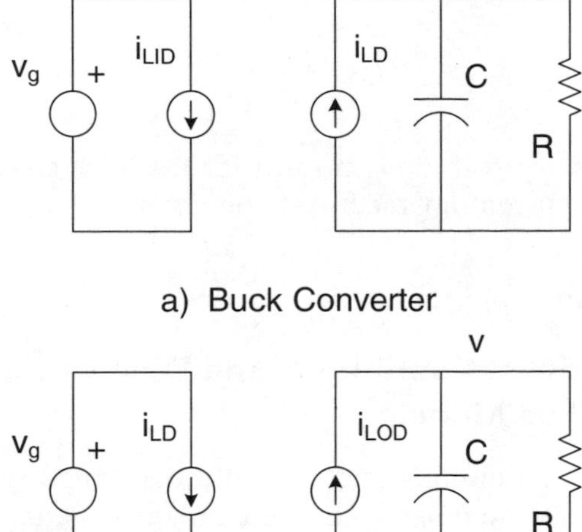

FIGURE 3.5
Discontinuous mode large signal models.

With this knowledge, the instantaneous or cyclical power into the converter is equal to the instantaneous power out.

$$p_{in} = p_{out} \tag{3.8}$$

For the buck converter,

$$p_{in} = v_g i_{LID} \tag{3.9}$$

$$p_{out} = v i_{LD} \tag{3.10}$$

and

$$i_{LID} = \frac{v}{v_g} i_{LD} \tag{3.11}$$

For the boost converter,

$$p_{in} = v_g i_{LD} \tag{3.12}$$

$$p_{out} = v i_{LOD} \tag{3.13}$$

and

$$i_{LOD} = \frac{v_g}{v} i_{LD} \tag{3.14}$$

Current i_{LID} is the correct input current for the buck converter and i_{LOD} is the correct output current for the boost converter.

3.3 Buck and Boost Continuous and Discontinuous Mode Unified Models

Now combine the continuous and discontinuous mode models into one encompassing unified model, which will emulate an actual converter going from one conduction mode to the other as the critical inductor current boundary is traversed. The schemes depicted in Figure 3.6 and Figure 3.7 provide a simple and straightforward way of implementing this model.

A detailed explanation of this topology is now in order. For the moment, ignore the center components of Figure 3.6 consisting of VM2, D1, D2, i_{LID}, and F1. The discontinuous mode current, i_{LD}, is calculated for all prevailing conditions of d, v_g, and v according to Equation 3.4. If the voltage, E, is sufficiently high to produce an inductor current larger than the value of i_{LD}, diode D3 will then conduct. (Diode D3 is modeled as an ideal diode with a forward voltage drop near 0 V. Section 3.10 discusses this further.) The converter is thus in the *continuous* mode with $i_L > i_{LD}$. D3 basically shorts out current source i_{LD}, thus making it appear as though it is not in the circuit.

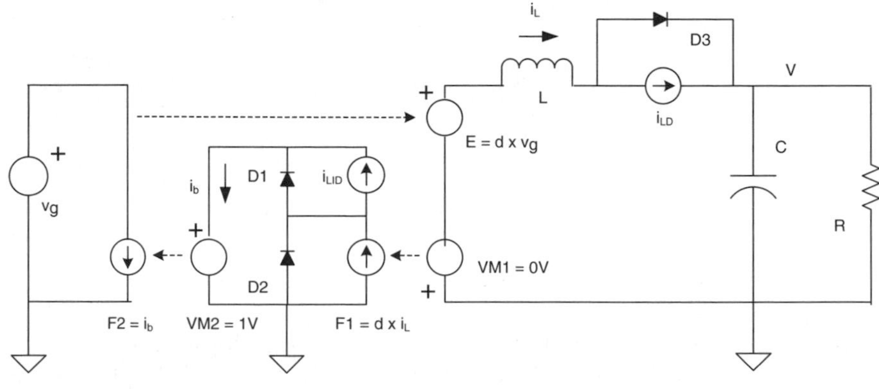

FIGURE 3.6
Buck converter combined continuous and discontinuous mode model.

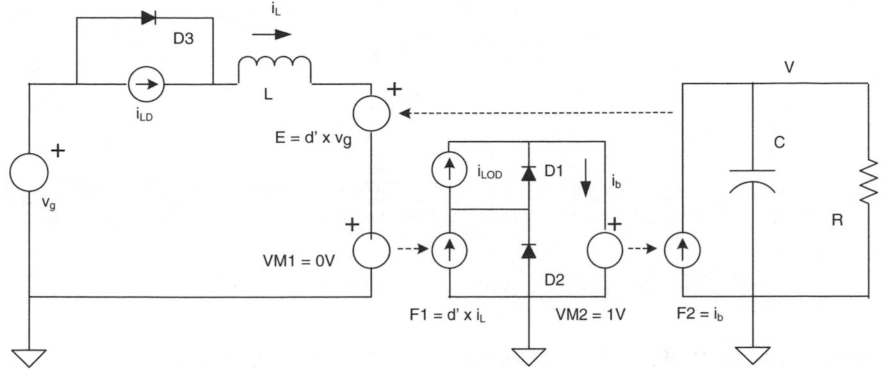

FIGURE 3.7
Boost converter combined continuous and discontinuous mode model.

This is appropriate because i_{LD} is a nonexistent fictitious quantity for i_L greater than any computed quantity greater than i_{LD}. If circuit conditions change to lower i_L so that i_L wants to be less than i_{LD}, D3 will be reversed biased and the actual inductor current will be i_{LD}. The converter is now in the discontinuous mode with $i_L = i_{LD}$.

Now consider the role of the components VM2, D1, D2, i_{LID}, and F1 that were ignored earlier. The circuit made from these components is necessary to provide the proper input current that is seen as a load on the power source v_g. The circuit is basically a current ORing circuit, with current i_b equal to the larger of the two currents F1 or i_{LID}. For example, if current F1 is larger than i_{LID}, nonideal diode D1 will conduct, effectively shorting out i_{LID}, and D2 is reversed biased with i_b = F1. This is the continuous mode case with F2 = i_b = F1 = $d \times i_L$, which is the correct converter input current in the continuous mode. If current F1 decreases below the computed discontinuous mode current, i_{LID}, then i_b will be equal to i_{LID}, which is the correct discontinuous mode current.

Voltage source VM2 is shown in Figure 3.6 as 1 V, but may be any arbitrary positive value. When a value greater than zero is used for VM2, the voltage at the cathode side of diode D2 will have basically one of two values, depending on the converter conduction mode. This circuit may now also be viewed as a conduction mode detector circuit when monitoring the voltage at the D2 cathode.

Thus, the complete unified continuous and discontinuous mode model for the buck converter is shown in Figure 3.6. It can be recognized from Figure 3.3 and Figure 3.5 that the buck and boost topologies are in essence the duals of each other; therefore, it can be easily deduced that the model shown in Figure 3.7 is the correct unified model for the boost topology by applying the corresponding line of development as for the buck converter.

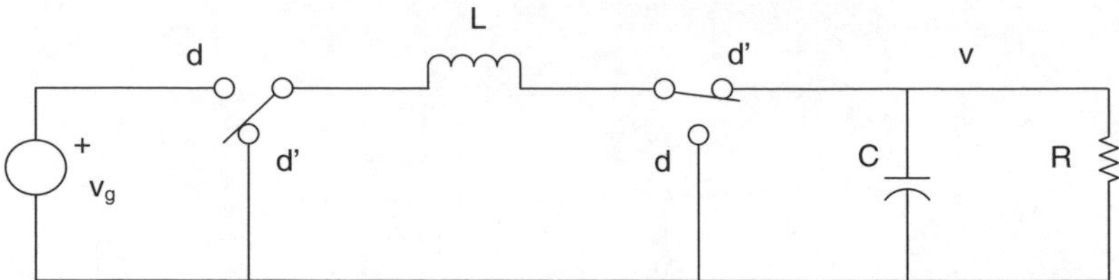

FIGURE 3.8
Basic buck–boost (flyback) converter continuous mode topology.

3.4 Buck–Boost (Flyback) Converter Continuous Mode Large Signal Model

The buck–boost converter is in essence a cascaded combination of the buck and boost converters (Figure 3.1a and Figure 3.1b, respectively). Figure 3.8 shows an equivalent representation of this combination. (See Cuk and Middlebrook[4] for more detail on this.) Replacing the duty ratio controlling switches with their equivalent circuit models as developed in Section 3.1, the equivalent continuous mode model is shown in Figure 3.9.

3.5 Buck–Boost (Flyback) Converter Discontinuous Mode Large Signal Model

From the inductor current waveform of Figure 3.4, Equation 3.1, and the following equations, the average discontinuous mode current, i_{LD}, is determined from Figure 3.8.

$$I_P = \frac{v_g dT_s}{L} \tag{3.15}$$

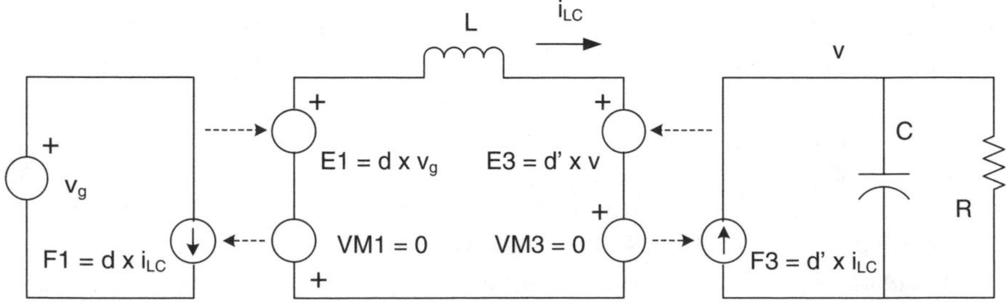

FIGURE 3.9
Buck–boost (flyback) continuous mode model.

Also,

$$dv_g = d_2 v$$

or

$$d_2 = d\frac{v_g}{v} \tag{3.16}$$

Substituting Equation 3.15 and Equation 3.16 into Equation 3.1 yields the desired control equation for the flyback converter:

$$i_{LD} = d^2 v_g \left(1 + \frac{v_g}{v}\right)\frac{T_s}{2L} \tag{3.17}$$

When the flyback converter is modeled only in the discontinuous mode, the average inductor current, i_{LD}, as indicated in Equation 3.17 is not essential, as shown in Figure 3.10. However, its necessity will be noted when developing the unified model. The equations for i_{LID} and i_{LOD} are easily noted from Figure 3.4 and Equation 3.15 and Equation 3.16.

$$i_{LID} = \frac{I_P}{2} d$$

$$i_{LID} = d^2 v_g \frac{T_s}{2L} \tag{3.18}$$

and

$$i_{LOD} = \frac{I_P}{2} d_2$$

$$i_{LOD} = d^2\left(\frac{v_g^2}{v}\right)\frac{T_S}{2L} \tag{3.19}$$

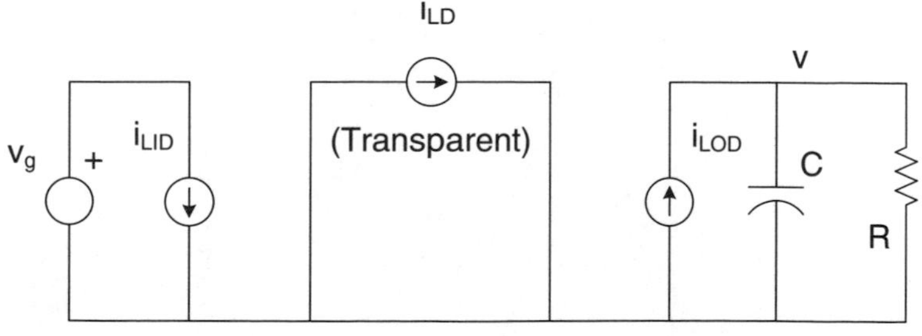

FIGURE 3.10
Flyback discontinuous mode model.

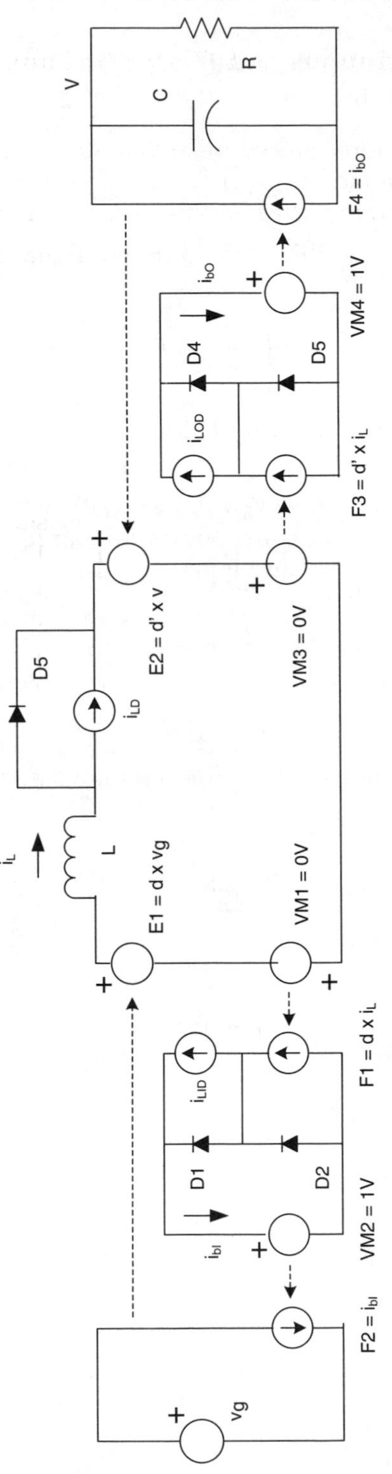

FIGURE 3.11
Flyback converter continuous and discontinuous mode model.

3.6 Buck–Boost Continuous and Discontinuous Mode Unified Model

Following the same unification procedure as that described in Section 3.3, the basic flyback unified model is shown in Figure 3.11. Note the necessity of including i_{LD} here, although it was unnecessary when the discontinuous mode model was considered alone.

3.7 Cuk Converter Continuous Mode Large Signal Model

The continuous mode Cuk converter is in essence a cascaded combination of the boost and buck converters shown in Figure 3.1b and Figure 3.1a, respectively. This is true for the continuous mode, but not necessarily true for the discontinuous mode. Figure 3.12 shows the equivalent representation of this combination. (See Cuk and Middlebrook[4] for more details on this.) Shown in Figure 3.12 is the noninverting equivalent of the classical Cuk converter (Figure 1.8).

It is interesting to note from Cuk[6] that the inductor currents i_{L1} and i_{L2} are simultaneously both continuous or both discontinuous and that in the general case these currents do not individually become zero for the discontinuous part of their cycles, but rather the sum of i_{L1} and i_{L2} becomes zero at this point of discontinuity (see Figure 3.14). Replacing the duty ratio controllers with their equivalent circuit models as developed in Section 3.1, the equivalent continuous mode model is shown in Figure 3.13. The output part of the model may be inverted if desired to reflect the more classical negative output.

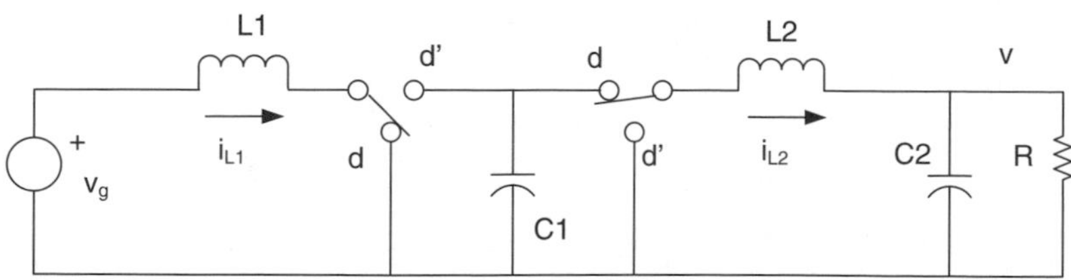

FIGURE 3.12
Basic boost–buck (Cuk) continuous mode topology.

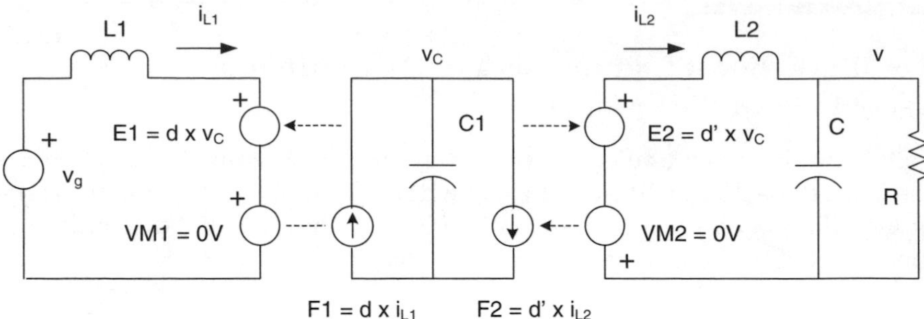

FIGURE 3.13
Boost–buck (Cuk) continuous mode topology.

3.8 Cuk Converter Discontinuous Mode Large Signal Model

From the inductor current waveforms of Figure 3.14 and the following equations, the average discontinuous mode inductor currents i_{L1D} and i_{L2D} are determined from the circuit of Figure 3.12:

$$i_{L1D} = \frac{I_{P1}}{2}(d + d_2) + i \tag{3.20}$$

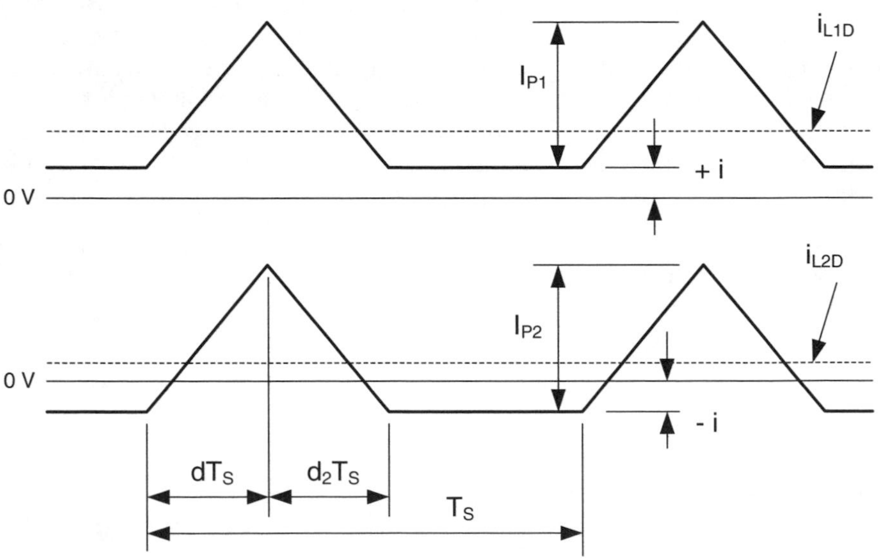

FIGURE 3.14
Cuk converter discontinuous mode inductor currents.

$$i_{L2D} = \frac{I_{P2}}{2}(d + d_2) - i \tag{3.21}$$

where

$$I_{P1} = \frac{v_g d T_s}{L_1} \tag{3.22}$$

and

$$I_{P2} = \frac{v d_2 T_s}{L_2} \tag{3.23}$$

For cyclical volt-second balance across L_1 and L_2,

$$d v_g = d_2(v_c - v_g) \tag{3.24}$$

and

$$d_2 v = d(v_c - v) \tag{3.25}$$

Eliminating v_c from Equation 3.24 and Equation 3.25 yields

$$d_2 = d\frac{v_g}{v} \tag{3.26}$$

Now, combining Equation 3.20 through Equation 3.26 yields

$$i_{L1D} = v_g d^2\left(1 + \frac{v_g}{v}\right)\frac{T_s}{2L_1} + i \tag{3.27}$$

and

$$i_{L2D} = v_g d^2\left(1 + \frac{v_g}{v}\right)\frac{T_s}{2L_2} - i \tag{3.28}$$

From Cuk and Middlebrook,[3]

$$i = i_{L2D}\frac{\frac{v_g}{v}L_1 - L_2}{L_1 + L_2} \tag{3.29}$$

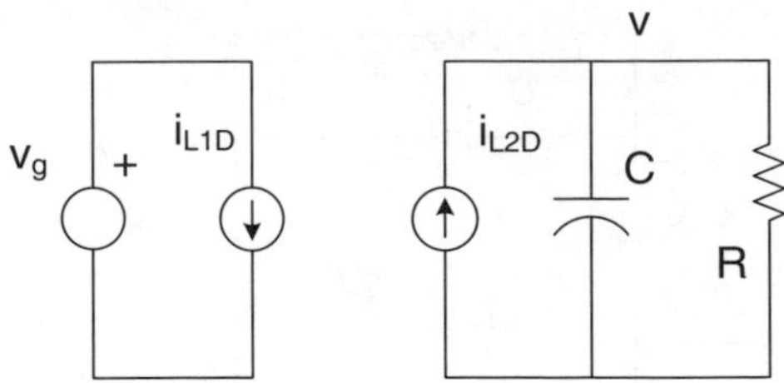

FIGURE 3.15
Cuk converter (noninverting) discontinuous mode large signal model.

and with cyclical input power equal to cyclical output power,

$$i_{L2D} = i_{L1D} \frac{v_g}{v} \tag{3.30}$$

and

$$i = i_{L1D} \frac{L_1 - \frac{v_g}{v} L_2}{L_1 + L_2} \tag{3.31}$$

Substituting Equation 3.31 into Equation 3.25 and solving for i_{L1D} yields

$$i_{L1D} = v_g d^2 \frac{T_s}{2L_e} \tag{3.32}$$

where

$$L_e = \frac{L_1 L_2}{L_1 + L_2}$$

The discontinuous mode noninverting Cuk converter model is then simply shown in Figure 3.15 with i_{L1D} indicated by Equation 3.32 and i_{L2D} indicated by Equation 3.30.

3.9 Cuk Converter Continuous and Discontinuous Mode Unified Model

Again, following the same unification procedure as described in Section 3.3, the basic Cuk converter unified model is shown in Figure 3.16. Note that the output stage may be inverted for a positive or a negative output.

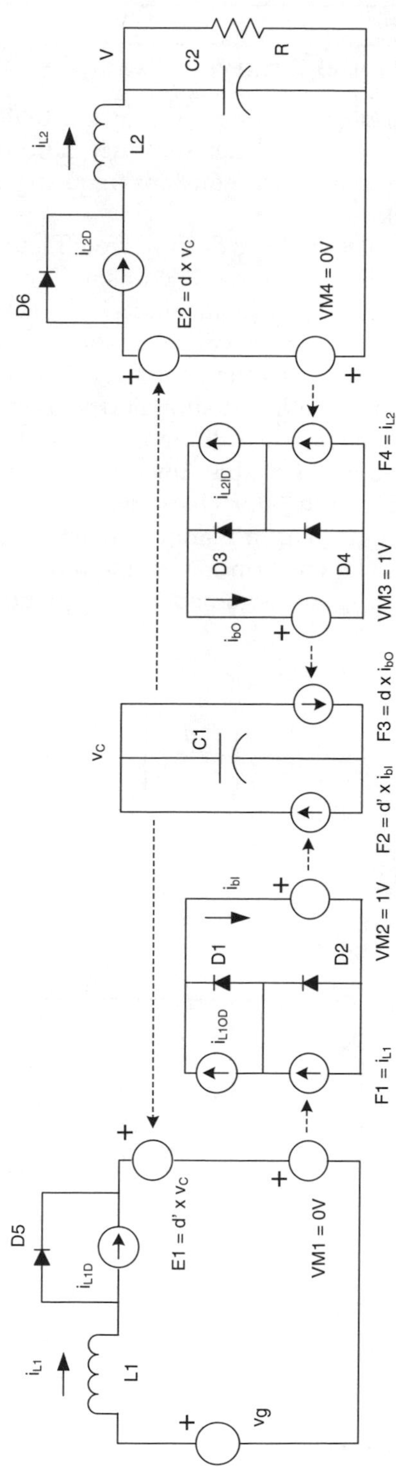

FIGURE 3.16
Cuk converter continuous and discontinuous mode model.

3.10 Boost Converter Model Analysis Example

Now create a model of a boost converter as an example to illustrate the ease and simplicity of the technique. Consider the duty ratio controlled boost converter of Figure 3.17 operating in the constant frequency mode of 75 kHz (a period of $T_s = 13.33$ μsec).

An equivalent unified PSPICE model is derived from Figure 3.7 and shown in Figure 3.18. The diode, DIDEAL, is modeled as an ideal diode with the characteristics shown in Figure 3.19. This ideal diode model uses $V_F = -1$ μV as opposed to a desired value of zero because, in some applications, a zero value could be divided into other factors, thus causing the simulation to crash. This model does not allow this situation and V_F is small enough that the diode is considered ideal. The ideal diode is here simply generated by a dependent voltage generator and a table function. (See the circuit netlists in Figure 3.22 and Figure 3.23.) In later chapters, netlists will show other ways of creating this ideal diode. If a simulation convergence problem exists, trying a different type of model creation may sometimes help.

The equations for i_{LD} and i_{LOD} are obtained from Equation 3.7 and Equation 3.14, respectively.

$$i_{LD} = 0.017d^2 \left(\frac{v}{\frac{v}{v_g} - 1} \right) \tag{3.33}$$

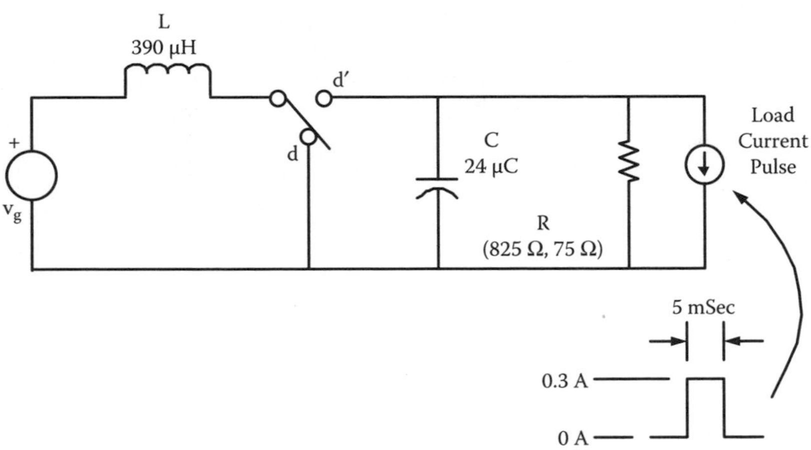

FIGURE 3.17
Duty ratio controlled boost converter example.

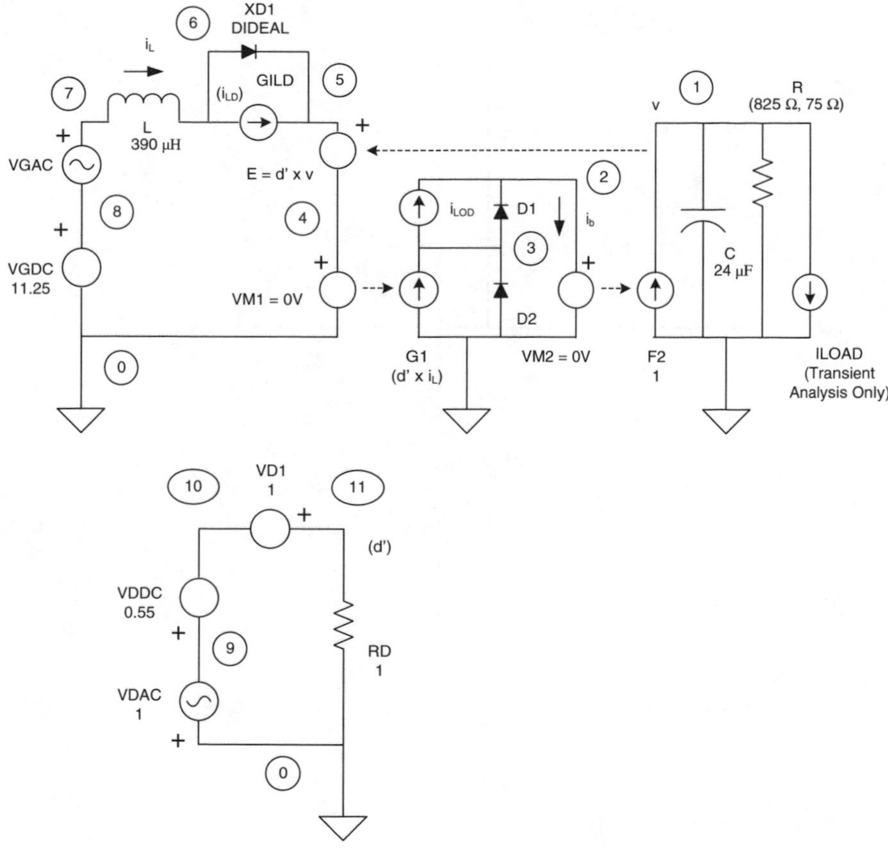

FIGURE 3.18
Boost converter PSPICE unified model.

and

$$i_{LOD} = \frac{v_g}{v} i_{LD} \tag{3.34}$$

Care must be taken in this simulation that v does not drop below v_g because a division by zero may result in Equation 3.33 and cause the simulation to crash. Although not used here, a LIMIT function may be used in the model equation for GILD to prevent this if convergence trouble is encountered. (See the netlists in Figure 3.22 and Figure 3.23).

The steady state critical inductor current (average inductor current at the boundary between the continuous and discontinuous modes of operation)

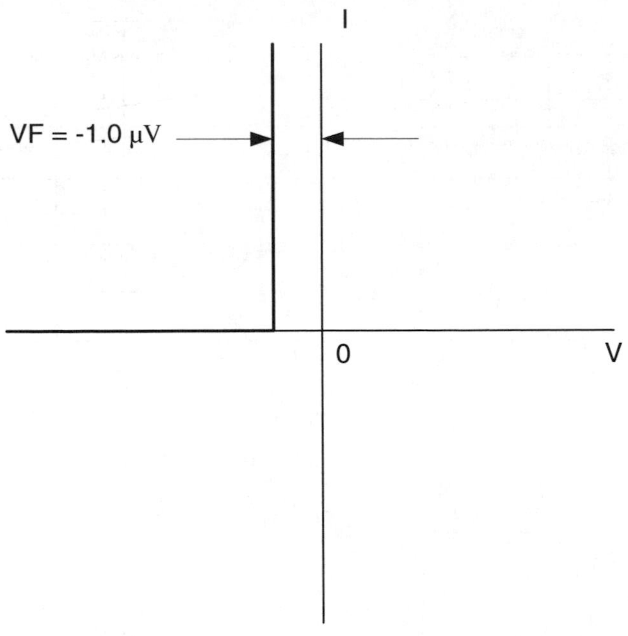

FIGURE 3.19
"Ideal" diode characteristics.

for this example occurs at

$$I_{CRIT} = \frac{V_g D T_s}{2L} = \frac{(11.25V)(0.55)(13.33\mu S)}{2(390\mu H)} = 0.106A \tag{3.35}$$

or

$$R_{CRIT} = \frac{V}{D' I_{CRIT}} = \frac{25V}{(0.45)(0.106A)} = 524\Omega \tag{3.36}$$

Three tests will be conducted on the model to reveal some of its characteristics. The first will show the AC response from the duty controller, *d*, to the output voltage, *v*, with the converter operating in the continuous mode ($R < R_{CRIT}$) for $R = 75\ \Omega$. Then, for the second test, the load resistor will be increased from 75 to 825 Ω, placing the converter in the deep discontinuous mode and examining the AC response under that condition. The third test will be to examine the effect on output voltage of a transient load current

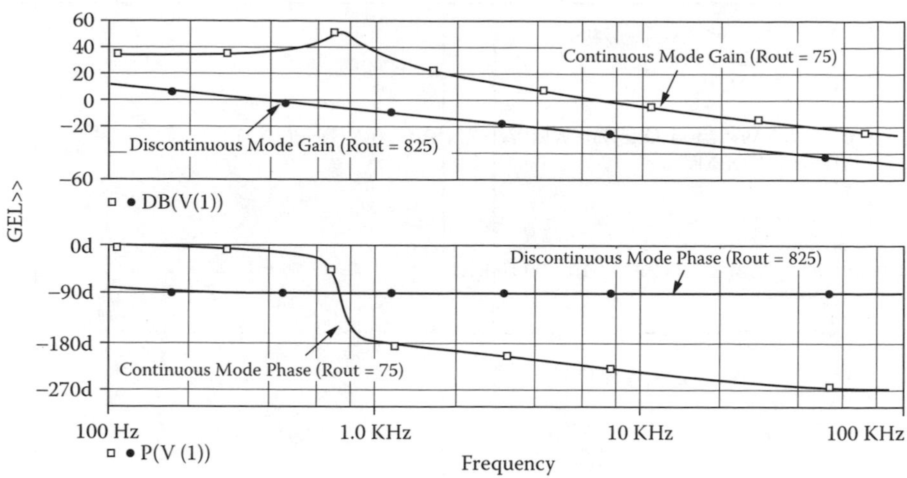

FIGURE 3.20
Boost converter AC analysis example.

pulse on the converter forcing it to move from the discontinuous mode ($R = 825\ \Omega$) to the continuous mode and back again with a load current pulse of 0.3 A lasting for 5 msec.

The results of these tests are shown in Figure 3.20 and Figure 3.21. Note the characteristic right-half plane zero for the continuous mode in Figure 3.20. Note that the discontinuous mode AC response of Figure 3.20 is of considerably lower bandwidth with a single pole roll-off and also the absence of

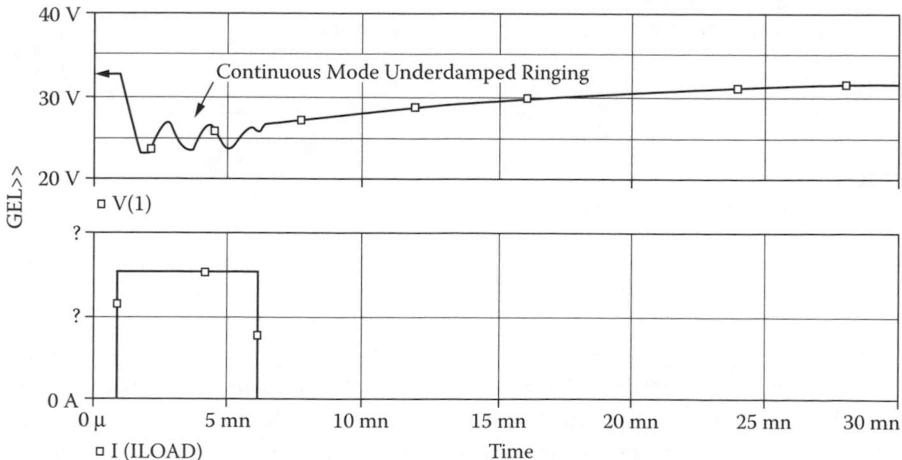

FIGURE 3.21
Boost converter load transient analysis example.

```
Boost Converter AC Analysis
VGDC 8 0 DC 11.25
VGAC 7 8 AC 0
L 7 6 390U
GILD 6 5 VALUE = {(0.017)*PWR(-V(10),2)*(V(1)/((V(1)/V(7))-1))}
XD1 6 5 DIDEAL
*
*Ideal Diode Model
.SUBCKT DIDEAL 1 2
EID 3 1 TABLE [V(3,2)] = (-1,1U)(0,1U)(1,1)
DIO 3 2 D
.ENDS
*
E 5 4 VALUE = {V(11)*V(1)}
VM1 4 0 DC 0
G1 0 3 VALUE = {V(11)*I(VM1)}
GILOD 3 2 VALUE = {V(7)/V(I)*I(VMI)}
D1 0 3 D
D2 3 2 D
VM2 2 0 DC 0
F2 0 1 VM2 1
C 1 0 24U
*
.PARAM RVAL = 1
ROUT 1 0 {RVAL}
.STEP PARAM RVAL 75 825 750
*
VD1 11 10 DC 1
VDDC 9 10 DC .55
VDAC 0 9 AC 1
RD 11 0 1K
*
.MODEL D D IS = 1N
.AC DEC 40 100 100K
.PROBE
.END
```

FIGURE 3.22
Boost converter AC analysis example netist.

the right-half plane zero. Now observe the results of the load transient test in Figure 3.21. Note that the predictable large signal result of the output voltage, v, varies considerably when the inductor current is less than its critical value of 0.106 A. Also note the expected underdamped ringing around the continuous mode output voltage value of 25 V.

Numerous other tests — possibly more practical ones — could be easily devised and conducted on the model to examine easily any large signal transient or small AC characteristic desired. For reference, the PSPICE™ netlists for the AC and transient analysis are shown in Figure 3.22 and Figure 3.23, respectively.

```
Boost Converter Load Transient Analysis
VGDC 8 0 DC 11.25
VGAC 7 8 AC 0
L 7 6 390U
GILD 6 5 VALUE = {(0.017)*PWR(-V(10),2)*(V(1)/((V(1)V(7))-1))}
XD1 6 5 DIDEAL
*
*Ideal Diode Model
.SUBCKT DIDEAL 1 2
EID 3 1 TABLE {V(3,2)} = (-1,1U)(0,1U)(1,1)
DIO 3 2 D
.ENDS
*
E 5 4 VALUE = {V(11)*V(1)}
VM1 4 0 DC 0
G1 0 3 VALUE = {V(11)*I(VM1)}
GILOD 3 2 VALUE = {V(7)/V(1)*I(VM1)}
D1 0 3 D
D2 3 2 D
VM2 2 0 DC 0
F2 0 1 VM2 1
C 1 0 24U
ROUT 1 0 825
ILOAD 1 12 PULSE(0.3 1M 10U 10U 5M)
VM3 12 0
VD 1 11 10 DC 1
VDDC 9 10 DC .55
VDAC 0 9 AC 1
RD 11 0 1K
*
.MODEL D D IS = 1N
.TRAN 10U 30M
.PROBE
.END
```

FIGURE 3.23
Boost converter load transient analysis example netist.

3.11 Summary

This chapter has developed the fundamental conceptual computer models for the four major pulse width modulated (PWM) power converter topologies. Many unique converter topologies exist, but in most instances they can be reduced to one in these four major categories. These models accept input power, v_g, and PWM control, d, to provide output voltage, v, with the controlling variable, d, controlled directly by some control signal. The converters, as depicted, are commonly known as "voltage mode" or "duty ratio-controlled" converters. These models may be used directly as shown in this chapter to obtain first-order analysis results for single output voltage SMPSs. In subsequent chapters, these models will be embedded within other control schemes to allow for current mode control analysis along with more practical expansions. These include multiple outputs and macromodels of commercially available PWM integrated circuit controllers.

4

Analyzing the Fundamental SMPS Model

Chapter 3 developed the basic switch mode power converter circuit-averaged models that could easily be adapted for use in a switch mode power supply (SMPS) computer simulation circuit analysis. These models provide the capability of analyzing the large signal (DC and transient) and small signal (AC) properties of a particular power converter. At the next higher level, these converter models can be enfolded into a total SMPS model and an analysis of the total power supply performance can make. In this chapter, only the most basic or first-order aspects of the SMPS will be considered by analyzing some of its fundamental DC and AC characteristics. Chapter 6 will continue with the general theme of this chapter and provide the more advanced or comprehensive detailed analysis methods necessary when a higher level of complexity is involved.

4.1 Buck Converter SMPS Analysis (Voltage Mode)

Consider the switching regulator (SMPS) depicted in Figure 4.1. This is a very elementary voltage regulator with a single output voltage and no transformer isolation separating the input from the output. Despite its simplicity, many fundamental analysis concepts with a broad range of applications can be learned by conducting a thorough first-order analysis on such a regulator. A first-order computer model will be set up for the voltage mode switching regulator and some of its most basic properties will be observed as it is analyzed.

4.1.1 Voltage Mode Converter Model Setup

A power converter model of the type shown in Figure 3.6 is determined to be of the type required for this voltage mode application. First, a feedback control circuit consisting of the error amplifier, voltage reference, and pulse width modulation block is added. The pulse width modulator is a circuit-averaged block appearing simply as a linear voltage-to-duty ratio, d, converter. This d is the control input to the power converter. For the purposes of this exercise, the control circuit blocks are all simplified functional blocks

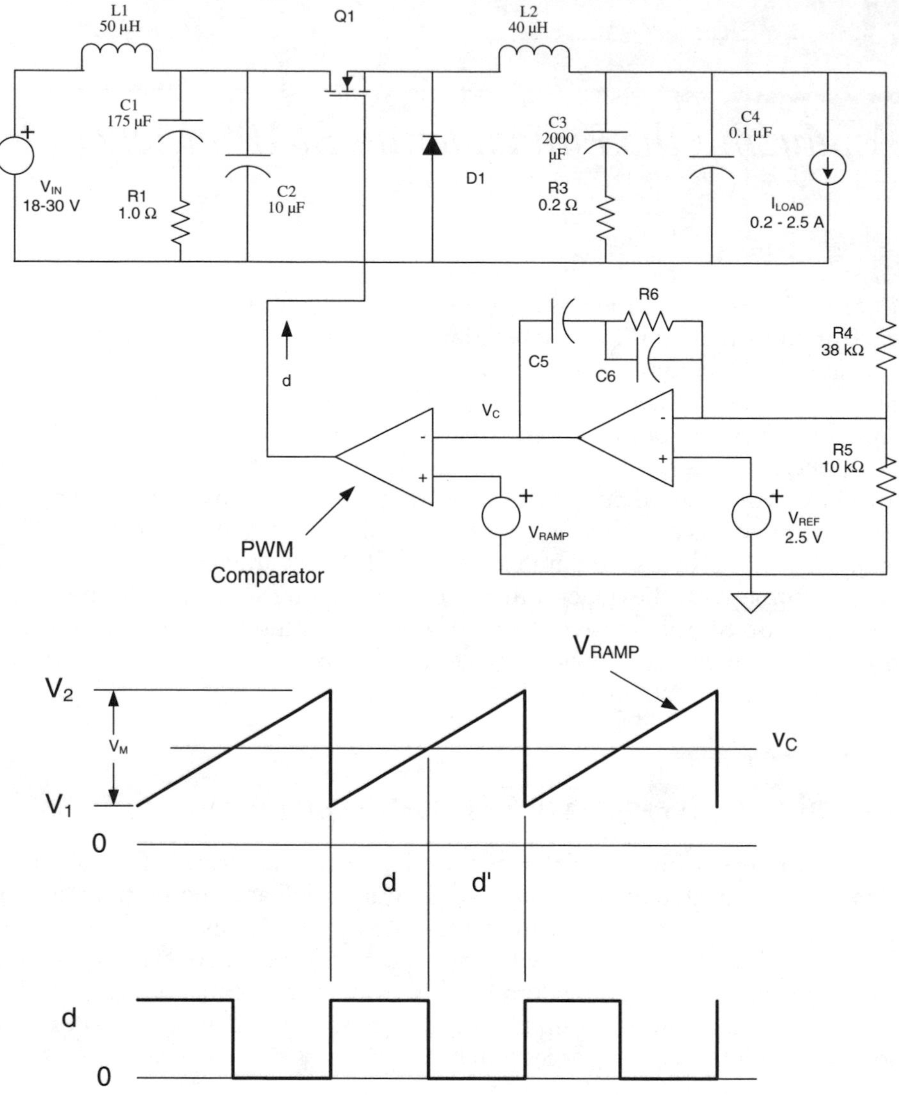

FIGURE 4.1
Simple buck voltage mode switching regulator (conceptual).

with near ideal characteristics. The converter model to be analyzed is of the type shown in Figure 4.2.

4.1.2 Open Loop Analysis (Continuous and Discontinuous Modes)

As a point of departure, the power supply will be analyzed initially by looking at the converter open loop characteristics. Then the feedback control circuit

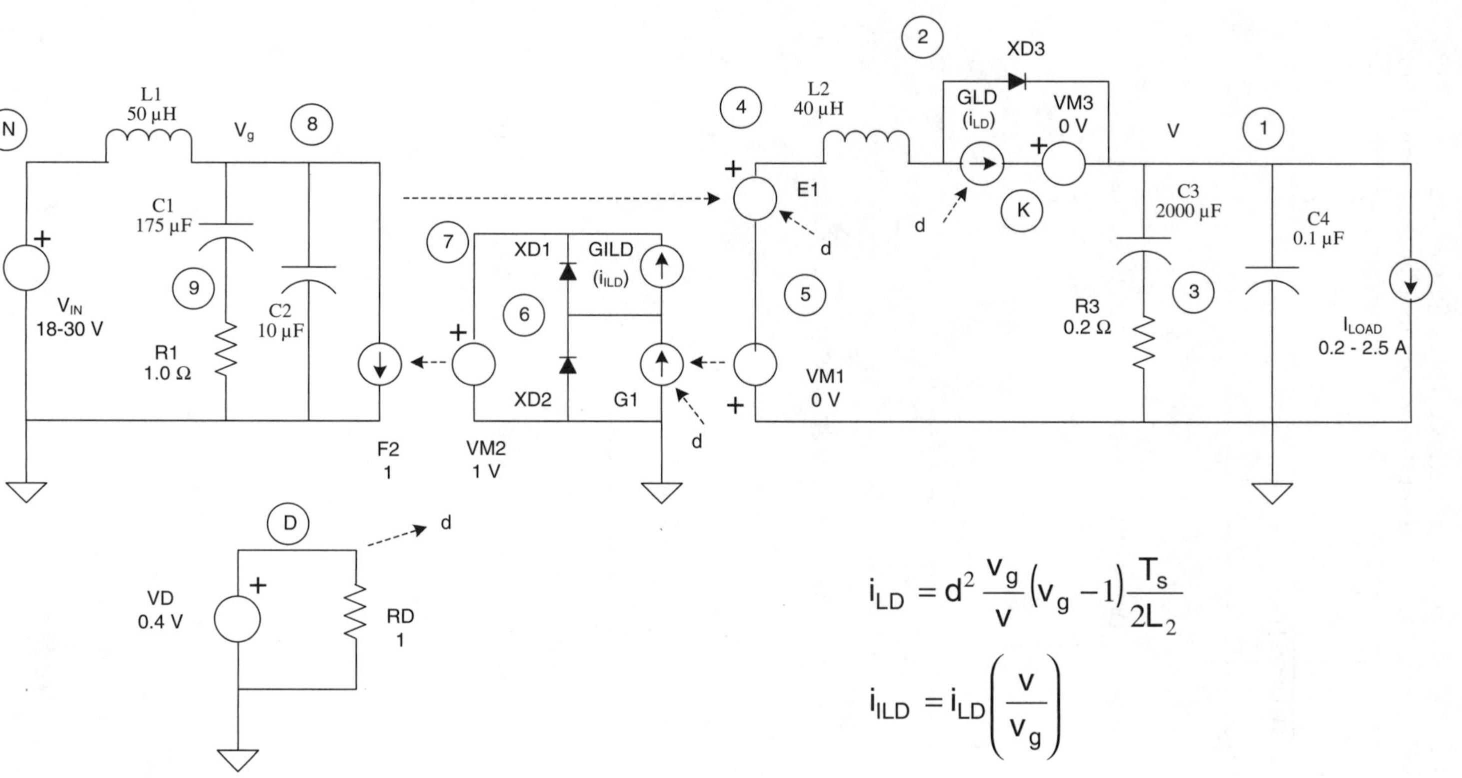

FIGURE 4.2
Buck converter model example.

will be added and the final performance characteristics noted. A PSPICE netlist, labeled Netlist 4.1, is used for these open loop analysis simulations.

NETLIST 4.1

```
BUCK CONVERTER OPEN LOOP ANALYSIS
*
**************************************************************
*
** SOURCE-LOAD CONFIGURATION FOR
** FIGURE 4.3
** BUCK CONVERTER DC FORWARD TRANSFER FUNCTION WITH D = 0.4
*
VIN N 0 DC 30
ILOAD 1 0 DC 2
.DC ILOAD 0 2.5 .01 VIN 20 30 5
VD D 0 DC .4 AC 0
*
**************************************************************
*
** SOURCE-LOAD CONFIGURATION FOR
** FIGURE 4.4 AND 4.5
** BUCK CONVERTER AC FORWARD TRANSFER FUNCTION
*
*VIN N 0 DC 30 AC 1
*ILOAD 1 0 DC 2
*.STEP ILOAD LIST .2 .7 1.2 2.5
*VD D 0 DC .4 AC 0
*.AC DEC 20 1 100K
*
**************************************************************
*
** SOURCE-LOAD CONFIGURATION FOR
** FIGURE 4.6
** BUCK CONVERTER OUTPUT IMPEDANCE (VOLTAGE AND CURRENT MODE)
*
*VIN N 0 DC 30
*ILOAD 1 0 DC 2
*.STEP ILOAD LIST .2 .7 1.2 2.5
*VD D 0 DC .4 AC 0
*IAC 1 0 AC 1
*.AC DEC 20 1 100K
*
**************************************************************
*
** SOURCE-LOAD CONFIGURATION FOR
** FIGURE 4.7
** BUCK CONVERTER DC "CONTROL "D" TO OUTPUT" TRANSFER FUNCTION
*
*VIN N 0 DC 30 AC 0
*ILOAD 1 0 DC 2
*.DC ILOAD 0 2.5 .01 VD 0 1 .1
*VD D 0 DC .4 AC 1
*
```

```
**************************************************************
*
** SOURCE-LOAD CONFIGURATION FOR
** FIGURE 4.8
** BUCK CONVERTER AC "D TO OUTPUT" TRANSFER FUNCTION (VOLTAGE MODE)
*
*VIN N 0 DC 30 AC 0
*ILOAD 1 0 DC 2
*.STEP ILOAD LIST .2 .7 1.2 2.5
*VD D 0 DC .4 AC 1
*.AC DEC 20 1 100K
*
**************************************************************
*
** SOURCE-LOAD CONFIGURATION FOR
** FIGURE 4.18
** BUCK CONVERTER AC FORWARD TRANSFER FUNCTION
** WITH FEEDFORWARD ADDITION (VOLTAGE MODE)
*
*VIN N 0 DC 30 AC 1
*ILOAD 1 0 DC 2
*.STEP ILOAD LIST .2 .7 1.2 2.5
*ED D 0 VALUE = {12/V(8)}
*.AC DEC 20 1 100K
*
**************************************************************
*
** BUCK CONVERTER, VOLTAGE MODE, NETLIST
*
RD D 0 1
C4 1 0 .1U
C3 1 3 2000U
R3 3 0 .2
GLD 2 K VALUE = {V(D)**2*(V(8)/V(1))*(V(8)-V(1))*.125}
VM3 K 1
XD3 2 1 DIDEAL
L2 4 2 40U
E1 4 5 VALUE = {V(8)*V(D)}
VM1 0 5
GILD 6 7 VALUE = {I(VM3)*V(1)/V(8)}
G1 0 6 VALUE = {I(VM1)*V(D)}
XD1 6 7 DIDEAL
XD2 0 6 DIDEAL
VM2 7 0 DC 1
F2 8 0 VM2 1
C2 8 0 10U
C1 8 9 175U
R1 9 0 1
L1 N 8 50U
*
.SUBCKT DIDEAL 1 2
EID 3 1 TABLE {V(3,2)} = (-1,1U) (1U,1U) (1,1)
DIO 3 2 D
.MODEL D D IS=1E-12
.ENDS DIDEAL
```

```
*
.PROBE
.END
```

4.1.2.1 Forward Transfer Function

The converter is set up with an input voltage of 30 Vdc and an output voltage of 12 Vdc. This will necessitate a steady state duty ratio, D, of

$$D = \frac{12v}{30v} = 0.4 \tag{4.1}$$

for the no feedback forward transfer analysis. With a fixed frequency of 100 kHz (T_S = 10 μS) and a value of 40 μH for the filter inductor, L2, the critical load current for continuous conduction mode operation is:

$$I_{CRIT} = \frac{VT_S(1-D)}{2L_2} = \frac{12V \times 10\,\mu S \times (1-0.4)}{2 \times 40\,\mu S} = 0.9A \tag{4.2}$$

The forward transfer function with continuous conduction mode currents of 1.2 and 2.5 amps (greater than the critical current of 0.9 amps) and then with discontinuous mode currents of 0.2 and 0.7 amps (less than the critical current of 0.9 amps) will be considered. The lowest value of 0.2 amps represents a deep discontinuous conduction mode. Figure 4.2 shows the converter model to be analyzed.

As a sidelight, Figure 4.3 shows how the output voltage of this constant frequency, fixed duty ratio, buck converter varies with a sweep of load current from 0 to 2.5 amps and input voltages of 20, 25, and 30 V. It is interesting to note that a fixed critical load resistance, R_{CRIT} (equal to 13.33 Ω in this case),

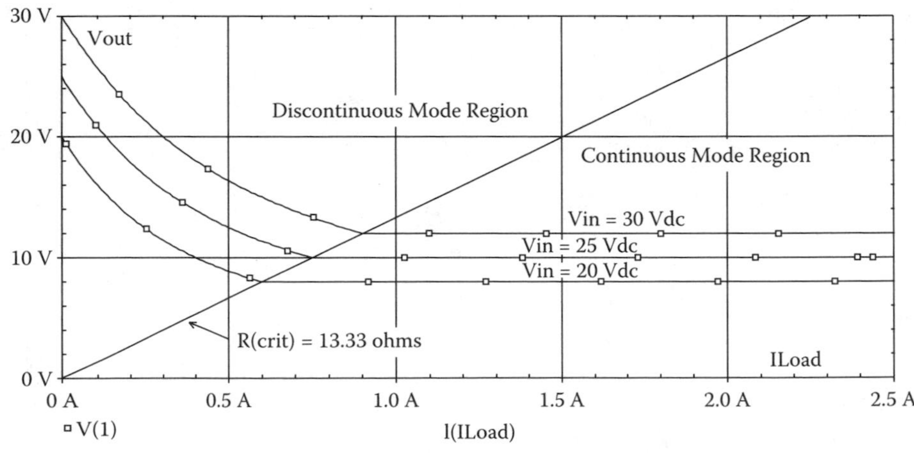

FIGURE 4.3
Buck converter DC forward transfer function with D = 0.4.

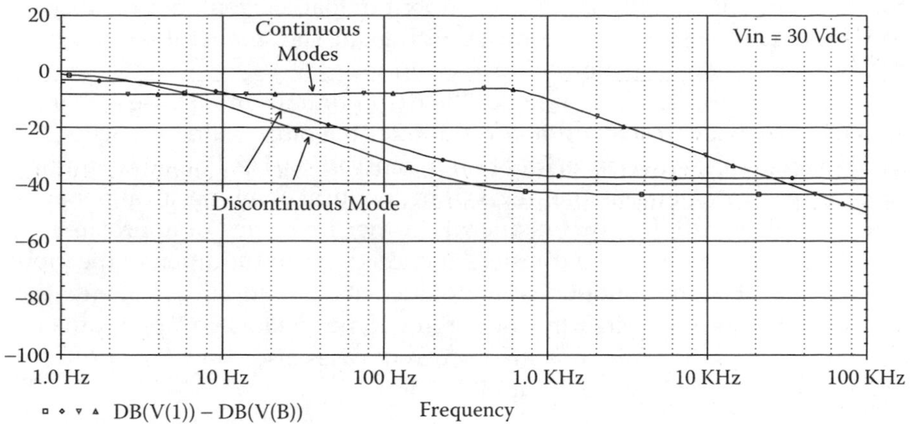

FIGURE 4.4
Buck converter AC forward transfer function.

exists as noted by the straight line intersection of all the critical conduction points separating continuous and discontinuous conduction modes.

Now consider the AC forward transfer characteristics. A constant input voltage of 30 Vdc, $D = 0.4$, and the same two pair of continuous and discontinuous mode DC load current are used here also. For this example, the corner frequencies of the input and output filter are purposefully separated from each other in order to show their possible individual effects. Figure 4.4 shows the transfer function for just the converter portion (node 8 to node 1 in the simulation), and Figure 4.5 shows the overall transfer function, including the input filter. Notice that both continuous mode current transfer

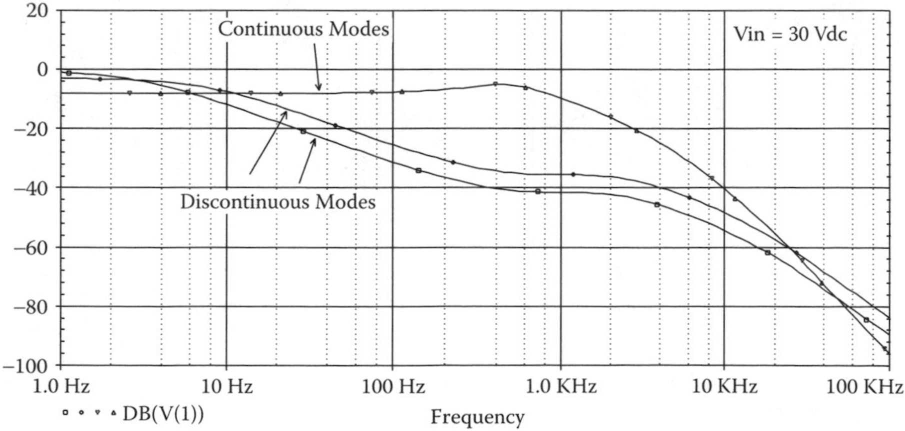

FIGURE 4.5
Buck converter AC forward transfer function with input filter added.

functions are the same and are independent of load current, but the discontinuous mode transfer function varies with load. The smaller the load is, the lower the initial low frequency break point is.

Another interesting observation for the discontinuous mode case is that the converter attenuation starts to flatten out after the output filter capacitor, C3, breaks with it series resistance, R3, at a frequency of 400 Hz. For the continuous mode case with the output filter (L2, C3) near critical damping, an attenuation slope of only 20 dB/Dec occurs after its corner frequency of approximately 600 Hz. Figure 4.5 shows the transfer functions with the addition of the input filter section. For the continuous mode currents, the additional attenuation and steeper slope occur after the input filter corner frequency (approximately 5 kHz). The two discontinuous mode current cases also show the additional attenuation provided by the input filter addition.

4.1.2.2 Output Impedance (Voltage Mode)

Now look at the AC output impedance of the converter for the same conditions as those set up in the previous paragraph. (The input filter section remains in the circuit unless otherwise noted.) Figure 4.6 shows the AC output impedance plots. (0 dB is equivalent to 1 Ω.) The discontinuous mode output impedance is relatively high and also a function of the load current prior to breaking with the output filter capacitor, C3, at frequencies near 10 Hz. At the low frequencies, the output impedance for the continuous mode is essentially the inductive reactance of inductor L2 in series with the reactance of L2 reflected by the duty ratio squared (D^2).

4.1.2.3 Control to Output (Voltage Mode)

Examine the DC transfer functions for the control, *D*, to output voltage, *v*. The input voltage will be maintained constant at 30.0 Vdc; in this case, the

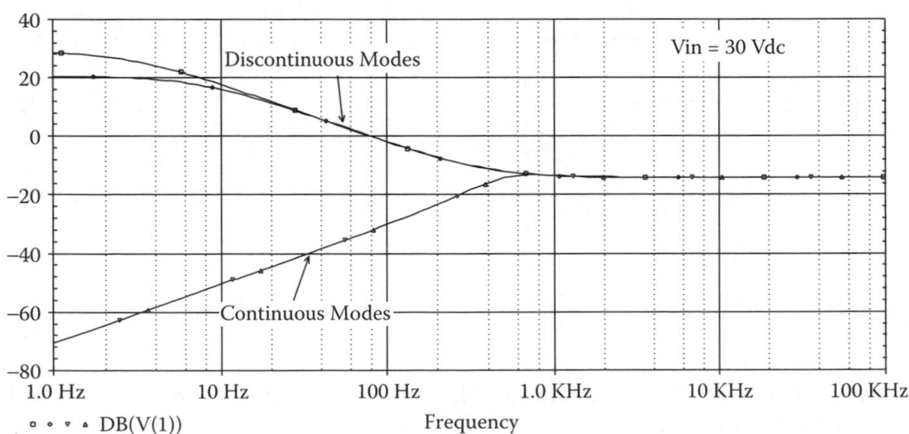

FIGURE 4.6
Buck converter output impedance (voltage mode).

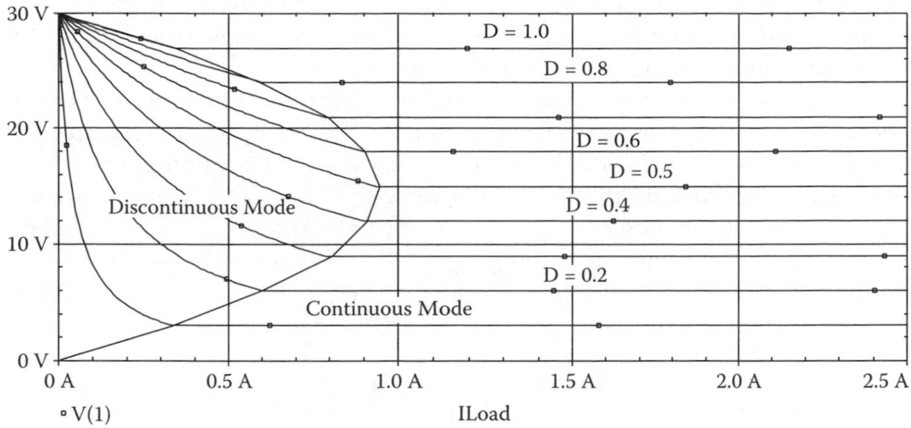

FIGURE 4.7
Buck converter DC "control to output" transfer function with *Vin* = 30 Vdc (voltage mode).

duty ratio, *D*, will be varied over its range of 0 to 1. Figure 4.7 shows the DC transfer functions for values of *D* stepped from 0 to 1 in increments of 0.1. Load currents were swept from 0 to 2.5 amps as was the case for the forward transfer function analysis. Note that the output voltage is independent of load current for continuous mode conduction and varying with load current for discontinuous modes. Of course, the duty ratio must vary considerably with load current in the discontinuous mode to regulate the output voltage, but theoretically may remain fixed for continuous mode currents. It is interesting to note the parabolic loci generated by connecting the critical current points for each of the stepped values of *D*. Also, note the maximum critical current occurring at the duty ratio of *D* = 0.5.

Now consider the AC control to output transfer functions. The same two pairs of continuous and discontinuous mode DC load currents will be used as were used for the forward voltage transfer analysis when analyzing the AC forward transfer function (continuous conduction mode currents of 1.2 and 2.5 amps and discontinuous mode currents of 0.2 and 0.7 amps). Also, as was the case for the forward transfer function, the input will be set to a constant 30 Vdc and the control, *D*, to a DC biased value of 0.4 while this is modulated with the small signal AC stimulus.

Figure 4.8 shows the results. Note that the discontinuous mode gains rolling off at lower corner frequencies are very much proportional to load current. The continuous mode gains roll off at higher frequencies and are seen to be fairly independent of load current. A very small, almost imperceptible dip in gain is noted around 2 kHz. This secondary effect is produced by the finite output impedance of the input filter. When analyzed with the input filter removed (not shown here), this small dip disappears. On the other hand, with an improperly designed input filter, this gain dip may increase and cause such negative effects as reduced loop gain; reduced

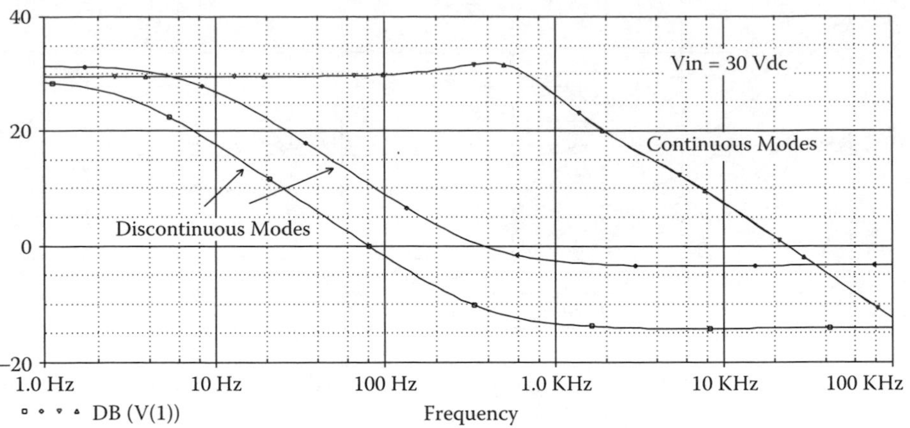

FIGURE 4.8
Buck converter AC "*d* to output" transfer function (voltage mode).

forward transfer attenuation; and higher output impedance; in severe cases, an input filter oscillating instability may occur.[2,13]

4.1.3 SMPS Closed Loop Analysis (Continuous and Discontinuous Modes)

Now add the feedback control circuit, which regulates the output voltage to 12 Vdc, and consider the SMPS characteristics and performance with this addition. Figure 4.9a shows the basic configuration. (A PSPICE netlist, labeled Netlist 4.2, is used for these voltage mode closed loop analysis simulations.) Using the same buck converter, an error amplifier has been added, along with a 2.5-V reference. Then the PWM gain block is inserted and combined with the AC-only loop opening, ultralow-pass filter circuit. It comprises LOL, COL, and VAC. This allows one to maintain the normal closed loop DC operating points while examining the small signal AC open loop characteristics at these operating points. This technique will be used repeatedly in many of the future analyses. VAC is the signal injection stimulus necessary for any AC gain measurement or analysis. The PWM gain is:

$$d = \frac{V(13)}{V_M} \tag{4.3}$$

where V_M is the PWM ramp amplitude ($V_M = 1$ V for these simulations). The input voltage and load currents are then varied over the same ranges as was done for the preceding open loop analysis. (Figure 4.9b shows the extremely simple op amp model used for illustration purposes.)

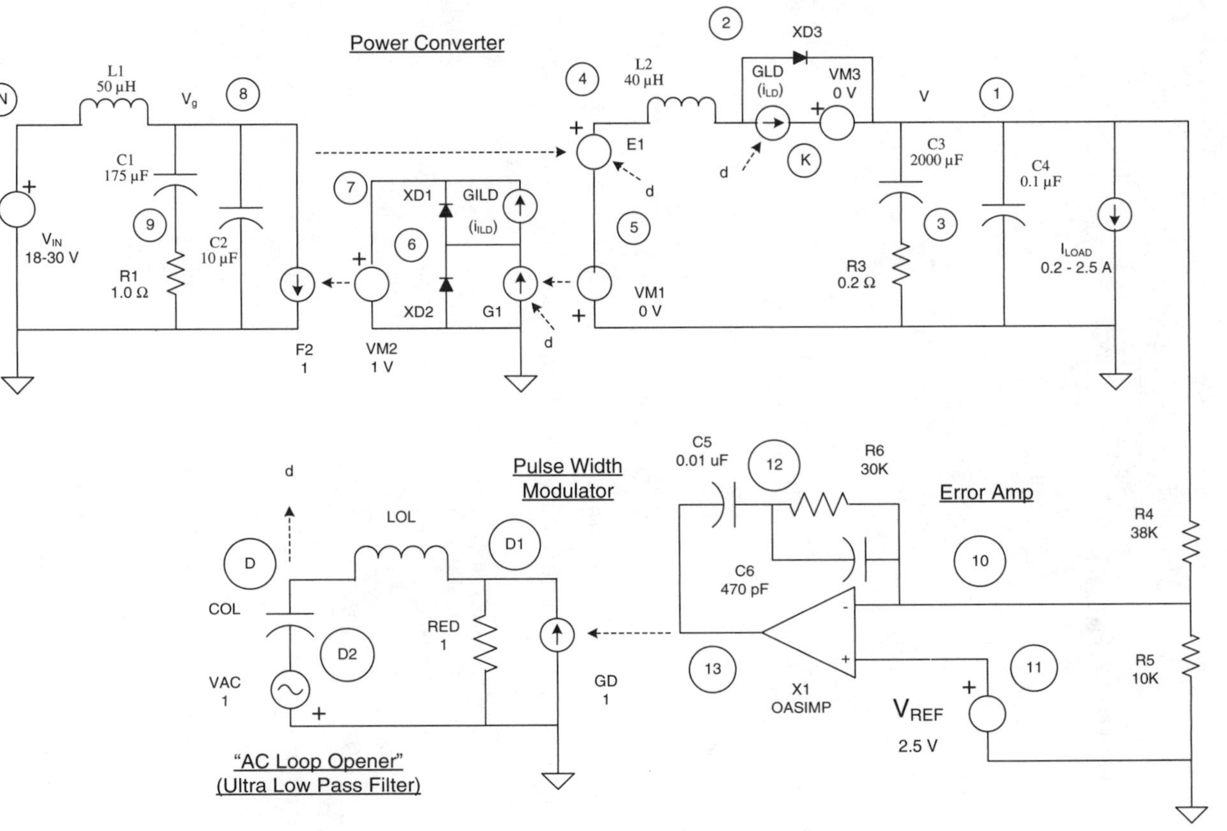

FIGURE 4.9a
Buck converter switching regulator (voltage mode).

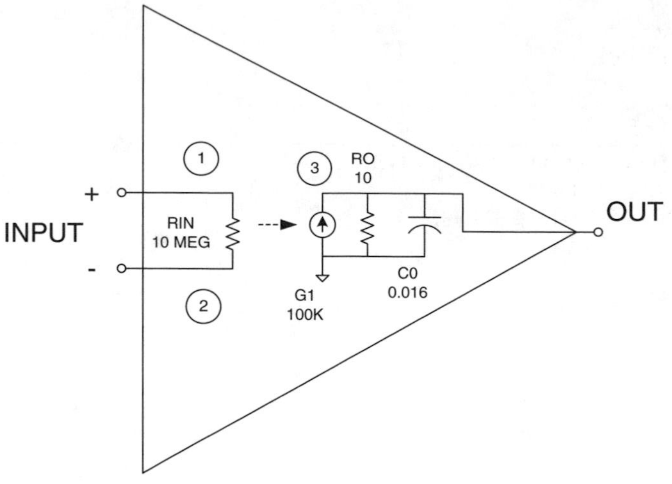

FIGURE 4.9b
Simple 1.0-MHz bandwidth op amp model.

NETLIST 4.2

```
BUCK CONVERTER SMPS, VOLTAGE MODE, CLOSED LOOP ANALYSIS
*
************************************************************
*
** SOURCE-LOAD CONFIGURATION FOR
** FIGURES 4.10 AND 4.11
** BUCK SMPS AC LOOP GAIN AND PHASE
** WITH LOAD CHANGE (VOLTAGE MODE)
*
VIN N 0 DC 30 AC 0
ILOAD 1 0 DC .2
.STEP ILOAD LIST .2 .7 1.2 2.5
LOL D1 D 1K
COL D D2 1K
VAC 0 D2 AC 1
.AC DEC 20 1 100K
*
************************************************************
*
** SOURCE-LOAD CONFIGURATION FOR
** FIGURES 4.12 AND 4.13
** BUCK SMPS AC LOOP GAIN AND PHASE
** WITH INPUT VOLTAGE CHANGE (VOLTAGE MODE)
*
*ILOAD 1 0 DC 2.5
*ILOAD 1 0 DC .2
*VIN N 0 DC 18
*.STEP VIN LIST 18 30
*LOL D1 D 1K
```

```
*COL D D2 1K
*VAC 0 D2 AC 1
*.AC DEC 20 1 100K
*
*************************************************************
*
** SOURCE-LOAD CONFIGURATION FOR
** FIGURES 4.14 AND 4.15
** BUCK SMPS AC OUTPUT IMPEDANCE
** WITH INPUT VOLTAGE CHANGE (VOLTAGE MODE)
*
*ILOAD 1 0 DC 2.5
*ILOAD 1 0 DC .2
*VIN N 0 DC 18
*.STEP VIN LIST 18 30
*LOL D1 D 1P
*COL D D2 1P
*VAC 0 D2 AC 0
*IAC 1 0 AC 1
*.AC DEC 20 1 100K
*
*************************************************************
*
** SOURCE-LOAD CONFIGURATION FOR
** FIGURES 4.16 AND 4.17
** BUCK SMPS AC FORWARD ATTENUATION
** WITH INPUT DC VOLTAGE CHANGE (VOLTAGE MODE)
*
*ILOAD 1 0 DC .2
*ILOAD 1 0 DC 2.5
*VIN N 0 DC 30 AC 1
*.STEP VIN LIST 18 30
*LOL D1 D 1P
*COL D D2 1P
*VAC 0 D2 AC 0
*.AC DEC 20 1 100K
*.NODESET V(1)=12 V(8)=30 V(13)=1
*
*************************************************************
*
** SOURCE-LOAD CONFIGURATION FOR
** FIGURES 4.19 AND 4.20
** BUCK SMPS AC FORWARD ATTENUATION WITH FEEDFORWARD COMPENSATION
** WITH INPUT DC VOLTAGE CHANGE (VOLTAGE MODE)
*
*ILOAD 1 0 DC .2
*ILOAD 1 0 DC 2.5
*VIN N 0 DC 30 AC 1
*.STEP VIN LIST 18 30
*LOL D1 D 1P
*COL D D2 1P
*VAC 0 D2 AC 0
*.AC DEC 20 1 100K
*.NODESET V(1)=12 V(8)=30 V(13)=1
```

```
*.OPTIONS GMIN=1E-10
*
** NOTE: THE FOLLOWING GD REMAINS IN THE CIRCUIT FOR ALL OTHER ANALYSIS
** BUT MUST ALTERNATE WITH THE SECOND GD BELOW FOR THIS
** BUCK SMPS AC FEEDFORWARD ATTENUATION ANALYSIS ONLY
GD 0 D1 13 0 1
*GD 0 D1 VALUE = {LIMIT((12*V(13))/V(8),0,1)}
*
*************************************************************
*
**VOLTAGE MODE SMPS NETLIST
*
RED D1 0 1
C4 1 0 .1U
C3 1 3 2000U
R3 3 0 .2
GLD 2 K VALUE = {V(D)**2*(V(8)/V(1))*(V(8)-V(1))*.125}
VM3 K 1
XD3 2 1 DIDEAL
L2 4 2 40U
E1 4 5 VALUE = {V(8)*V(D)}
VM1 0 5
GILD 6 7 VALUE = {I(VM3)*V(1)/V(8)}
G1 0 6 VALUE = {I(VM1)*V(D)}
XD1 6 7 DIDEAL
XD2 0 6 DIDEAL
VM2 7 0 DC 1
F2 8 0 VM2 1
C2 8 0 10U
C1 8 9 175U
R1 9 0 1
L1 N 8 50U
R4 1 10 38K
R5 10 0 10K
VREF 11 0 DC 2.5
X1 11 10 13 OASIMP
C5 13 12 .01U
R6 12 10 30K
C6 12 10 470P
*
.SUBCKT OASIMP 1 2 3
RIN 1 2 10MEG
G1 0 3 1 2 100K
RO 3 0 10
CO 3 0 .016
.ENDS OASIMP
*
.SUBCKT DIDEAL 1 2
VAS 1 3 DC -1U
D1 3 2 D
D2 3 4 D
D3 4 2 D1
FAS 4 2 VAS 1
.MODEL D D IS=1E-6
.MODEL D1 D
```

```
CC 1 2 .1P
.ENDS DIDEAL
*
.PROBE
.END
```

4.1.3.1 AC Loop Gain (Voltage Mode)

Figure 4.10 shows the AC loop gain and phase plots for the two continuous mode loads of 1.2 and 2.5 amps, and Figure 4.11 shows the AC loop gain and phase plots for the two discontinuous mode loads of 0.2 and 0.7 amps. The continuous mode gains in Figure 4.10 are independent of load current variations, as might be expected; however, the gains in the discontinuous mode cases of Figure 4.11 do vary with load current, as also might be expected. Note that the unity gain crossover frequency is reduced for the discontinuous mode cases and continues to decrease for further decreasing loads.

To ensure that the feedback control loop maintains AC stability, the phase shift must not approach or exceed a magnitude of –180° at the unity gain crossover frequency. The actual phase difference, or "phase margin," from –180° at the unity gain crossover frequency is conventionally considered the most important figure of merit for determining the degree of AC stability of a feedback control system. The accompanying figure of merit is the gain margin. This is the reduction in loop gain magnitude below unity (0 dB) corresponding to the –180° phase shift point. Typical numbers desired for the phase and gain margin are 45° and 10 dB. See Kuo[10] and D'Azzo and Houpis[11] for more fundamental information on feedback control stability.

Now consider the effects of input DC voltage variations on loop gain by stepping V_{in} from 18 to 30 Vdc with the load cases of 2.5 amps (continuous)

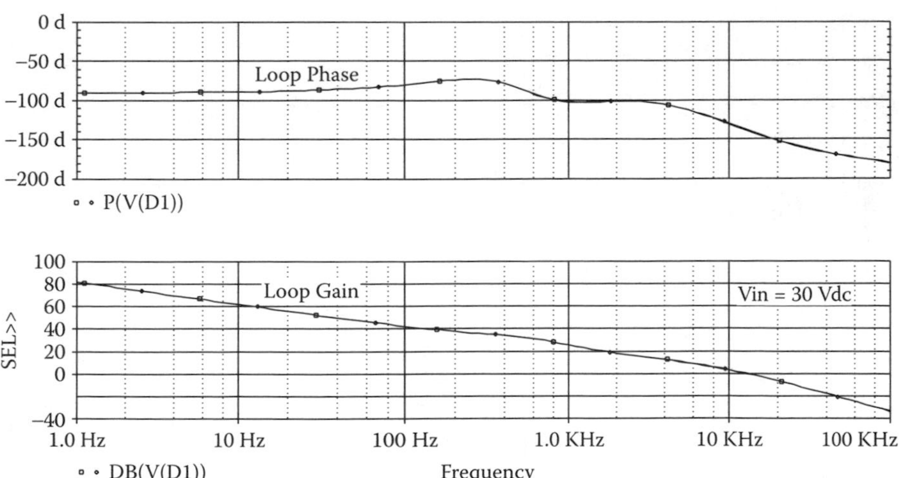

FIGURE 4.10
Buck SMPS continuous mode AC loop gain and phase with load change (voltage mode).

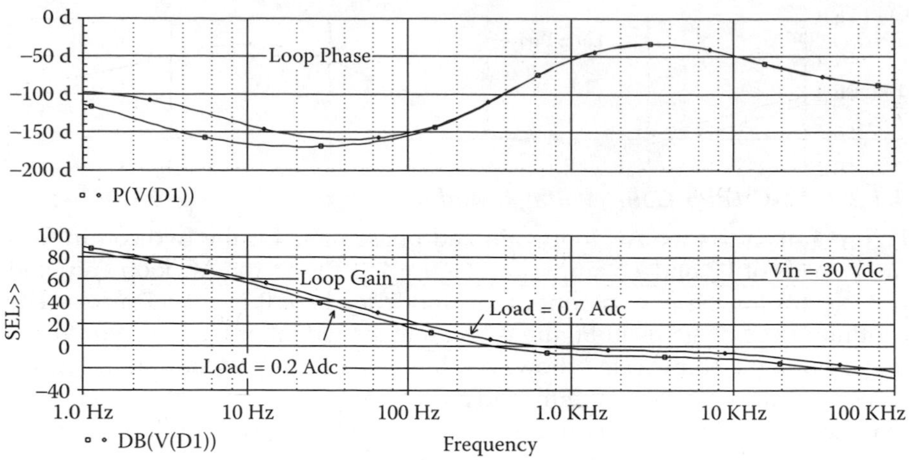

FIGURE 4.11
Buck SMPS discontinuous mode AC loop gain and phase with load change (voltage mode).

and 0.2 amps (discontinuous). Figure 4.12 and Figure 4.13 show the respective plots. Both plots indicate a shift in gain that is somewhat proportional to the input voltage change ratio.

4.1.3.2 SMPS Output Impedance (Voltage Mode)

Now look at the SMPS output impedance for extreme variations in line voltage and load currents that were considered in the preceding loop gain analyses. By examining the plots of Figure 4.14 and Figure 4.15, AC impedance

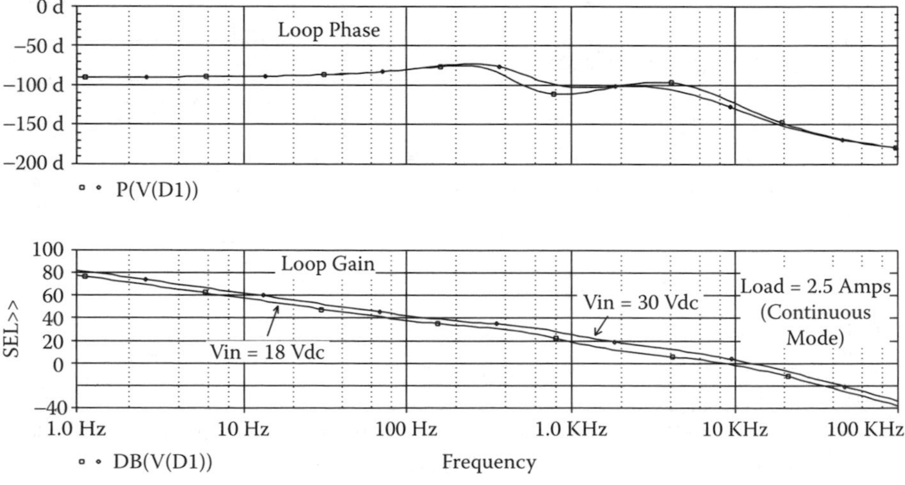

FIGURE 4.12
Buck SMPS continuous mode AC loop gain and phase with input voltage change (voltage mode).

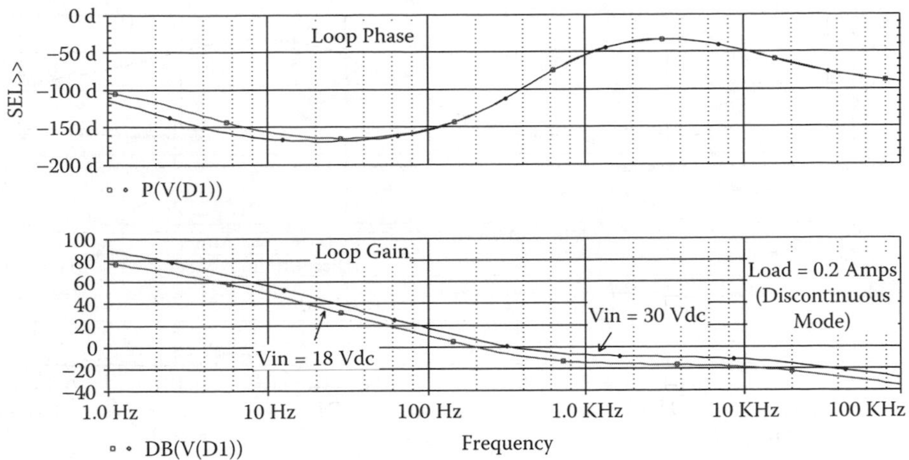

FIGURE 4.13
Buck SMPS discontinuous mode AC loop gain and phase with input voltage change (voltage mode).

magnitudes for input voltages of 18 and 30 Vdc, respectively, can be observed. The 30-Vdc cases provide higher loop gains, which produce the lower output impedances that might be expected (see Equation 2.2). To illustrate specifically for the 30-Vdc input case, the impedance plots of Figure 4.14 and Figure 4.15 could be essentially derived by dividing the open loop impedance of Figure 4.6 by one plus the corresponding loop gains of Figure 4.12 and Figure 4.13 (see Equation 2.2).

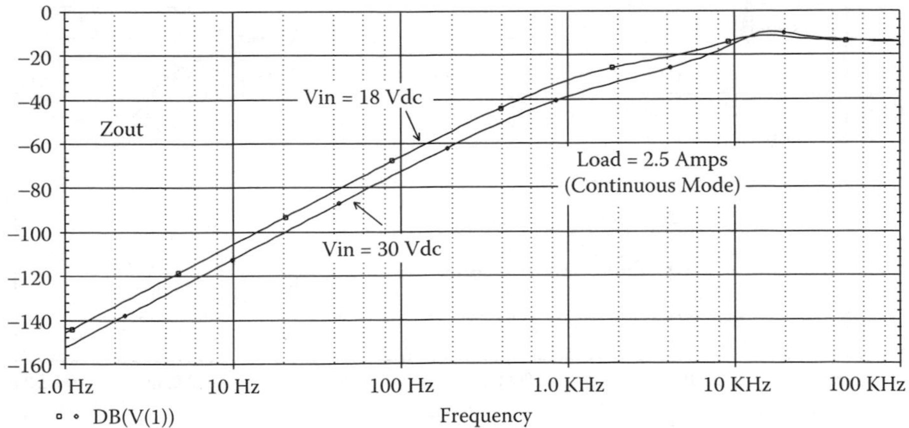

FIGURE 4.14
Buck SMPS continuous mode AC output impedance with input voltage change.

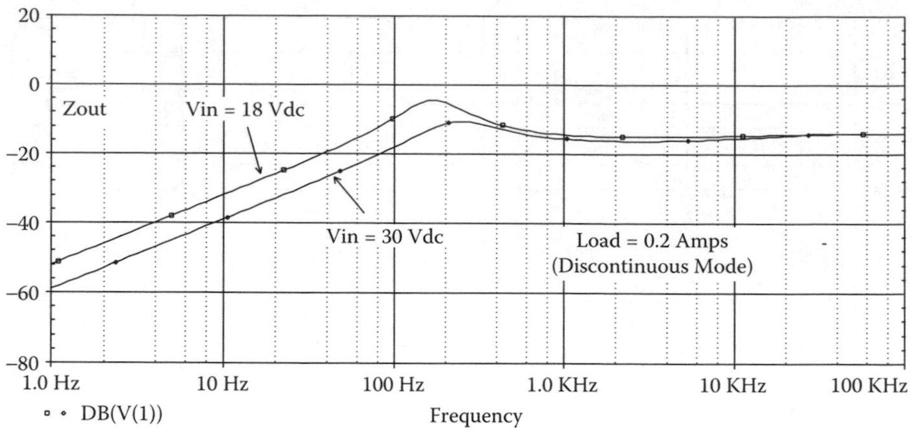

FIGURE 4.15
Buck SMPS discontinuous mode AC output impedance with input voltage change.

4.1.3.3 SMPS Line Regulation (Voltage Mode)

Consider the AC line rejection characteristics of the closed loop configuration. This is sometimes called "audio susceptibility." Figure 4.16 and Figure 4.17 show the AC attenuation of the SMPS for the indicated operating conditions of line and load. By examining the plots of these two figures, it is possible to observe the AC forward attenuation at input voltages of 18 and 30 Vdc for discontinuous and continuous modes, respectively. The 30-Vdc cases with the higher loop gains provide the greater attenuations, as one might expect (see Equation 2.1). To illustrate specifically for the 30-Vdc input case, the attenuation plots of Figure 4.16 and Figure 4.17 could be essentially derived

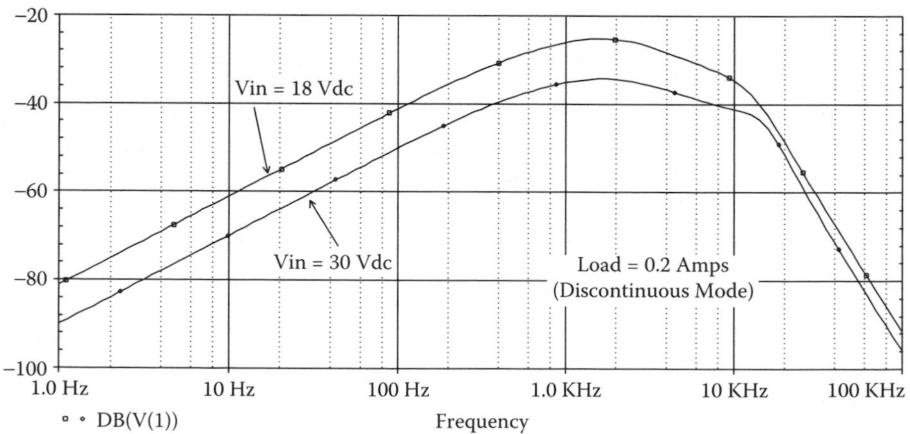

FIGURE 4.16
Buck SMPS continuous mode AC forward attenuation with input DC voltage change.

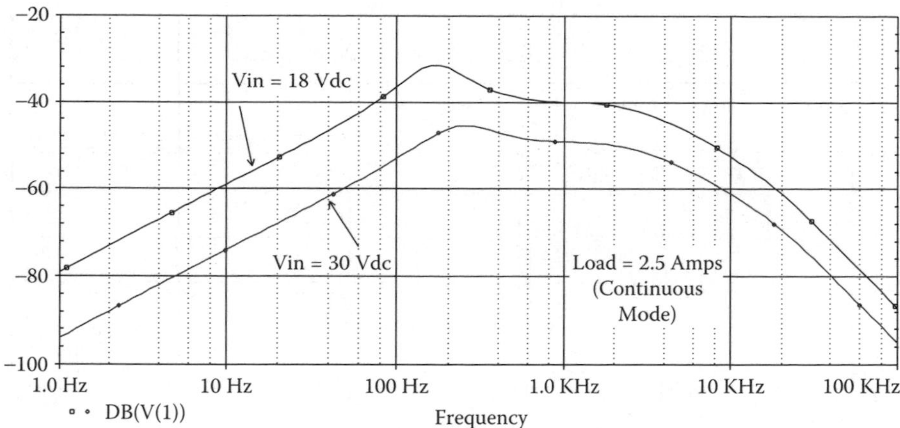

FIGURE 4.17
Buck SMPS discontinuous mode AC forward attenuation with input DC voltage change.

by dividing the open loop attenuations of Figure 4.8 by one plus the corresponding loop gains of Figure 4.12 and Figure 4.13 (Equation 2.1).

4.1.3.4 SMPS Feedforward Analysis

In many applications that utilize voltage mode control, the AC line rejection requirements are sometimes difficult to satisfy because of the difficulty in achieving the required AC loop gain and maintaining AC loop stability in the continuous mode. One technique of improving this situation is to provide a type of input voltage feedforward. This is implemented by letting the slope of the pulse width modulator (PWM) control ramp (Figure 4.1) be proportional to the input line voltage. For fixed-frequency converters, this translates into a control ramp amplitude that is proportional to input voltage.

Since the PWM gain is inversely proportional to this ramp amplitude, it has the effect of *reducing* loop gain with increasing input voltage. On the other hand, an increasing input voltage will increase the converter control to output gain, thereby *increasing* loop gain. With these offsetting effects, the feedback control loop can effectively be fixed and "linearized" by being almost if not completely independent of input voltage variations. Although this gain stabilizing effect simplifies the control loop, the AC line rejection improvement is in reality achieved by the immediate duty ratio adjustment being made with the immediate change in input voltage. Perfect feedforward compensation is generally not practical, but considerable improvements can be realized.

Take the open loop converter that produced the results of Figure 4.5 and, instead of a fixed $D = 0.4$, ideally add some feedforward by letting d have the following inverse relationship to V_g:

$$d = \frac{1}{\left(\frac{v_g}{V}\right)} = \frac{1}{\left(\frac{v_g}{12}\right)} = \frac{12}{v_g} \tag{4.4}$$

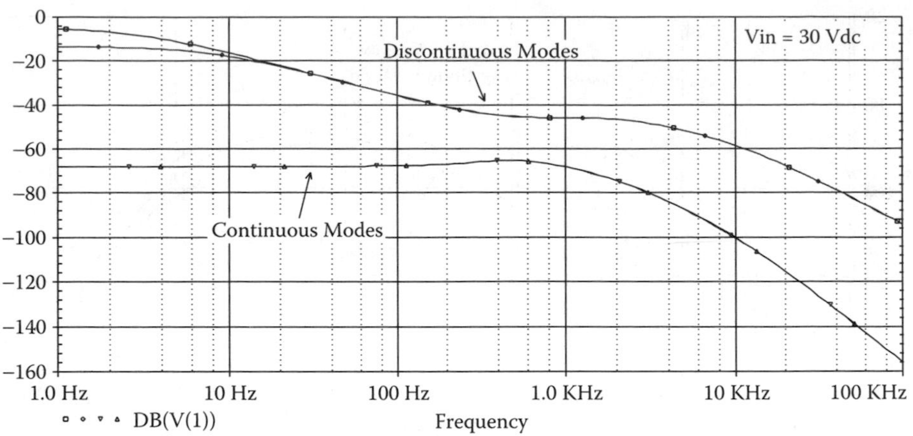

FIGURE 4.18
Buck converter AC forward transfer function with feedforward.

Comparing the results of Figure 4.5 to those in Figure 4.18 makes it possible to see the dramatic increase in line to output attenuation for the continuous mode cases at the lower frequencies (~60 dB improvement). The discontinuous modes are only slightly affected.

4.1.4 SMPS Combined Feedback and Feedforward Analysis (Line Regulation)

Now take the closed loop feedback regulated circuit of Figure 4.9 and add the feedforward implementation of Section 4.1.3.4 to it and compare the AC line rejection characteristics to those obtained with feedback only, as shown in Figure 4.16 and Figure 4.17. With the PWM control ramp amplitude simply modulated by the input voltage, the implementation is simply to multiply the circuit model control voltage, *d*, by a function inversely proportional to the input voltage. A function as expressed by Equation 4.5 might be a possible idealistic implementation:

$$d = \frac{v_e}{\left(\frac{v_g}{12}\right)} = 12\left(\frac{v_e}{v_g}\right) \tag{4.5}$$

The results are shown in Figure 4.19 and Figure 4.20. Compare these figures with Figure 4.16 and Figure 4.17, respectively, and note the improvement in forward attenuation with the addition of the feedforward compensation. About 50 dB of additional attenuation was noted for the

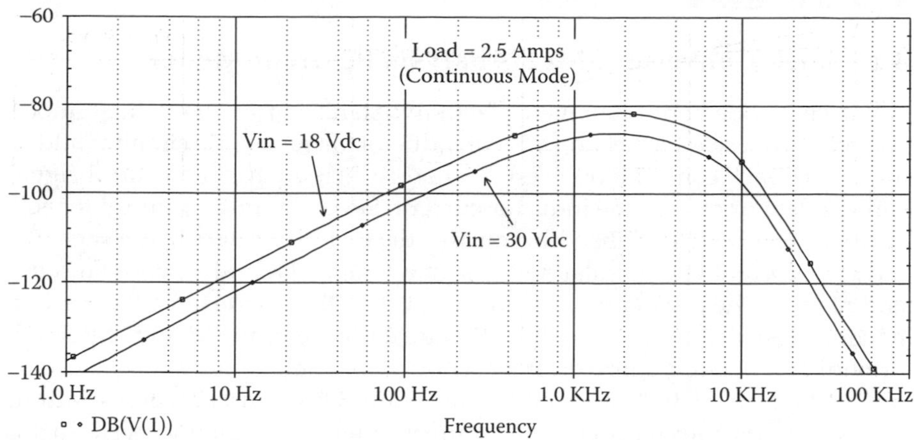

FIGURE 4.19
Buck SMPS with feedforward compensation. Continuous mode AC forward attenuation with input DC voltage change.

continuous mode cases, but practically no noticeable increase was noted for the discontinuous mode cases. The idealistic feedforward addition considered here, combined with the feedback, provides somewhat idealistic results that may not be achievable in actual practice. Nevertheless, the method of analysis and the general expected results have been shown.

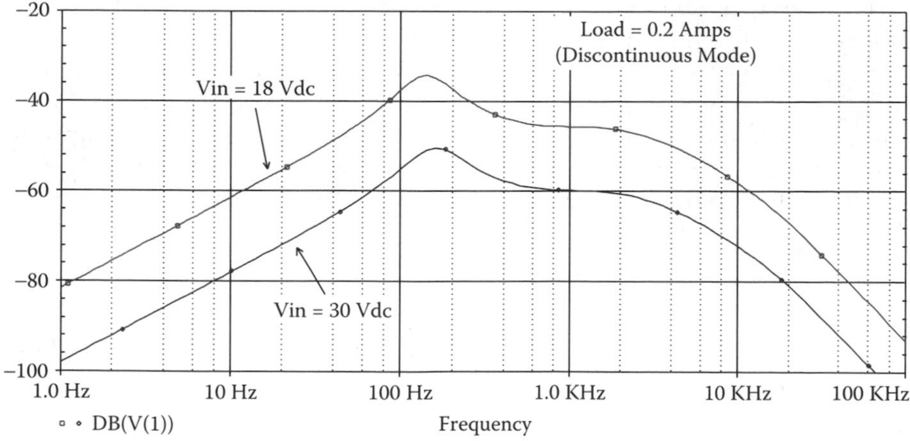

FIGURE 4.20
Buck SMPS with feedforward compensation. Discontinuous mode AC forward attenuation with input DC voltage change.

4.2 Buck Converter SMPS Analysis (Current Mode)

A current mode control converter actually starts out as the voltage mode converter described in Section 4.1. An inductor current sense circuit is added and a feedback control loop is established so that an applied control signal can regulate or control the inductor current. Several types of control for this current are possible, depending on the desired characteristics. When the inductor current is triangular and has an averaged DC value, one can sense and regulate the peak current or, in some cases, the converse valley current. In some designs, the average inductor current is determined and regulated, providing a possibly more desirable control for some applications. For output voltage regulation, a second outer control loop circuit senses output voltage and its output provides the control signal for regulating the inductor current. Controlling the inductor current subsequently allows regulation of the output voltage.[5,12] (A PSPICE netlist, labeled Netlist 4.3, is used for these current mode analysis simulations.)

NETLIST 4.3

```
BUCK CONVERTER SMPS, CURRENT MODE, ANALYSIS
*
****************************************************************
*
** SOURCE-LOAD CONFIGURATION FOR
** FIGURES 4.24 AND 4.25
** BUCK SMPS AC LOOP GAIN AND PHASE
** WITH LOAD CHANGE (CURRENT MODE)*
VIN N 0 DC 30
ILOAD 1 0 DC .2
.STEP ILOAD LIST .2 .7 1.2 2.5
LOL 14 16 1K
COL 16 15 1K
VAC 0 15 AC 1
.AC DEC 20 1 100K
.OPTIONS ITL2=200
*
****************************************************************
*
** SOURCE-LOAD CONFIGURATION FOR
** FIGURES 4.26 AND 4.27
** BUCK SMPS AC LOOP GAIN AND PHASE
** WITH INPUT VOLTAGE CHANGE (CURRENT MODE)*
*ILOAD 1 0 DC 2.5
*ILOAD 1 0 DC .2
*VIN N 0 DC 30
*.STEP VIN LIST 18 30
*LOL 14 16 1K
*COL 16 15 1K
*VAC 0 15 AC 1
*.AC DEC 20 1 100K
```

```
*.OPTIONS STEPGMIN ITL2=200
*.NODESET  V(1)=12 V(8)=30 V(D)=.6 V(17)=1 V(13)=1 V(14)=1
*
***************************************************************
*
** SOURCE-LOAD CONFIGURATION FOR
** FIGURE 4.28
** OUTPUT IMPEDANCE OF THE "OPEN VOLTAGE LOOP" CURRENT MODE
** CONVERTER*
*ILOAD 1 0 DC .2
*.STEP ILOAD LIST .2 .7 1.2 2.5
*VIN N 0 DC 30
*LOL 14 16 1K
*COL 16 15 1K
*VAC 0 15 AC 0
*IAC 1 0 AC 1
*.AC DEC 20 1 100K
*.OPTIONS STEPGMIN ITL2=200
*
***************************************************************
*
** SOURCE-LOAD CONFIGURATION FOR
** FIGURE 4.29 AND 4.30
** AC FORWARD TRANSFER FUNCTION OF THE "OPEN VOLTAGE LOOP"
** CURRENT MODE CONVERTER WITHOUT THE INPUT FILTER.*
*ILOAD 1 0 DC .2
*.STEP ILOAD LIST .2 .7 1.2 2.5
*VIN N 0 DC 30 AC 1
*LOL 14 16 1K
*COL 16 15 1K
*VAC 0 15 AC 0
*.AC DEC 20 1 100K
*.OPTIONS STEPGMIN ITL2=200
*
***************************************************************
*
** SOURCE-LOAD CONFIGURATION FOR
** FIGURE 4.31
**BUCK CONVERTER AC "D TO OUTPUT" TRANSFER FUNCTION (CURRENT MODE).*
*ILOAD 1 0 DC .2
*.STEP ILOAD LIST .2 .7 1.2 2.5
*VIN N 0 DC 30
*LOL 14 16 1K
*COL 16 15 1K
*VAC 0 15 AC 1
*.AC DEC 20 1 100K
*.OPTIONS STEPGMIN ITL2=200
*
***************************************************************
*
** SOURCE-LOAD CONFIGURATION FOR
** FIGURE 4.32
** INNER INDUCTOR CURRENT CONTROL LOOP GAIN*
*ILOAD 1 0 DC .2
*.STEP ILOAD LIST 1.2 2.5
```

```
*VIN N 0 DC 30
*LOL 14 16 1K
*COL 16 15 1K
*VAC 0 15 AC 0
*HI 100 0 VM1 1
*LOL2 100 101 1000
*COL2 101 102 1000
*VAC2 0 102 AC 1
*.AC DEC 20 1 100K
*.OPTIONS STEPGMIN ITL2=200
*
** NOTE: THE FOLLOWING EDC REMAINS IN THE CIRCUIT FOR ALL ANALYSIS
** BUT MUST ALTERNATE WITH THE SECOND EDC BELOW FOR THIS
** INNER INDUCTOR CURRENT CONTROL LOOP GAIN ANALYSIS ONLY
EDC 17 0 VALUE = {(V(16)/1-I(VM1))/(.125*(24+V(8)-V(1)))}
*EDC 17 0 VALUE = {(V(16)/1-V(101))/(.125*(24+V(8)-V(1)))}
*
**************************************************************
*
** SOURCE-LOAD CONFIGURATION FOR
** FIGURE 4.33
** OUTPUT IMPEDANCE OF THE CURRENT MODE SMPS*
*ILOAD 1 0 DC .2
*.STEP ILOAD LIST .2 .7 1.2 2.5
*VIN N 0 DC 30
*LOL 14 16 1P
*COL 16 15 1P
*VAC 0 15 AC 0
*IAC 1 0 AC 1
*.AC DEC 20 1 100K
*.OPTIONS STEPGMIN ITL2=200
*
**************************************************************
*
** SOURCE-LOAD CONFIGURATION FOR
** FIGURE 4.34 AND 4.35
** AC LINE REJECTION OF THE CURRENT MODE SMPS*
*VIN N 0 DC 30 AC 1
*VIN N 0 DC 18 AC 1
*ILOAD 1 0 DC .2
*.STEP ILOAD LIST .2 .7 1.2 2.5
*LOL 14 16 1P
*COL 16 15 1P
*VAC 0 15 AC 0
*.AC DEC 20 1 100K
*.OPTIONS STEPGMIN ITL2=200
*.NODESET  V(1)=12 V(8)=30 V(D)=.6 V(17)=1 V(13)=1 V(14)=1
*
**************************************************************
*
**CURRENT MODE SMPS NETLIST
*
C4 1 0 .1U
```

```
C3 1 3 2000U
R3 3 0 .2
GLD 2 K VALUE = {V(D)**2*(V(8)/V(1))*(V(8)-V(1))*.125}
VM3 K 1
XD3 2 1 DIDEAL
L2 4 2 40U
E1 4 5 VALUE = {V(8)*V(D)}
VM1 0 5
GILD 6 7 VALUE = {I(VM3)*V(1)/V(8)}
G1 0 6 VALUE = {I(VM1)*V(D)}
XD1 6 7 DIDEAL
XD2 0 6 DIDEAL
VM2 7 0 DC 1
F2 8 0 VM2 1
C2 8 0 10U
C1 8 9 175U
R1 9 0 1
L1 N 8 50U
R4 1 10 38K
R5 10 0 10K
VREF 11 0 DC 2.5
X1 11 10 13 OASIMP
C5 13 12 .01U
R6 12 10 300K
C6 12 10 470P
EVC 14 0 13 0 1
*
.SUBCKT OASIMP 1 2 3
RIN 1 2 10MEG
G1 0 3 1 2 100K
RO 3 0 10
CO 3 0 .016
.ENDS OASIMP
*
X2 D 17 DIDEAL
EDD 18 0 VALUE = {(V(16)/1)/(.25*(12+V(8)-V(1)))}
X3 D 18 DIDEAL
IRM 0 D DC 1M
RM 0 D 10K
*
.SUBCKT DIDEAL 1 2
VAS 1 3 DC -1U
D1 3 2 D
D2 3 4 D
D3 4 2 D1
FAS 4 2 VAS 1
.MODEL D D IS=1E-6
.MODEL D1 D
CC 1 2 .1P
.ENDS DIDEAL
*
.PROBE
.END
```

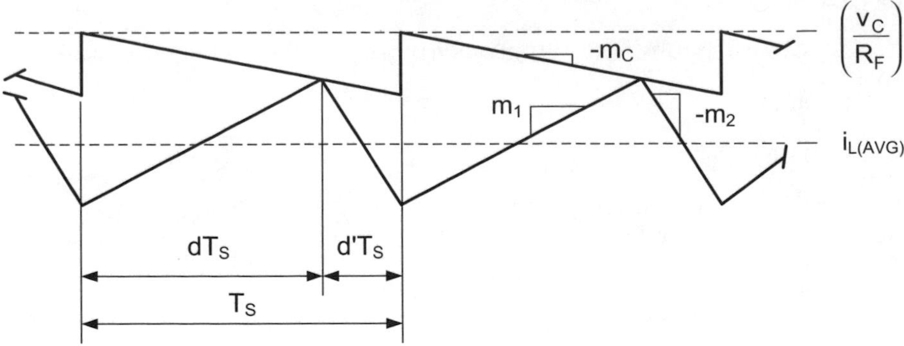

FIGURE 4.21
Peak current sensing current mode control waveforms (continuous conduction mode).

4.2.1 Current Mode Converter Model Setup

To maintain a sense of consistency, use the same converter of Figure 4.1 and its model (Figure 4.2) and implement a current mode control scheme with it for the current mode analysis. Figure 4.21 and Figure 4.22 show the fundamental inductor current and current control waveforms for the continuous and discontinuous modes of operation, respectively. From these waveforms, the following expressions for the control parameter, d, are determined. Slope m_1 is the ON time slope of the inductor current and $-m_2$ is the OFF time slope. Slope m_C is that of the stabilizing ramp added to the peak current control voltage, v_C/R_f, for stability.[12] For continuous conduction mode, d is derived from Figure 4.21 as:

$$d = \frac{\frac{v_C}{R_f} - i_{L2}}{\frac{T_S}{2L_2}(2m_C L_2 + v_g - v)} \tag{4.6}$$

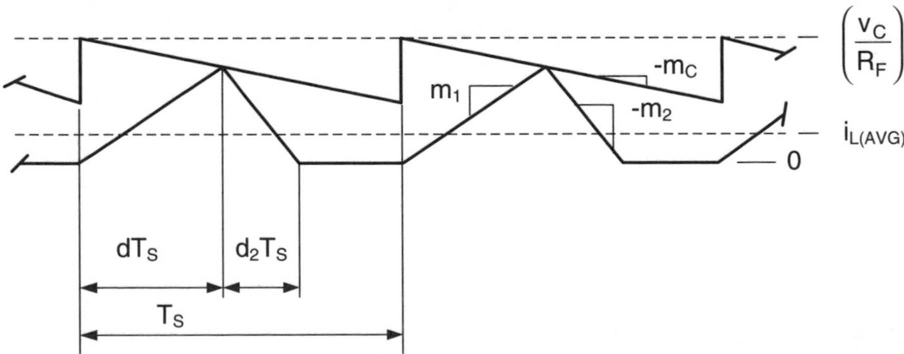

FIGURE 4.22
Peak current sensing current mode control waveforms (discontinuous conduction mode).

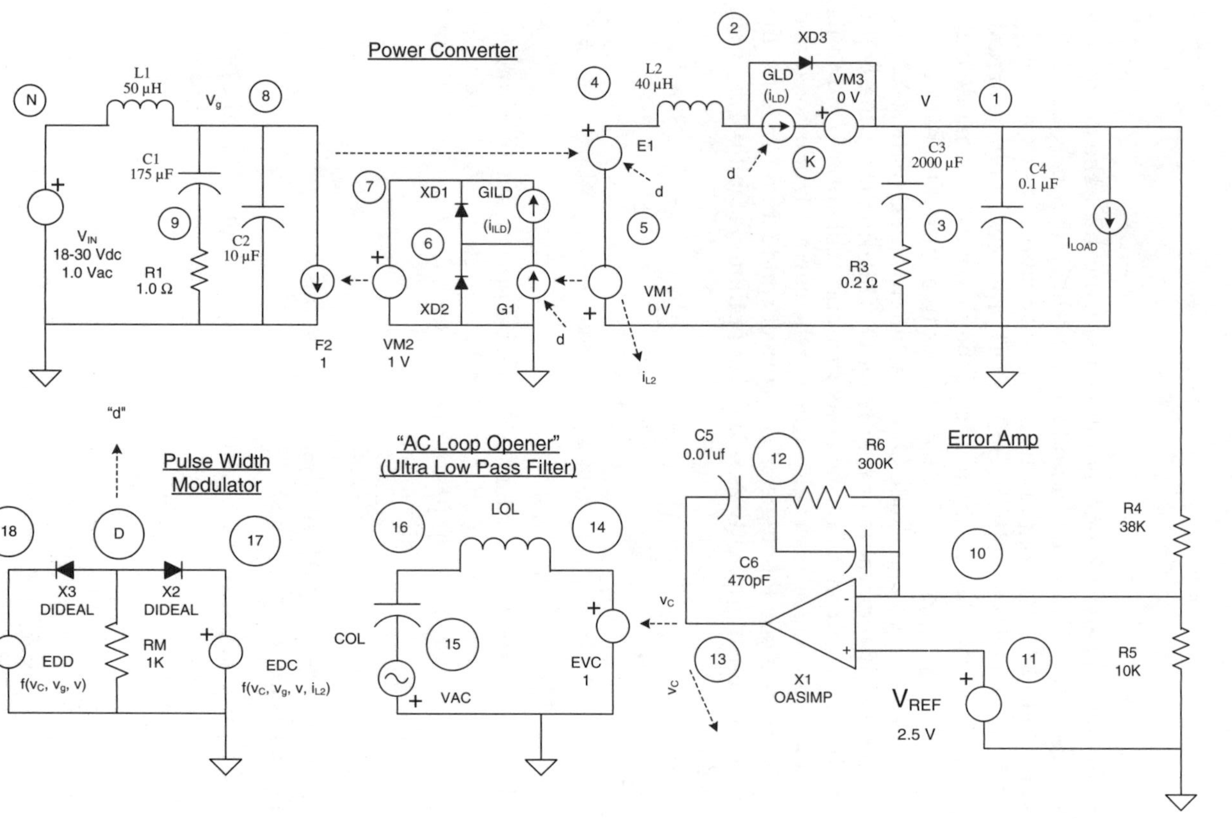

FIGURE 4.23
Buck converter switching regulator model (current mode).

and the discontinuous mode case from Figure 4.22 is:

$$d = \frac{\frac{v_C}{R_f}}{\frac{T_S}{L_2}(m_C L_2 + v_g - v)} \tag{4.7}$$

The current mode analysis model is shown in Figure 4.23. The dependent voltage source EDC will produce a voltage at node 17 that is equal to the calculated magnitude of the continuous mode duty ratio, *d*, as expressed by Equation 4.6. The dependent voltage source EDD produces a voltage at node 18 representing the calculated magnitude of the discontinuous mode duty ratio, *d*, expressed by Equation 4.7. Each of these two duty ratio parameters is continuously calculated and diode ORed by the two ideal diodes, X2 and X3, producing the smaller value of *d* at node D. This smaller value is the correct *d* to be used for control of the converter and will correspond to its correct conduction mode. (This is also true for the current mode boost and buck–boost current mode topologies.) The correct conduction mode for the converter is determined implicitly as usual. For optimum stability,[5] m_C is set equal to $-m_2$ or

$$m_C = \frac{V}{L_2} = \frac{12V}{40\,\mu H} = 0.3\frac{V}{\mu S} \tag{4.8}$$

R_f will be 1 Ω because this scaling will be appropriate for this analysis (Chapter 6 contains more information about this selection). With the value of T_S still at 10 μS, and L_2 at 40 μH, equations for the values of EDC and EDD can now be inserted in the model.

4.2.2 Open Voltage Loop AC Analysis (Continuous and Discontinuous Modes)

The plot of Figure 4.24 shows the continuous mode loop gain and phase of the outer voltage control loop of the current mode switching regulator; Figure 4.25 shows the discontinuous mode cases. (The inner current control loop is closed via the equation for EDC.) The loads are the same as they were for the voltage mode regulator in Section 4.1.3. The error amplifier frequency response characteristic was modified from the voltage mode example to provide more practical results for the current mode analysis.

As was the case for the voltage mode converter, note that the continuous mode gains in Figure 4.24 are independent of load current variations (current sink loads), but that the gains in the discontinuous mode case of Figure 4.25 do vary with load currents. Note that the unity gain crossover frequency is

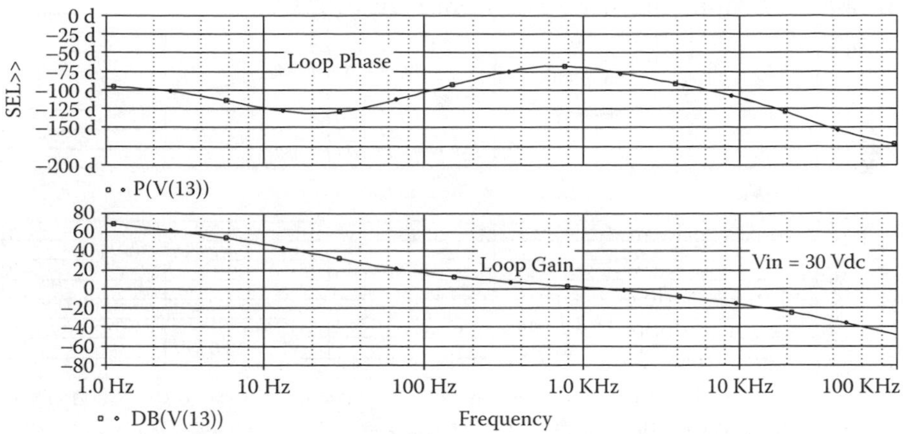

FIGURE 4.24
Buck SMPS continuous mode AC loop gain and phase with load change (current mode).

reduced for the discontinuous mode cases and continues to decrease for continuously decreasing loads. Also, to ensure that the feedback control loop maintains AC stability, the phase shift must not approach or exceed a magnitude of −180° at the unity gain crossover frequency. As stated earlier, typical desired numbers for the phase and gain margins are 45° and 10 dB, respectively. Kuo[10] and D'Azzo and Houpis[11] offer more fundamental information on feedback control stability.

Now consider the effects of input DC voltage variations on loop gain by stepping V_{in} from 18 to 30 Vdc with the load cases of 2.5 amps (continuous)

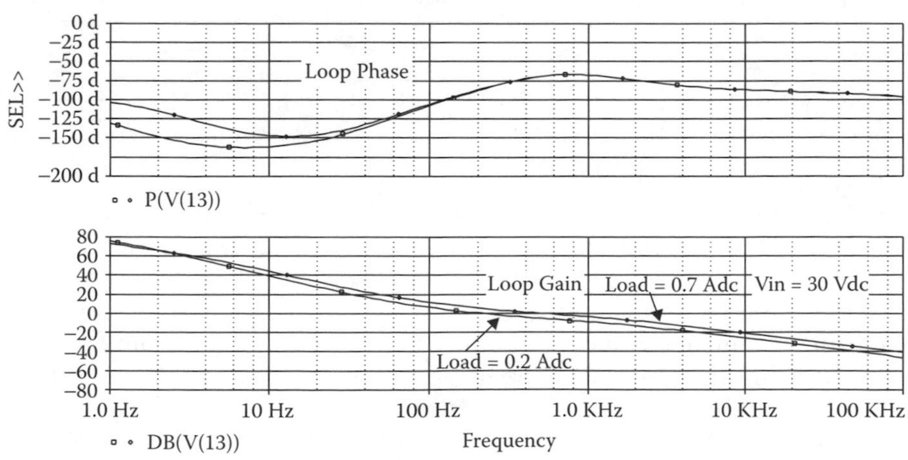

FIGURE 4.25
Buck SMPS discontinuous mode AC loop gain and phase with load change (current mode).

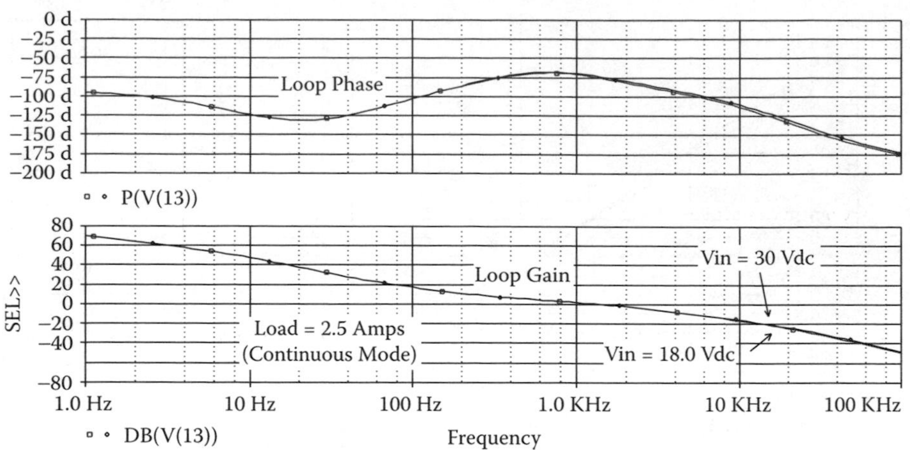

FIGURE 4.26
Buck SMPS continuous mode AC loop gain and phase with input voltage change (current mode).

and 0.2 amps (discontinuous). Figure 4.26 and Figure 4.27 show the respective plots. Only the discontinuous mode case plot indicates a noticeable shift in gain with the input voltage. This may not be true for all designs, but a small change will generally be noticed.

4.2.2.1 Open Voltage Loop AC Output Impedance (Current Mode)

Now look at the AC output impedance of the converter for the same conditions as those set up in the previous section. This is the output impedance of the

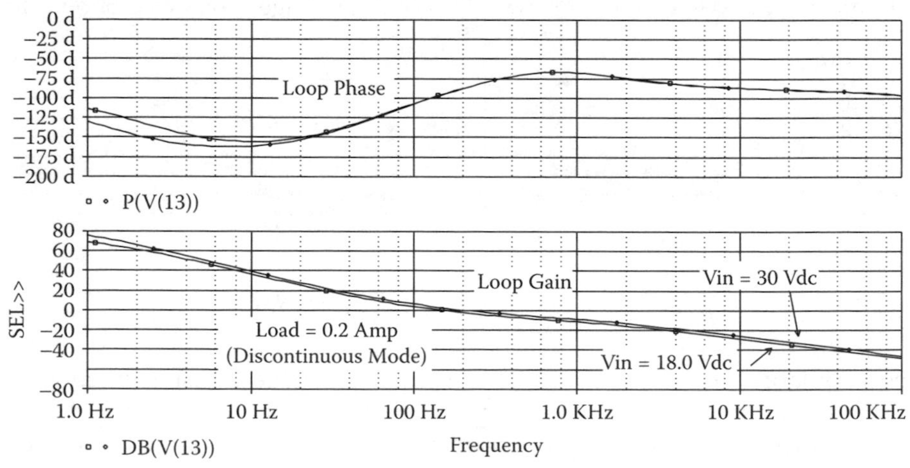

FIGURE 4.27
Buck SMPS discontinuous mode AC loop gain and phase with input voltage change (current mode).

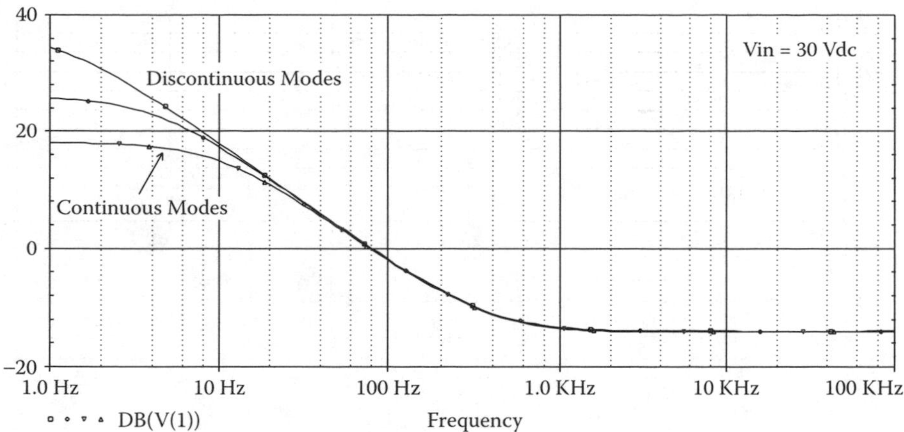

FIGURE 4.28
Output impedance of the open voltage loop current mode converter.

SMPS with the inner inductor current control loop closed and the outer output voltage control loop opened "AC wise" with the ultralow-pass filter circuit (see Section 4.1.3). Figure 4.28 shows the plot (0 dB is equivalent to 1 Ω). The low-frequency discontinuous mode output impedances are relatively high and also a function of the load current prior to breaking with the output filter capacitor, C3, at frequencies near 4 Hz. The continuous mode cases are theoretically independent of load current and have high impedance at low frequencies and even approaching that of the discontinuous mode cases. This is considerably higher than that of the voltage mode output impedance when comparing this same converter with that of the voltage mode case shown in Figure 4.6. This is characteristic of current mode converters.

4.2.2.2 Open Voltage Loop Forward Transfer Function (Current Mode)

The AC line rejection of the current mode open voltage loop configuration will now be examined. Figure 4.29 shows the results without the input filter. Figure 4.30 shows the result with the input filter added.

4.2.2.3 Control to Output Analysis (Current Mode)

With the voltage loop opened AC wise by letting LOL and COL (Figure 4.23) be very large values, the control to output AC voltage transfer function of the current mode converter is now considered. The load currents are 1.2 and 2.5 amps for continuous conduction mode and 0.2 and 0.7 amps for the discontinuous conduction mode cases. The input voltage is fixed at 30 Vdc and the output voltage is maintained at 12 Vdc by the voltage loop DC-only feedback control. Figure 4.31 shows the analysis results. Note that the continuous mode gains are independent of load and the discontinuous mode

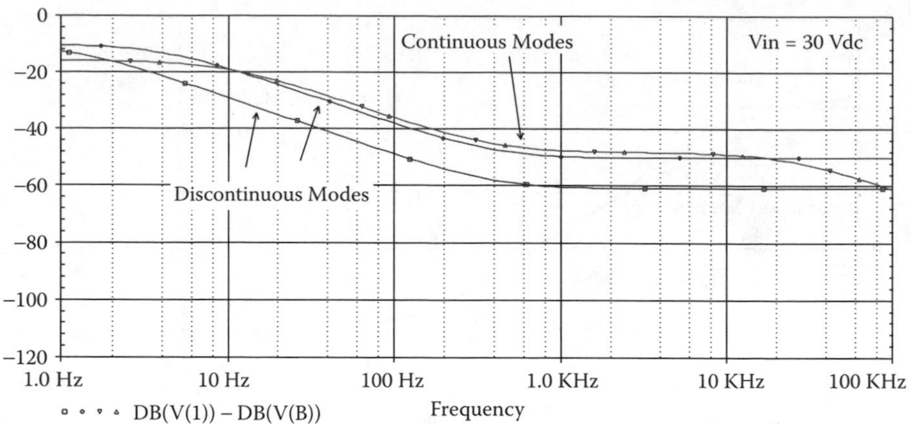

FIGURE 4.29
AC forward transfer function of the open voltage loop current mode converter without the input filter.

cases do vary with load current. The gain decreases at a slope of –1, which is generally characteristic of current mode converters in continuous and discontinuous modes of operation.

4.2.2.4 Inner Inductor Current Control Loop Gain

Occasionally, it may be desirable to examine the inner inductor current control loop to see if its AC characteristics are as expected. This is accomplished by opening the inductor current feedback signal AC wise (this current is sensed

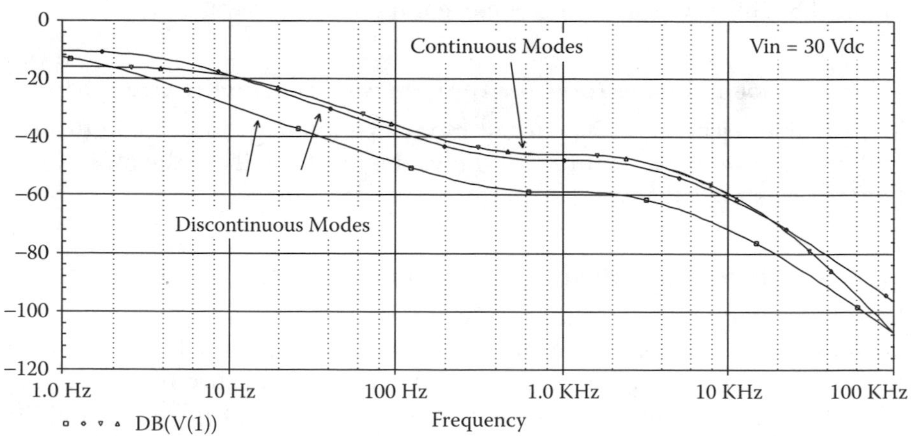

FIGURE 4.30
AC forward transfer function of the open voltage loop current mode converter with input filter added.

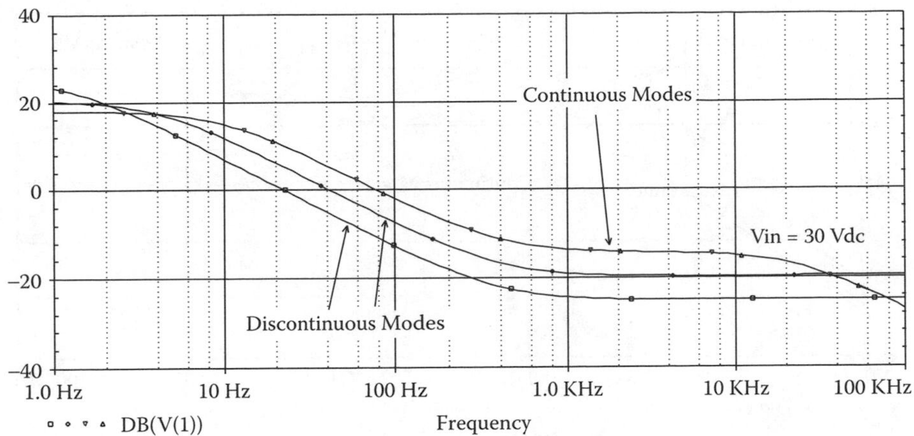

FIGURE 4.31
Buck converter AC "*d* to output" transfer function (current mode).

by VM1 in Figure 4.23). The outer voltage loop will also be opened for AC signals by maintaining the ultralow-pass filter with large values of LOL and COL; however, its VAC is now set to zero. Then another ultralow-pass filter is created for the inductor current signal to open its loop for AC analysis. (Note: the circuit for this second ultralow-pass filter is not shown on the model schematic of Figure 4.23, but it can be easily visualized from the current mode netlist, Netlist 4.3.) VAC for this new ultralow-pass filter insertion is conveniently set to 1.0 VAC as the AC stimulus for analysis for this loop. Figure 4.32 shows the results.

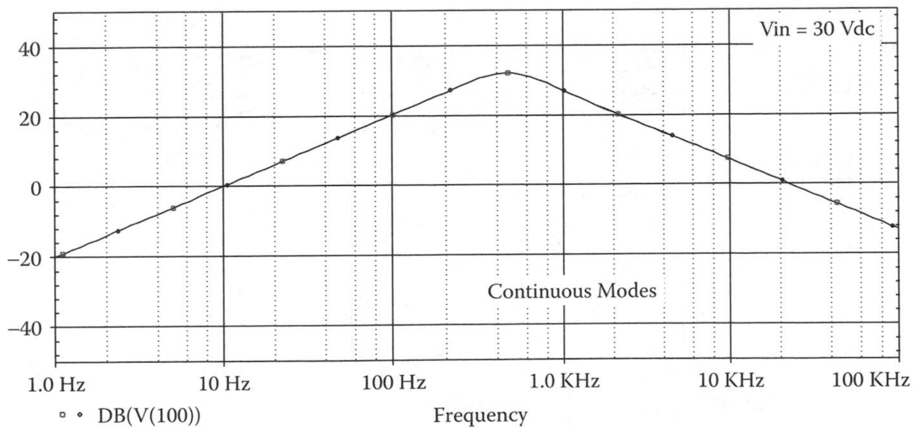

FIGURE 4.32
Inner inductor current control loop gain.

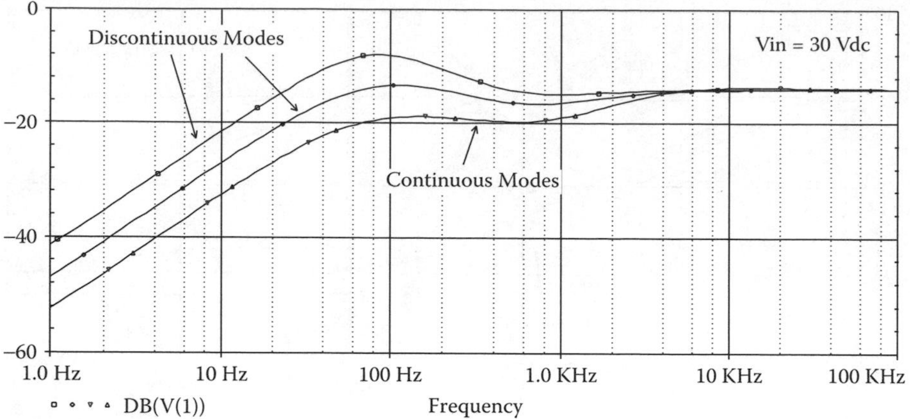

FIGURE 4.33
Output impedance of the current mode SMPS.

4.2.3 SMPS Closed Loop Analysis (Current Mode)

Now close the inner current loop and the outer voltage loop, configure the SMPS in its normal operating configuration, and then examine its properties of output impedance and line regulation. The addition of an input voltage feedforward analysis is not considered here. This implementation is generally not necessary with current mode control because of its superior line rejection characteristics in continuous as well as discontinuous modes.

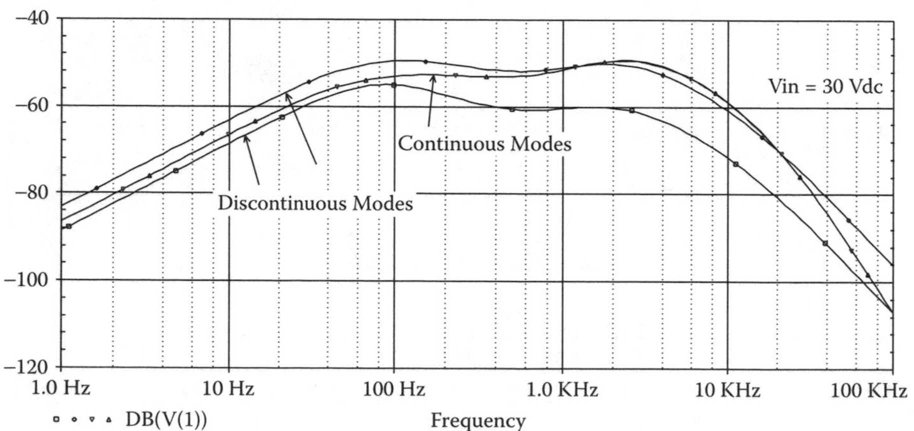

FIGURE 4.34
AC line rejection of the current mode SMPS (*Vin* = 30 Vdc).

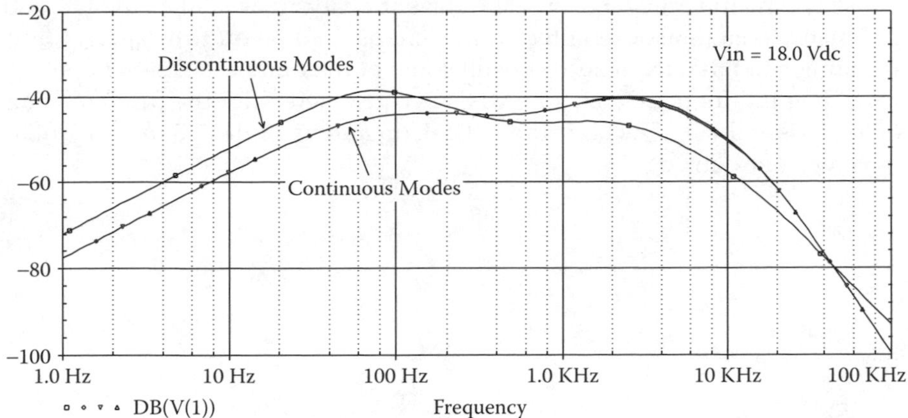

FIGURE 4.35
AC line rejection of the current mode SMPS (*Vin* = 18 Vdc).

4.2.3.1 SMPS Output Impedance (Current Mode)

Now look at the SMPS output impedance for the extreme variations in load currents considered in the open voltage loop gain analyses in Section 4.2.2.1. Input voltage was held at 30 Vdc. The plot of Figure 4.33 shows the AC impedance magnitudes. Notice that the impedances of Figure 4.28 are reduced to those of Figure 4.33 by the voltage loop gain factor of Equation 2.2.

4.2.3.2 SMPS Line Regulation (Current Mode)

Figure 4.34 and Figure 4.35 show the plots of AC line rejection for input voltages of 30 and 18 Vdc, respectively. Note the excellent line rejection for all conditions; this is one of the salient characteristics of current mode control.

4.3 Summary

This chapter has considered most of the fundamental kinds of DC and AC analysis in which one might be interested when analyzing a very basic switching regulator. These have been for the power converter with no duty ratio control; the voltage mode-controlled SMPS; and the current mode-controlled SMPS. Chapter 6 will continue with the general theme of this chapter and provide more advanced or comprehensive detailed analysis methods.

These will include the large signal transient analyses as well as the DC and AC analysis methods presented in this chapter. To be on firm ground, it is recommended that the reader consult some of the numerous references[14,24,26] regarding the fundamentals of SMPSs because here only the fundamentals of analysis and very few basic theoretical, operating, or design fundamentals have been presented.

5

SMPS Component Models

Chapter 3 provided the conceptual and fundamental development of switch mode power supply (SMPS) analysis models for computer simulations. This chapter provides information on constructing comprehensive and definitive circuit component models. These are required when it is desired to create more realistic SMPS analysis models containing higher than first-order effects. Real-world capacitor, inductor, and diode behavioral model concepts are considered. Comments are made on the desirability of developing circuit-averaging integrated circuit PWM controller macromodels. Also included are general comments relating to many peripheral circuit additions associated with power supplies such as postregulators, inrush current limiters, power factor correction circuits, etc.

5.1 Component Model Development

In this section the development of the sometimes complex real-world models of the passive components capacitors and inductors is included. A section on resistors is not included here because they are generally represented by a simple resistive element in SMPS analysis, with the exception of sometimes including the simple equivalent series inductance effect or a possible resistance tolerance variation with environment. Also included is the development of a real-world circuit-averaged model for a diode macro. Integrated circuit macros are also considered for some generic types of controllers in use today.

5.1.1 Real-World Capacitor Macro

Figure 5.1 shows the equivalent circuit of a capacitor with the parasitic effects of equivalent series resistance, ESR, equivalent series inductance, ESL, and the leakage resistance effects, R_P. All of these effects are, in general, not necessarily included simultaneously in a component model

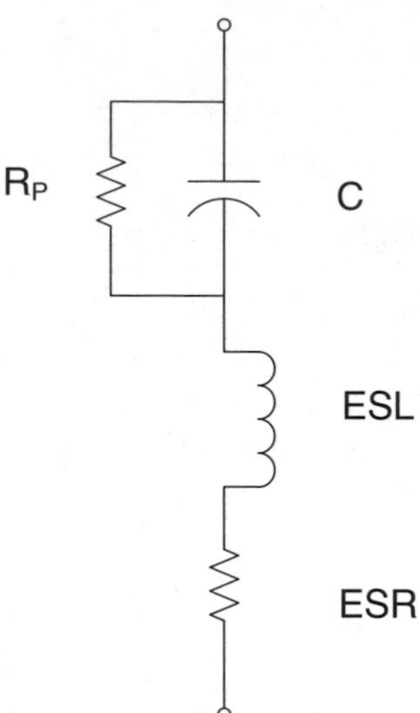

FIGURE 5.1
Capacitor equivalent circuit model.

when a circuit analysis is conducted. For example, the equivalent leakage resistance, R_P, need not be added to the model when another resistance of a much lower value, such as a power load on a filter capacitor, is in parallel with the capacitor. Also, the equivalent series parasitic components, ESL and ESR, need not be included if a low-frequency analysis is being conducted and it can be determined from the outset that the magnitudes of these effects are sufficiently low so as not to have any effect on circuit operation.

The ESL of a capacitor is generally a simple series inductor element added in series with the capacitance element of the model. Its magnitude is essentially a function of the physical geometry and lead connections to the device and is usually modeled as a fixed lumped inductance value. (It is very common also to lump circuit wiring inductances into the ESL value to produce a more realistic analysis result.)

The ESR of the capacitor is a more complex parameter. In many cases, it is more than a simple lumped resistance, although it may be modeled as such when a first-order analysis is performed or when only a limited range of operating conditions of frequency and temperature is considered. The particular capacitor technology (i.e., tantalum, aluminum, ceramic, film, etc.) also contributes to the unique complexity of this ESR impedance of a capacitor. The ESR in more complex capacitor structures such as tantalum and aluminum capacitors has the unique property of varying in a nonlinear fashion with temperature and frequency. Some details of how this effect may be modeled will now be considered.

Figure 5.2 shows a typical impedance plot of a wet slug tantalum capacitor at various temperatures. Figure 5.3 shows the +25°C case with the

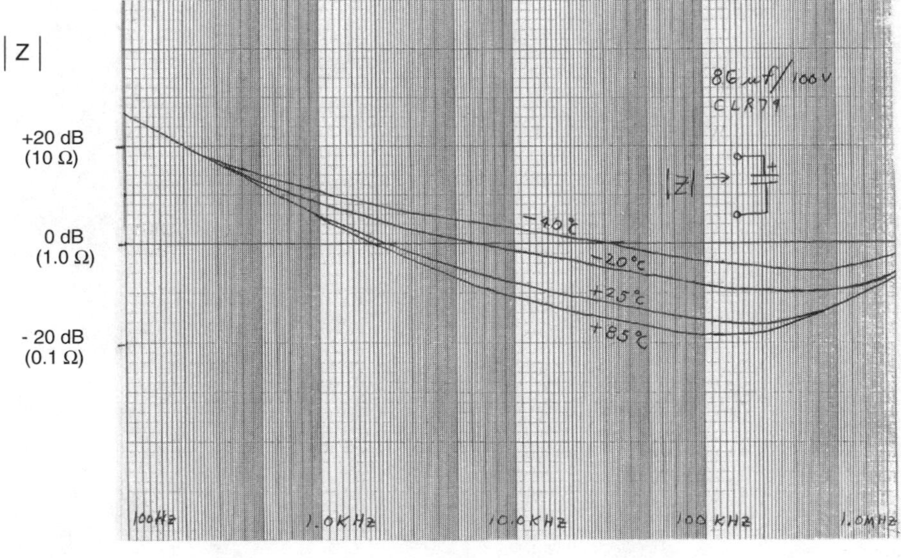

FIGURE 5.2
Typical impedance plot of a wet slug tantalum capacitor at various temperatures.

asymptotes sketched in. Note that the asymptotic slope of the ESR impedance in this case varies with frequency — unlike the familiar manner of an ideal resistor, which has a slope of 0 dB/decade. Also, it does not have the asymptote of an ideal capacitor, which has a slope of –20 dB/decade; it has a slope somewhere in between. (Calling this effect a "resistance" is a misnomer in this case because it does not behave like a pure resistance.) The slope of 0 dB/decade is defined in the literature as a slope of 0 and the slope of –20 dB/decade is defined as a slope of –1, so this "in between" slope will be defined as a fractional slope because its value lies between 0 and –1.

For these tantalum capacitors, a simple exponential relationship may be used to model this nonlinear ESR vs. frequency, f, effect. The graphical data shown in Figure 5.3 are needed to determine the constants of Equation 5.1, which will be used to define this effect.

$$ESR = R_1\left(\frac{f_1}{f}\right)^k \tag{5.1}$$

R_1 is the impedance at frequency f_1 and k is the fractional slope of the plot in this ESR region. After analyzing the capacitor model, an additional iterative

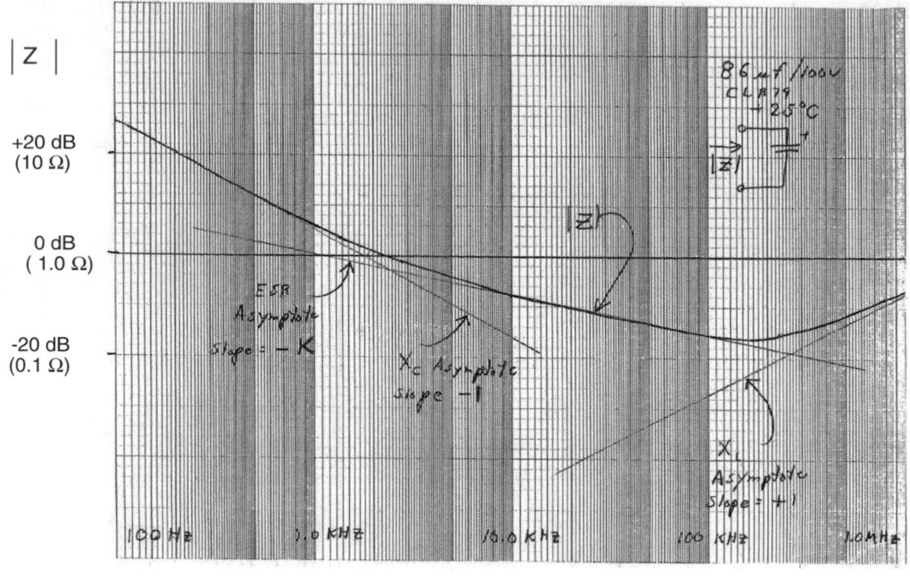

FIGURE 5.3
AC impedance plot of +25°C case showing asymptotes.

determination of these constants may sometimes be required if it is desired to have very close correlation between the model impedance and the actual component graphical impedance. This is because it is not always easy to determine an actual equivalent straight line value for the ESR asymptotic slope from the graphical data due to possible skin effects in the conductors and other nonideal factors. When a computer simulation is conducted, the ESR effect may be modeled in the Laplace equation form as

$$ESR = R_1\left(\frac{2\pi f_1}{s}\right)^k \tag{5.2}$$

This will provide a valid ESR effect for small signal AC as well as large signal transient analyses.

To illustrate, now determine the desired computer model for the capacitor of Figure 5.2. Figure 5.3 shows the impedance plot for the +25°C case along with the most accurate asymptotes that can initially be constructed visually. The low-frequency slope of –1 obviously represents the ideal capacitor of magnitude

$$C = \frac{1}{2\pi f X_C} \tag{5.3}$$

with X_C the capacitance reactance at a corresponding frequency f. For this case, select the impedance of 1 Ω (0 dB) occurring at a frequency of 1.85 kHz. This yields a capacitance value of

$$C = 86\ \mu\text{F} \tag{5.4}$$

Similarly, the ESL is determined from

$$L = \frac{X_L}{2\pi f} \tag{5.5}$$

with X_L the inductive reactance at a corresponding frequency f. In this case, again select the impedance of 0.1 Ω (–20 dB) occurring at a frequency of 240 kHz. This yields an inductance of

$$L = 0.066\ \mu\text{H} \tag{5.6}$$

Now determine the ESR using the impedance plot and Equation 5.1. Select a value of ESR (R_2) at a corresponding frequency f_2 that is different from f_1. Rearranging Equation 5.1, one can solve for the value of k:

$$k = \frac{\log\left(\frac{R_2}{R_1}\right)}{\log\left(\frac{f_1}{f_2}\right)} \tag{5.7}$$

Using the values of R_1, R_2, f_1, and f_2 from Figure 5.3, the value of k can now be calculated and the desired model for the ESR obtained. In this case, the value of k is:

$$k = \frac{\log\left(\frac{0.1\Omega}{1.0\Omega}\right)}{\log\left(\frac{1.1kHz}{360kHz}\right)} = 0.40 \tag{5.8}$$

With the value of the capacitance, inductance, and the ESR expressed by Equation 5.2, the desired three-element model of the capacitor at +25°C has been obtained.

Now look at the temperature aspects of the components. The capacitor specification data sheet and/or the experimental graph of Figure 5.2 will provide a temperature coefficient of capacitance, which may be implemented into the model with an appropriate equation. This may be linear or nonlinear.

From Figure 5.1, the ESR vs. temperature for this capacitor is apparently nonlinear. Using some curve-fitting technique for the ESR variation with temperature at the geometric mean frequency of f_1 and f_2 will generally provide a good useable representation of the ESR variation with temperature over the desired temperature range. The temperature-modifying equations may simply be a multiplying factor to the capacitance and the ESR equation (Equation 5.2).

The approach presented here seems to work for most types of tantalum capacitors, but other expressions for ESR may be necessary when modeling other capacitor technologies.

5.1.2 Real-World Inductor Macro

Inductive components have several complex parasitic effects that should be considered when SMPS second-order effects are to be modeled. Figure 5.4 shows a lumped equivalent circuit of an inductor showing the second-order parasitic effects. In any inductor, the magnetic flux generated by the current flowing in the conductor has two components: the flux contained in the magnetic core material and the flux not in the magnetic material — or the leakage flux, as it is more commonly known.

The inductance attributed to the flux in the core is generally called the magnetizing inductance, L_M, and the inductance attributed to the leakage flux is represented by the leakage inductance, L_l. The AC flux in the core is capable of producing core losses, which are represented by losses in a parallel resistance, R_P. Capacitor, C_P, represents the lumped equivalent capacitance existing across the windings of the inductor. The series resistance, R_S, represents the effective resistance of the winding. This resistance is equivalent to the DC resistance of the winding but also may include effective resistance increases caused by the AC skin and proximity effects.[16,17] These resistances may also need to be adjusted for temperature effects caused not only by the environment, but also by self-heating.

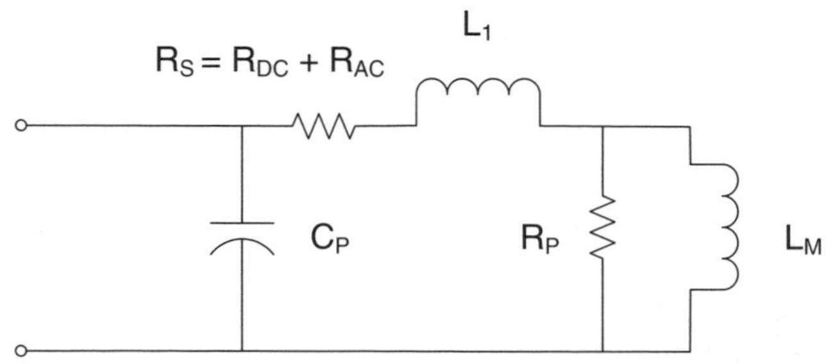

FIGURE 5.4
Inductor equivalent circuit model.

An experimental determination of the actual component temperature rise can generally be made by operating the device under the desired operating conditions, suddenly remove operating power, and then measuring the DC winding resistance. Knowing the winding material resistivity, one can then determine the actual temperature rise in the device.

Now look at some ways of determining the expressions for these five circuit elements that comprise the inductor macro. Once these have been determined, a behavioral computer model can be derived. Some of these parameters are best obtained by measured data; others are more easily derived from calculations or possibly a combination of both methods. It generally takes a certain amount of experience with magnetic components to do this accurately. If the device is a purchased item, one might request that the equivalent circuit model or some part of it be provided by the manufacturer.

Consider one way to develop the inductor model in a systematic and admittedly idyllic manner. First, consider a measurement technique by adding an auxiliary winding on the inductor with a number of turns — preferably, at least 10% of the number of turns of the inductor. (This approach may not be possible for some preassembled magnetics and thus other provisions must be made.) Use a wire gauge that is sufficiently small to be accommodated by the allowable excess window area, but not so small as to be unable to handle the expected test current. (In most cases, this current will be relatively small.) The number of turns and wire size of this auxiliary winding are generally not individually exacting parameters, so some iteration may be required here to provide acceptable data.

First use a low-resistance measuring ohmmeter to measure the DC resistance of the inductor, R_{DC}, accurately. Next, the AC component factors are to be determined. First apply a short to the auxiliary winding. A commercially available network analyzer is a great aid at this point. With the analyzer, make an impedance plot vs. frequency of the inductor. The plot might have a shape similar to the one shown in Figure 5.5. From this plot, one can calculate the leakage inductance, L_l, from the inductive reactance, as shown in Figure 5.5. The low-frequency resistance is also shown; this should be approximately equal to the previously measured R_{DC}.

As the frequency increases, the resistance increases approximately proportional to the square root of the frequency. This is the AC skin and proximity effect resistances, R_{AC}. As the frequency is increased further, these AC resistances are masked by the higher impedance of the leakage inductance reactance, X_{Ll}, but the phase angle readout can help determine the AC resistance component if desired. The more advanced network analyzers sometimes have a synthesis capability, which can create a series R–L model at the various frequencies; this can aid in determining the AC resistance at the higher frequencies.

Now remove the short from the auxiliary winding and determine the magnetizing inductance for various operating conditions of DC bias current. One can insert a DC current imposed on a small AC current component to

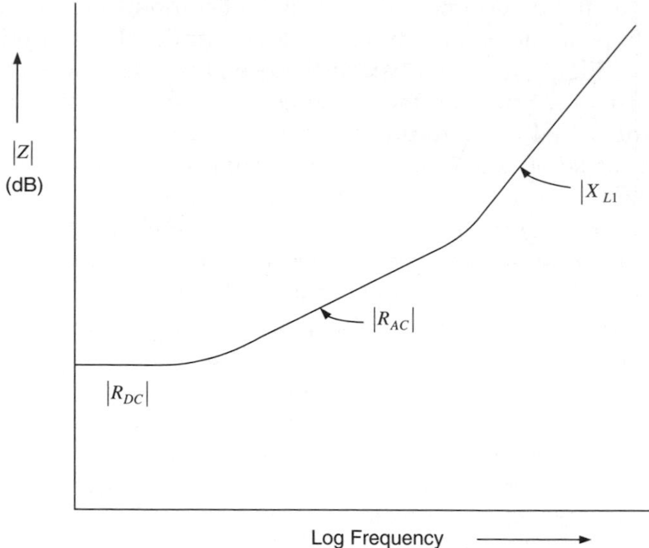

FIGURE 5.5
Inductor impedance plot with shorted auxiliary winding.

determine the incremental inductance at a particular DC bias — or magnetizing current, as it is sometimes called. Figure 5.6 shows a typical ungapped inductor and the same inductor containing an air gap. Note that the incremental inductances decrease with DC current slightly at first and then drop rapidly with the core approaching magnetic saturation. The ungapped inductor is highly variable with DC bias current; the gapped inductor is much more stable with a much smaller inductance existing over its larger magnetizing current operating range. Gapped cores are always used when a large DC current is present.

Using the data from these inductance vs. current plots, one can use curve-fitting techniques to develop a mathematical expression for these plots if desired. A general expression of the form shown in Equation 5.9 might be appropriate, although a more accurate curve fit might be obtained by a ratio of polynomials; this could be used if the computer analysis software has the capability.

$$L = \frac{L_O}{1 + ai + bi^2 + ci^3 + \cdots} \tag{5.9}$$

Now determine the winding capacitance, C_P. This can be determined by using a simple capacitance measurement with a capacitance meter operating at the higher frequencies (greater than 100 kHz) or it might be more accurately

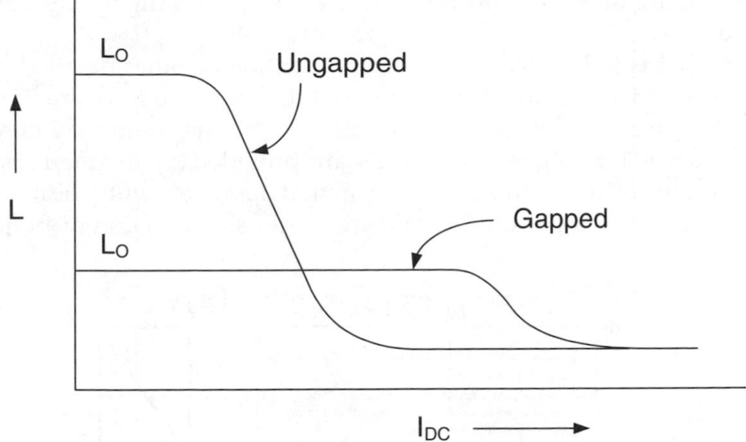

FIGURE 5.6
Inductance vs. current an ungapped core and with an air gap inserted (gapped).

determined by using the network analyzer and making an impedance vs. frequency plot that extends past the parallel resonant peak of the inductor, as shown in Figure 5.7. After this peak is reached, it is a simple calculation to determine C_P from the capacitance reactance, X_{CP}. The peak parallel resonant impedance magnitude is the core loss resistance, R_P, at the operating condition imposed by the analyzer at this point. This may be used as a fixed

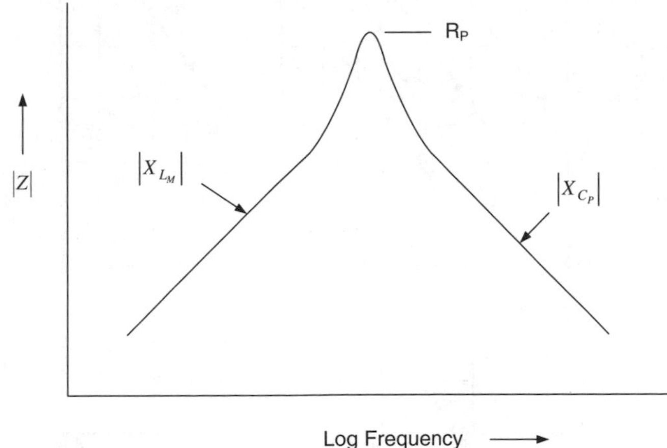

FIGURE 5.7
Inductor impedance plot showing parallel resonance.

resistance figure of merit if desired, but in reality R_P is highly variable and more complex.

If the model is to be used for power calculation or other detailed assessments, the following equations may be used to determine an operating point value of R_P. The magnetic parameters must be brought into play here. The core loss curves for magnetic materials are provided by the manufacturer. Figure 5.8 shows an example of the actual loss per unit volume (it is sometimes given in a per-unit-weight quantity) as a function of frequency,

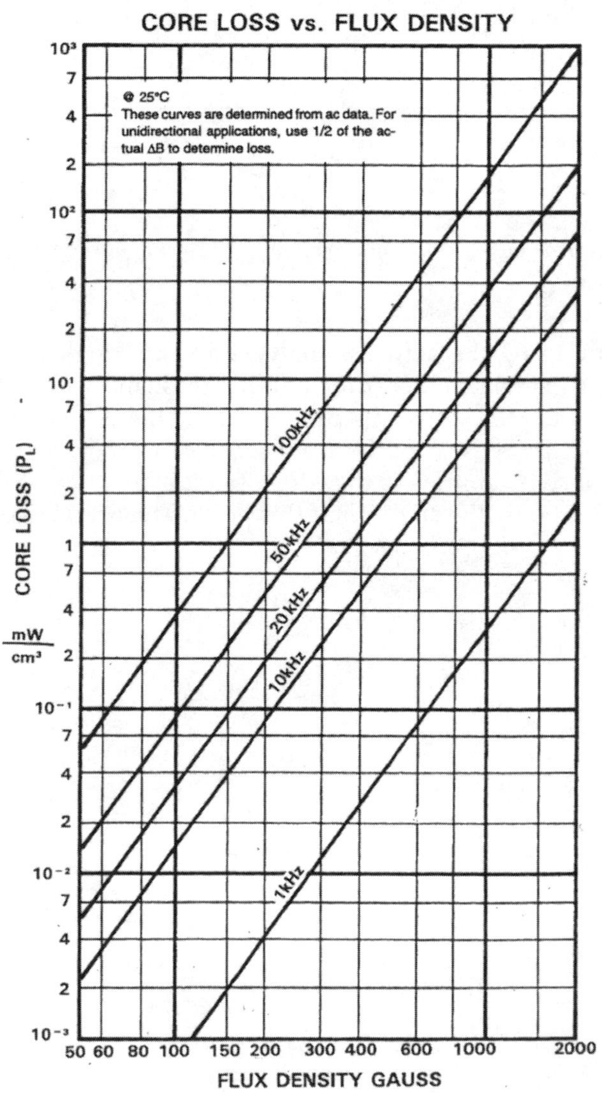

FIGURE 5.8
Typical magnetic core loss curves. (Magnetics Inc., W. Material.)

f, and the peak magnetic AC flux density variation, *B*. A fairly good acceptable expression for core loss can be approximated by Equation 5.10:

$$\frac{P_{LOSS}}{Volume} = Kf^{x}B^{y} \tag{5.10}$$

where *K*, *x*, and *y* are constants that can be obtained from the plot or, in some instances, provided by the core manufacturer. (See Schmit[18] for additional information.) From these curves, the large signal nonlinear value of the parameter R_P may be calculated for a particular inductor design with the following equations. Using Faraday's law to make the flux density, *B*, to voltage and frequency conversion,

$$B = \frac{V_{avg}}{4fNA_c} \tag{5.11}$$

N is the number of turns; A_c is the core cross sectional area; and V_{avg} is the average voltage of the AC waveform across the inductor. The value of R_P is noted as

$$R_P = \frac{V_{rms}^2}{P_{LOSS}} \tag{5.12}$$

The form factor, *FF*, of the particular wave shape provides the relationship between V_{rms} and V_{avg} as shown in Equation 5.13:

$$FF = \frac{V_{rms}}{V_{avg}} \tag{5.13}$$

Combining Equation 5.10 through Equation 5.13 yields the expression for R_P:

$$R_P = C_1 f^{(y-x)} V_{avg}{}^{(2-y)} (FF)^2 \tag{5.14}$$

where constant C_1 is

$$C_1 = \frac{(4NA_c)^y}{K(Volume)} \tag{5.15}$$

The form factor, *FF*, is a variable and must be computed for the different wave shapes; this will be a function of the voltage and converter duty ratio, *d*. Another core loss variable is temperature. Schmit[18] shows a polynomial relationship of core loss vs. temperature at a particular operating frequency

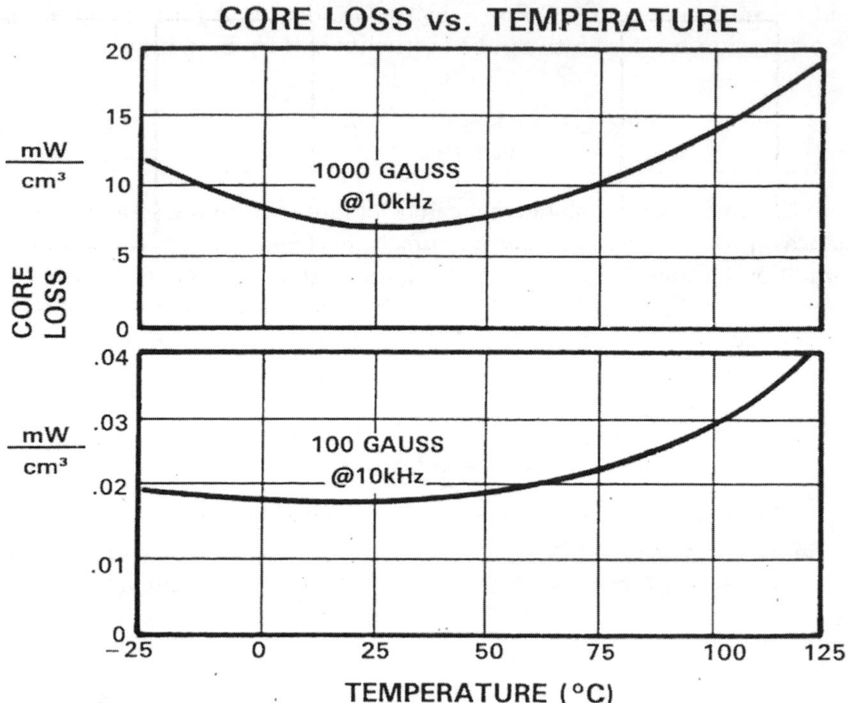

FIGURE 5.9
Core loss vs. temperature for differing operating conditions.

and flux density, but Figure 5.9 shows that core loss vs. temperature is different for differing operating conditions. Thus, a core loss resistance equation containing all three parameters of frequency, flux density, and temperature may be difficult to derive.

In some situations, a polynomial curve-fitted equation derived from a curve similar to Figure 5.9 may be used as a divider for Equation 5.14 if a core loss vs. temperature plot for operating conditions near those of the actual frequency and flux density operating conditions can be obtained. This rather involved method of determining R_P may seem impractical and may be so for most analysis; nevertheless, it is presented here as a possible method of doing so if desired. In most cases, a constant figure of merit value of R_P with a third- or fourth-order polynomial divider for temperature adjustments per Schmit[18] would be satisfactory.

5.1.3 SMPS Diode Macro

When simulating a power diode in an SMPS rectifier application, one of two conditions must be noted: whether the diode current is continuous or pulsed. If the current is continuous or smooth, it is simply a matter of using the diode model in the conventional manner. However, when the current is

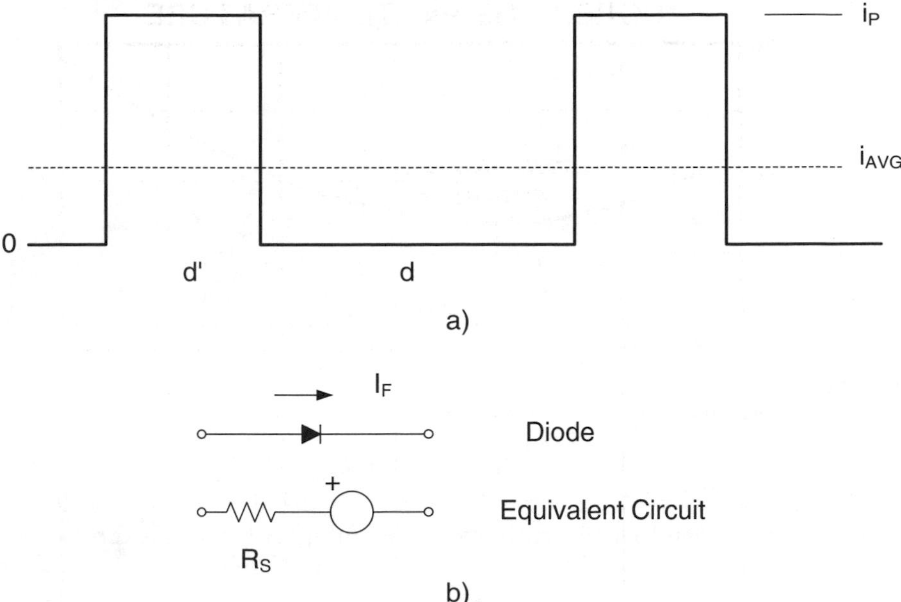

FIGURE 5.10
a) Typical diode pulsed current waveform. b) Typical diode forward voltage drop representation.

pulsed, as in a buck–boost topology, the actual operating point of the diode exhibits characteristics existing at the peak current and not those of the average current, which is the current seen to flow in the diode when one is working with a circuit-averaged SMPS model. The goal is therefore to force the diode macro to exhibit peak current characteristics while conducting average current.

Figure 5.10a shows a somewhat typical pulsed current waveform that might be seen in a power diode. Figure 5.10b shows a circuit model containing the ohmic resistance, R_S, along with a voltage source representing the classical diode equation voltage drop.[19]

$$V_F = \eta \frac{kT}{q} \ln\left(1 + \frac{I_F}{I_S}\right) \tag{5.16}$$

The term kT/q is approximately 26 mV at a temperature of 298.15°K (+25°C) and η is an assigned constant between 1 and 2. The saturation current, I_S, is a theoretical value of current, for which V_F approaches 0 when I_F is reduced to. An I_S variation with temperature is the most dominant contribution to diode temperature effects.

The goal is to make the diode conduct peak current artificially while averaged current flows in the diode macro. Figure 5.11 shows a typical macro for this

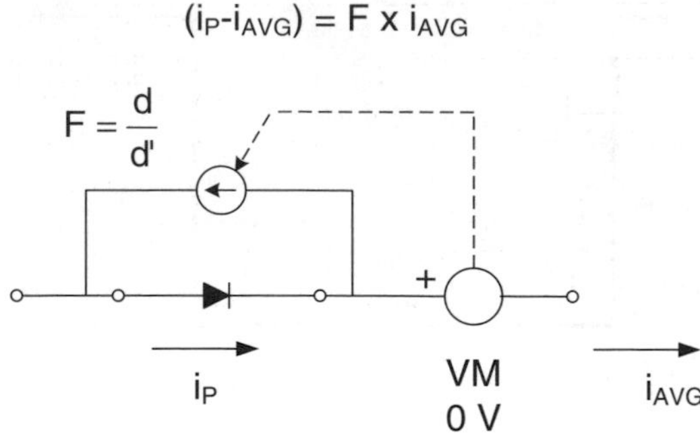

FIGURE 5.11
Typical diode macro for pulsed current applications.

application. The dependent current generator is equal in magnitude to the difference in peak and average currents. The F multiplier for i_{AVG} is expressed as:

$$F = \frac{i_p - i_{avg}}{i_{avg}} = \frac{d}{d'} \tag{5.17}$$

This will produce the correct desired results for dynamic and steady state conditions. In most simulations, the model of the actual diode used is simply inserted into the behavioral model macro.

5.1.4 Integrated Circuit Controllers

All PWM controllers have one main function: to create a desired duty ratio control of the power switches in the SMPS. There may be many conditions of control, such as normal output voltage regulation, output short circuit, or soft start-up conditions; in any case, a duty ratio control function is required. Figure 1.10 and Figure 1.11 show the simple general schemes required for voltage and current mode control, respectively. Figure 5.12 shows the block diagram of an early generation integrated circuit *voltage mode* PWM controller (UC1524A); Figure 5.13 shows it embedded in a typical application. Figure 5.14 shows the block diagram of the *current mode* controller (UC3825A,B) and Figure 5.15 shows this controller embedded in an example SMPS current mode design. As is noted each controller has several internal components. These examples are representative, but are just a sample of the many different ones available.

Appendix B provides information on constructing a circuit-averaging macro model of such a controller (UC1844) and Chapter 6 contains an SMPS example that uses this controller. One of the thrusts of this book is to advocate the development and use of these behavioral macromodels for commercially available and also custom PWM controllers. The macromodel developed in

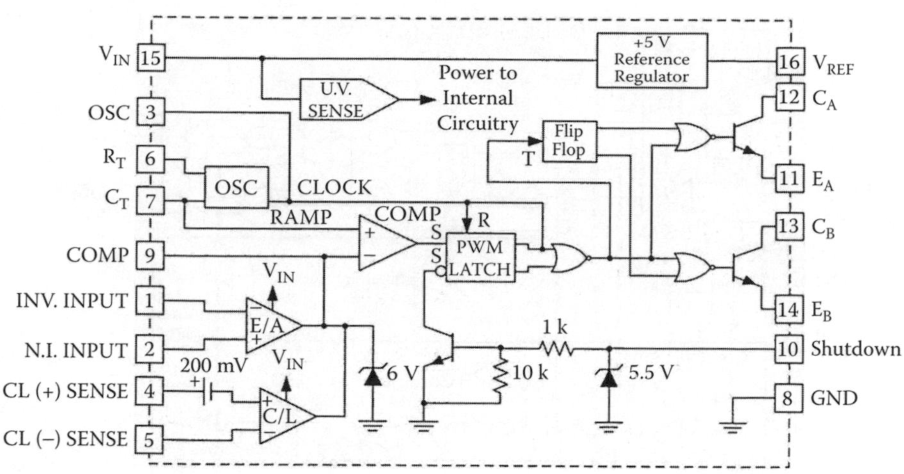

FIGURE 5.12
Block diagram of an early generation integrated circuit voltage mode PWM controller (UC1524A).

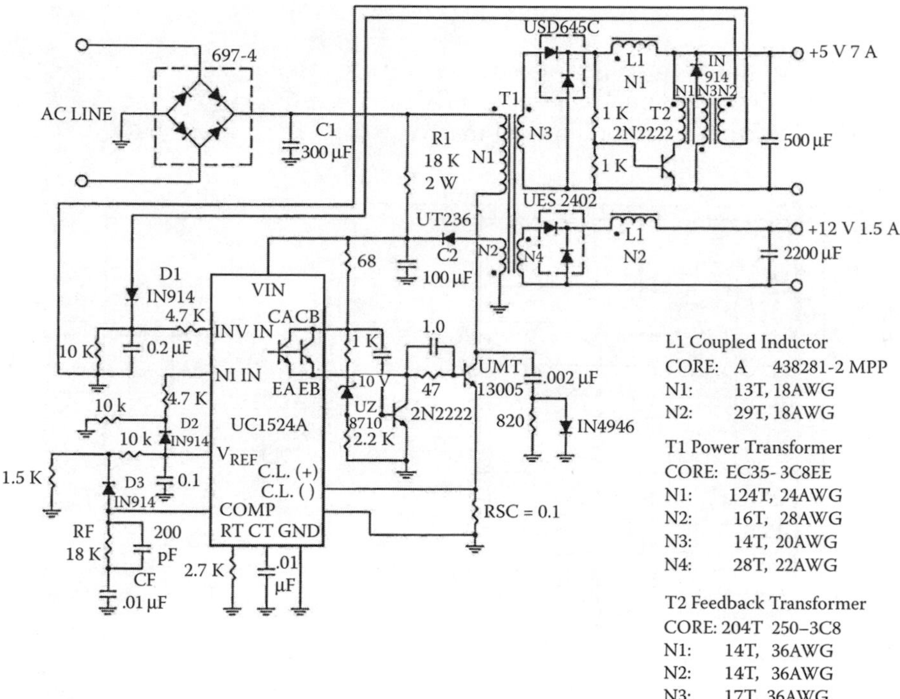

FIGURE 5.13
Early generation integrated circuit voltage mode PWM controller (UC1524A) embedded in a typical application.

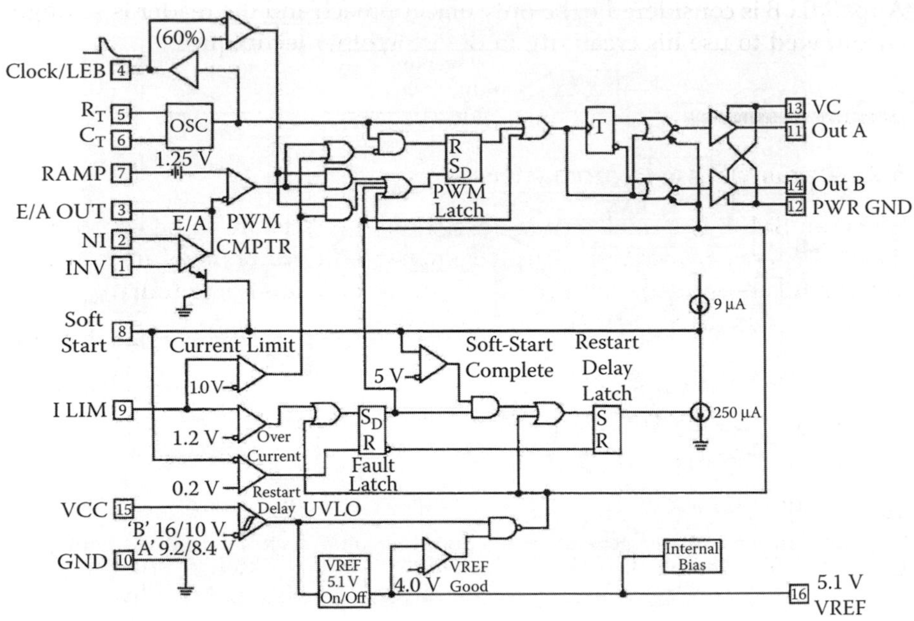

FIGURE 5.14
Block diagram of the current mode controller (UC3825A,B).

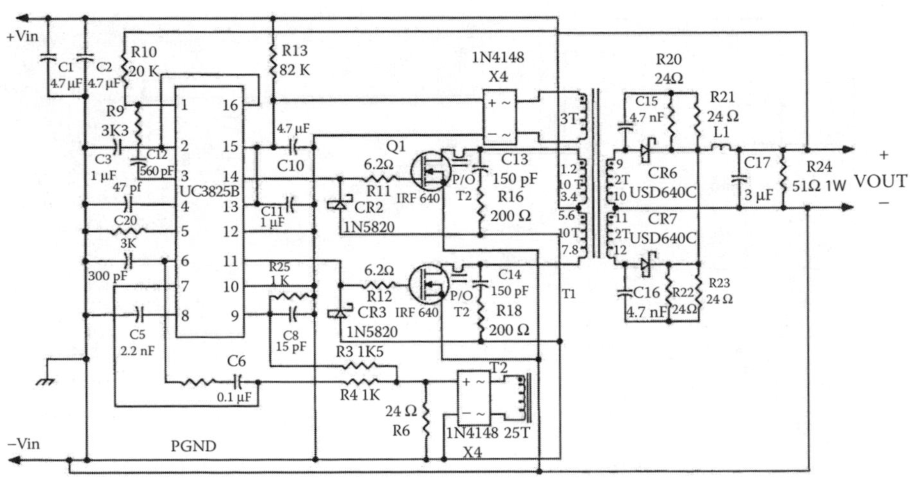

FIGURE 5.15
Current mode controller (UC3825A,B) embedded in an example SMPS current mode design.

Appendix B is considered to be only one approach and the reader is strongly encouraged to use his creativity in devising other techniques.

5.2 Parasitic Resistance Effects

When an SMPS is modeled and second-order effects are to be considered, the major parasitic resistance effects of most power components are required. These parasitic resistances have many sources; the following four are representative of the ones normally considered:

- Power switch resistance
- Power rectifier equivalent resistance
- Inductor resistance
- Input and output filter capacitor equivalent series resistances

Other component resistances, such as fuses and wiring, that are effectively in series with these resistances may be lumped in with them. Polivka et al.[20] created the seminal paper on the topic of parasitic resistances and this is recommended reading. The large signal models obtained using developments from this paper and the expressions relating these parasitic resistances for the buck, boost, and buck–boost topologies are shown in Figure 5.16 through Figure 5.18.

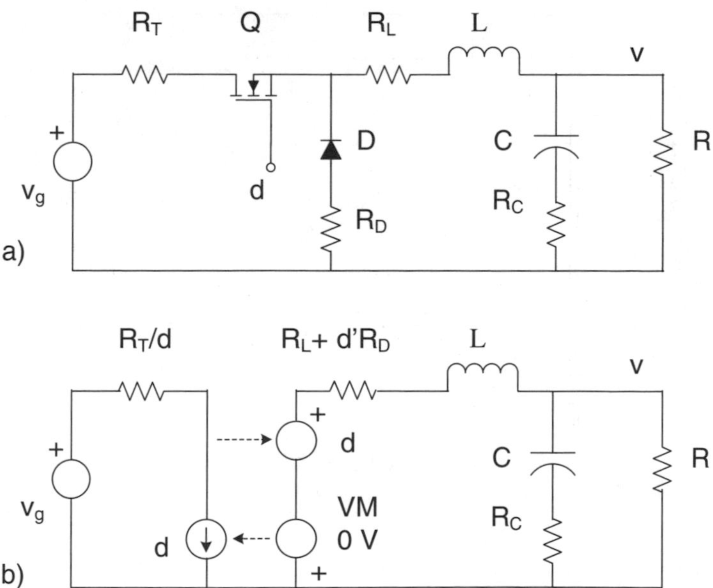

FIGURE 5.16
a) Buck converter with parasitics. b) Buck model with parasitics.

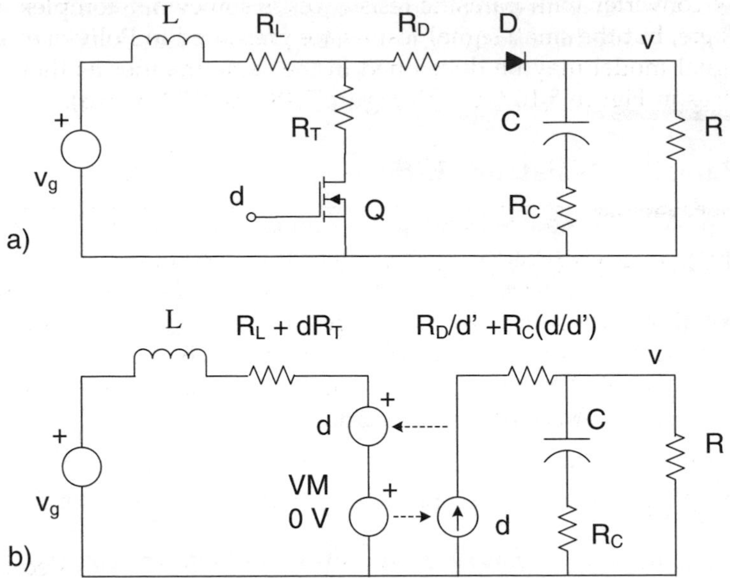

FIGURE 5.17
a) Boost converter with parasitics. b) Boost model with parasitics.

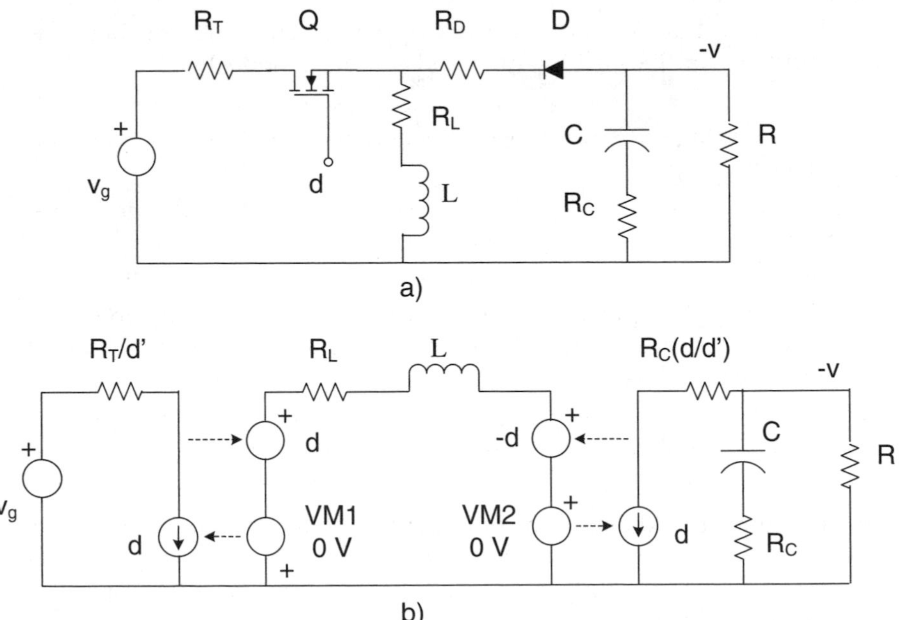

FIGURE 5.18
a) Buck–boost converter with parasitics. b) Buck–boost model with parasitics.

The Cuk converter with parasitic resistances is somewhat complex and not shown here, but the small signal results are presented in Polivka et al.[20] The large signal model may be developed in the same manner as those for the three cases in Figure 5.16 through Figure 5.18 (see Section 6.3).

5.3 Peripheral Circuit Additions

The next series of topics will provide general information on some of the main peripheral circuits associated with SMPSs. The list is by no means intended to be all inclusive, but it does show some of the main topics encountered when an SMPS is modeled.

5.3.1 Input Filter

When the many functions that an input filter of an SMPS is called upon to perform are considered, it soon becomes apparent that it is one of the most constrained circuits in the design. A detailed computer analysis is very essential in determining that all these constraints are satisfied. The following list contains many of these functions and constraints:

- The input filter limits conducted ripple current emissions onto the input power lines (differential and common modes).
- The input filter provides RF and some measure of required audio noise voltage susceptibility rejection (differential and common modes).
- The filter AC output impedance must be held to less than a necessary maximum value to ensure that the input filter instability problem does not exist. See Middlebrook and Cuk[2] and Middlebrook.[13]
- The filter AC input impedance must not be allowed to produce a series resonant dip that would violate any power source loading limitations.
- The filter should be able to limit the converter switching ripple voltage on the filter output to a specified level to ensure glitch-free power converter operation.
- The filter should limit audio susceptibility-produced resonant peaks and valleys so that: (1) undervoltage lockouts and/or regulation dropouts do not occur; (2) overvoltage (and overcurrent) component stresses do not occur; and (3) input fuses or circuit breakers are not tripped with the increased current.
- Inductive voltage spikes must be suppressed when input power is being switched off.

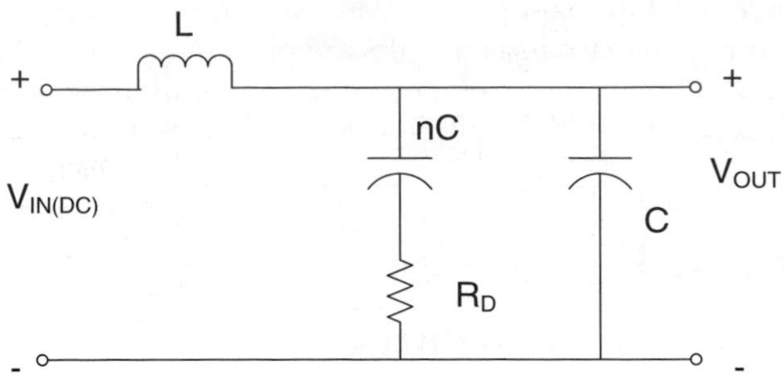

FIGURE 5.19
Generic SMPS low-pass input filter.

- Outrush surge current stresses must be controlled for line dropout conditions to protect switches, relays, fuses, etc.
- Low-frequency reflected load ripple current may need to be attenuated by the input filter in some cases.
- High-voltage line spikes may need to be suppressed or rejected.
- Proper filtering and rectification for off-line AC power sources are required.
- Size, weight, and cost limitations must be met.

Other custom requirements may need to be satisfied; however, this list provides the general idea of the many functions and constraints of the SMPS input filter. Figure 5.19 shows an example of an input filter that might be employed. Appendix A considers optimal design approaches to input filter design.

5.3.2 Inrush Current Limiter

The inrush current limiter, as its name implies, limits surge currents into the SMPS when the input line voltage is switched on. This surge current results from the initial charging of the filter capacitors of the input filter and, subsequently, the output filter capacitors of the converter and load. There are several implementations for this function. Typically, a surge-limiting series resistor is inserted that is switched out after near steady state operation is achieved. Figure 5.20 shows the concept.

This switch may be a mechanical or electronic switch and may be closed after some time delay, allowing for near steady state capacitor charge voltage to be achieved. Conversely, it may be closed at some particular charge voltage threshold of the capacitors involved. If the voltage threshold scheme is used, a substantial amount of hysteresis must be implemented in the voltage detector to prevent an ON/OFF oscillating limit cycle when this capacitor voltage sags with the converter starting to draw power. As may be surmised,

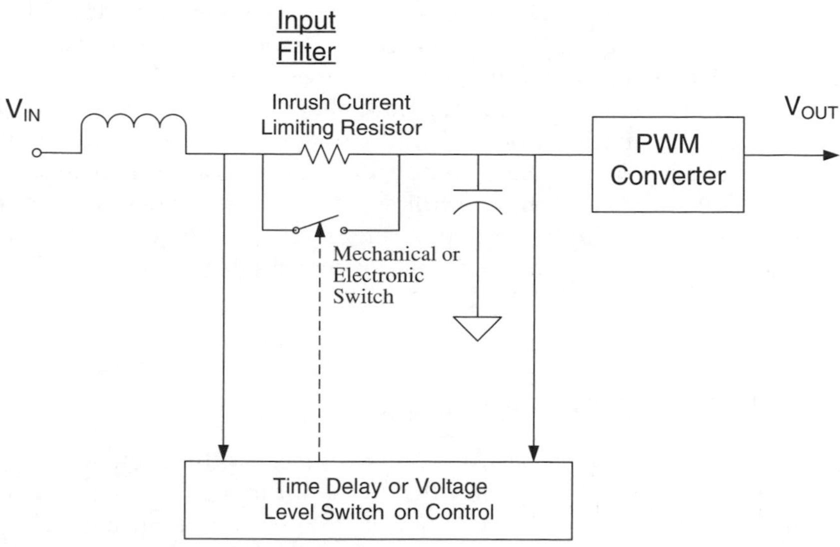

FIGURE 5.20
Concept of an SMPS inrush current limiter.

these peripheral circuits are tricky to design and a transient analysis using a large signal computer model is an invaluable tool in analyzing this start-up condition.

These inrush limiter circuits not only must function in close cooperation with the input filter circuit, as might be noted, but also must function in very close cooperation with the undervoltage lockout and soft start circuits discussed in the next section.

5.3.3 Undervoltage Lockout and Soft Start Circuits

An undervoltage lockout (UVLO) circuit functions by preventing the power converter circuit from switching until sufficient voltages are present to allow adequate control of the converter. A sufficient supply voltage must be available for the control circuit as well as the power circuits; however, in most designs the undervoltage lockout is implemented only for the control circuit power supply. This circuit is simply a voltage level sensing circuit with a prescribed hysteresis, which enables or disables the control circuit output to the converter power switches.

The function of the soft start circuit is to allow the converter to start up with a duty ratio that is near zero or very low. The duty ratio is allowed to increase slowly until steady state operation is achieved. This facilitates slow charging of the output filter capacitors and in essence provides a type of surge current limit through the converter in charging these capacitors. This prevents excessive component stresses, as well as excessive voltage sag on the input filter capacitors, which could result in resetting of the UVLO and inrush current limiting circuits. In such cases, chaotic and possibly destructive circuit operation could occur.

Soft start control circuits of this type are almost always implemented by slowly charging a capacitor from 0 V and having the pulse width modulator track this voltage. This signal is diode ORed with the normal feedback control and, at some point, the control signal takes over the PWM control and provides normal output regulation. Figure 2.4a and Figure 2.4b show a conceptual example of how a coordinated inrush limiter, input filter, UVLO, and soft start circuit may be implemented.

5.3.4 Power Factor Correction Circuits

When operating from an off line AC power source, a rectifier and a low-pass filter circuit are required to convert the AC input to a DC voltage that can be used by the SMPS power converter. The filter can be an inductive or capacitive input filter. In either case, the rectifier/filter distorts the input current waveform, resulting in a less than unity power factor and harmonic distortion of the input current waveform. Most SMPSs today use a capacitive input filter for a number of practical reasons, such as cost and weight; unfortunately, it generally produces the worst case of distortion and power factor.

The solution used to reduce these drawbacks is to include an active power factor correction, PFC, circuit in the design. The existing technology for PFC calls for a circuit to be inserted between the input voltage rectifier and what would otherwise be a capacitive input filter. Figure 5.21 shows the input current wave shapes for the uncorrected capacitive input filter and the input current corrected with a PFC circuit. From a power system point of view, it is desired that the input current waveform have the same and in-phase sine wave shape as the source voltage or, in other words, have a unity power factor. This results in the minimum RMS current and therefore the lowest transmission losses from the power source. To accomplish this, a uniquely controlled power converter is inserted to provide the desired controlled input current waveform.

The most widely used power converter topology for this application is the boost converter operating at a frequency much greater than that of the AC input power. The buck–boost converter and its variants such as the Cuk[4] and SEPIC[21] topologies have also been used successfully. When the boost converter is used, the output DC voltage should always be greater than the expected peak of the AC input voltage to maintain control. The buck–boost topologies, however, may operate with the output voltage greater or less than the peak AC input. Todd[22] provides a very detailed look at the concepts and design of a PFC circuit using the boost topology and is recommended reading.

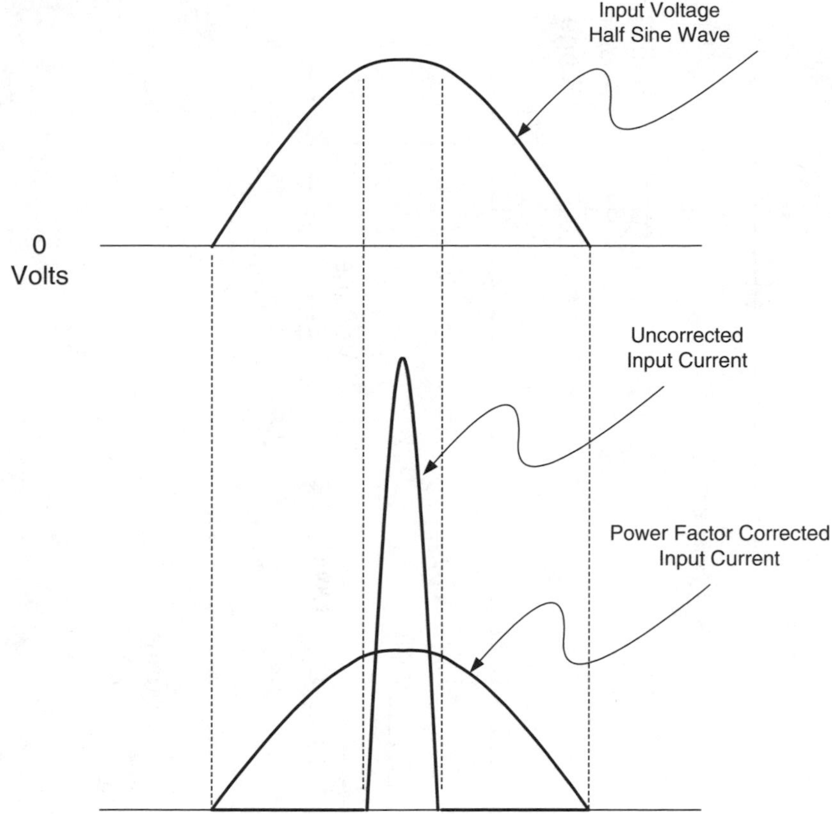

FIGURE 5.21
Rectified input currents for a capacitive input filter.

An active PFC circuit has two primary functions: to control the input current, forcing it to assume the same wave shape as the rectified sine wave, and to regulate its output DC voltage. Figure 5.22 shows a conceptual circuit using the boost topology. Shown are two basic control loops identified as Loop 1 and Loop 2. Loop 1 is the control loop for the inductor (and input) current and, when it is closed, it is embedded within the output voltage regulation control, Loop 2. At first glance, the control seems somewhat nonlinear with the multiplier, squarer, and divider circuits, but these nonlinear circuit effects are compensatory and tend to cancel out. Also, the bandwidths are sufficiently separated so that linear control can be assumed.

The multiplier accepts the current control, V_C, from the divider and multiplies (or modulates) it with the rectified sine wave shape from V_{IN}, which in turn forces I_{IN} to assume the desired sine wave shape. Loop 1 therefore

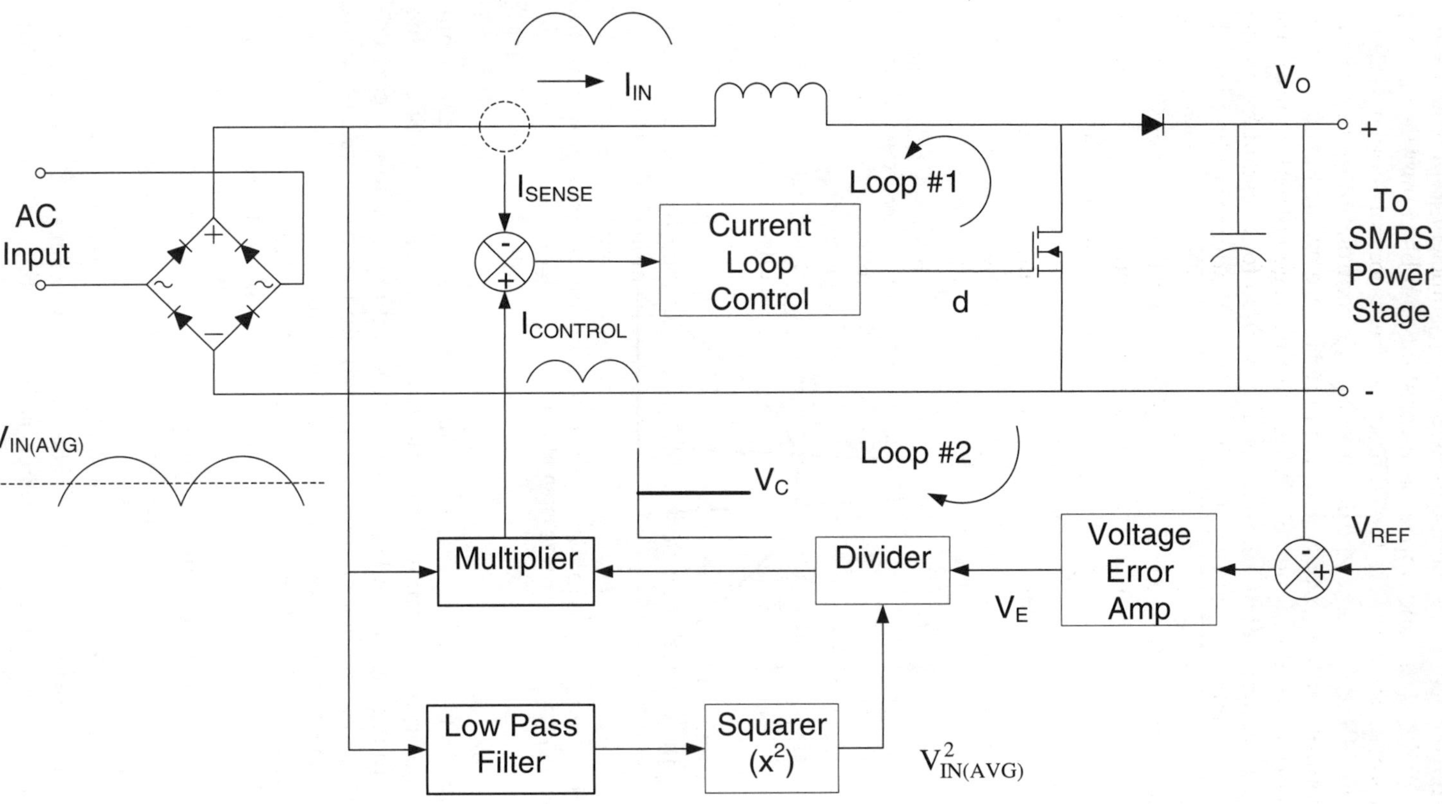

FIGURE 5.22
Conceptual power factor control using boost converter topology.

needs to have a relatively wide bandwidth and sufficient gain to track the control signal $I_{CONTROL}$ and appear as a constant gain block insofar as Loop 2 is concerned. Incidentally, the inductor current control loop may be a conventional fixed frequency peak current mode control, but more accurate current control and better power factor performance are achieved with an average current control scheme.[27] Also, with average current control, the current slope compensation ramp normally required for constant frequency peak current control is not required. This is a real advantage because the converter duty ratio is required to vary over a wide range when tracking V_{IN}. Ramp stability can sometimes be difficult under these circumstances.

This average current control loop assumes the properties of a voltage mode control and must be dealt with accordingly. Also, continuous and discontinuous modes of converter operation are acceptable. (In the continuous mode, do not forget the right-half plane zero.) For the moment, hold the control signal, V_C, constant and increase the input voltage, V_{IN}. Note that the input voltage has increased, and the input current has also increased by the higher modulated peak of $I_{CONTROL}$. In other words, the input power increases as the square of the input voltage. To prevent this input voltage change from influencing the output voltage, V_O, which has a constant power load, the control signal must be obtained by dividing the voltage error amplifier output signal, V_E, by the square of the *average* input voltage, V_{IN}. This is facilitated by the low-pass filter and the squarer circuits of Figure 5.22.

The bandwidth of the output voltage control loop, Loop 2, must be considerably lower than the line frequency so that it does not attempt to track and negate the desired modulation of $I_{CONTROL}$. On the other hand, it must be as wide as is practical to provide the maximum amount of transient line rejection. In practice, it appears that a bandwidth of about one-sixth of the rectified line frequency is typical, depending on the desired power factor and harmonic distortion specifications. Another scheme that can improve transient performance and permit an effective wider voltage loop bandwidth is to use a sample-and-hold scheme to update ${V_{IN(avg)}}^2$ at the end of every line-rectified cycle.

Several other practical considerations associated with this concept are noted in Todd.[22] Other PFC schemes have been used successfully; Andreycak[23] shows one that operates at critical inductor current and has a simpler control scheme using a variable frequency, controlled ON time converter. Unfortunately, component stresses generally prevent operation at power levels greater than 500 W for this scheme with today's technology.

The standards of power factor and total harmonic distortion can be varied depending on the usage circumstances. Kassakian[24] has shown a useful and interesting relationship between power factor, *PF*, and total harmonic distortion, *THD*, as:

$$PF = \sqrt{\frac{1}{1+THD^2}} \tag{5.18}$$

5.3.5 Multiple Outputs

When a power converter with multiple outputs is modeled, some special considerations must be noted. Figure 2.2 shows a typical example of a push–pull converter with split secondary multiple outputs. Of course, any number of multiple outputs can be added to the power transformer. Also, some or all of them may have coupled inductors.

First, consider a case of multiple outputs for a buck topology in which some outputs are in the continuous mode and some are in the discontinuous mode. Figure 5.23 shows a conceptual way of modeling the two outputs of the converter of Figure 2.2 when the output inductors are *not* coupled. (When output inductors *are* coupled together, all of these coupled loads theoretically function together as continuous or discontinuous, regardless of the individual loads. When these circuits are modeled, the outputs should be paralleled at the inductor input.) Note that the mode of each individual output is

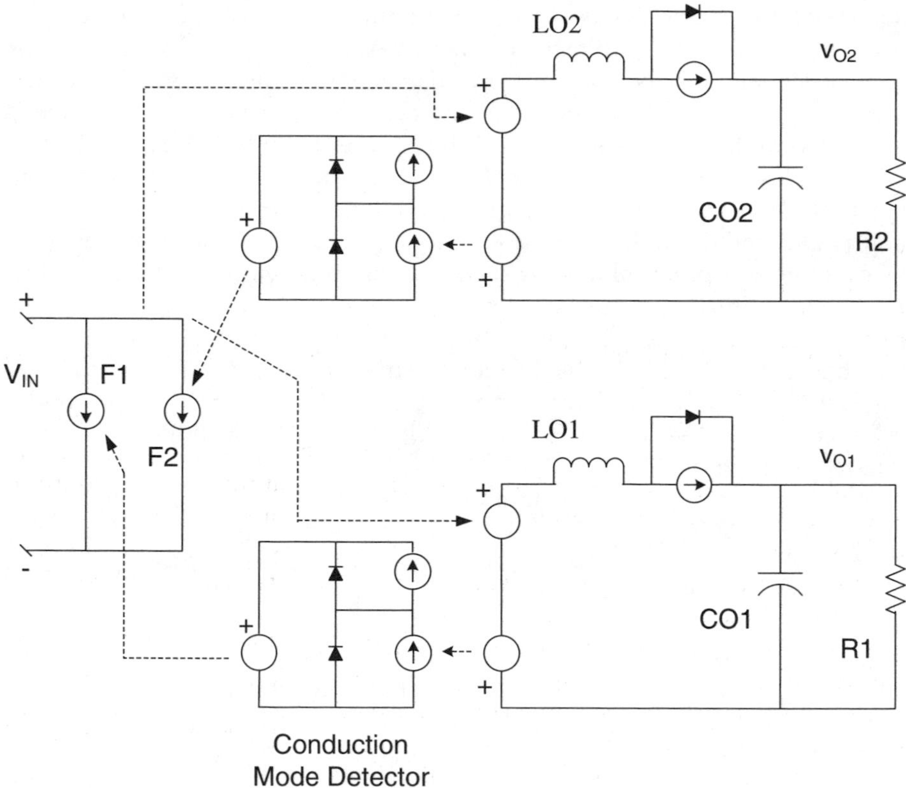

FIGURE 5.23
Modeling example of multiple output converter with continuous output or discontinuous conduction mode.

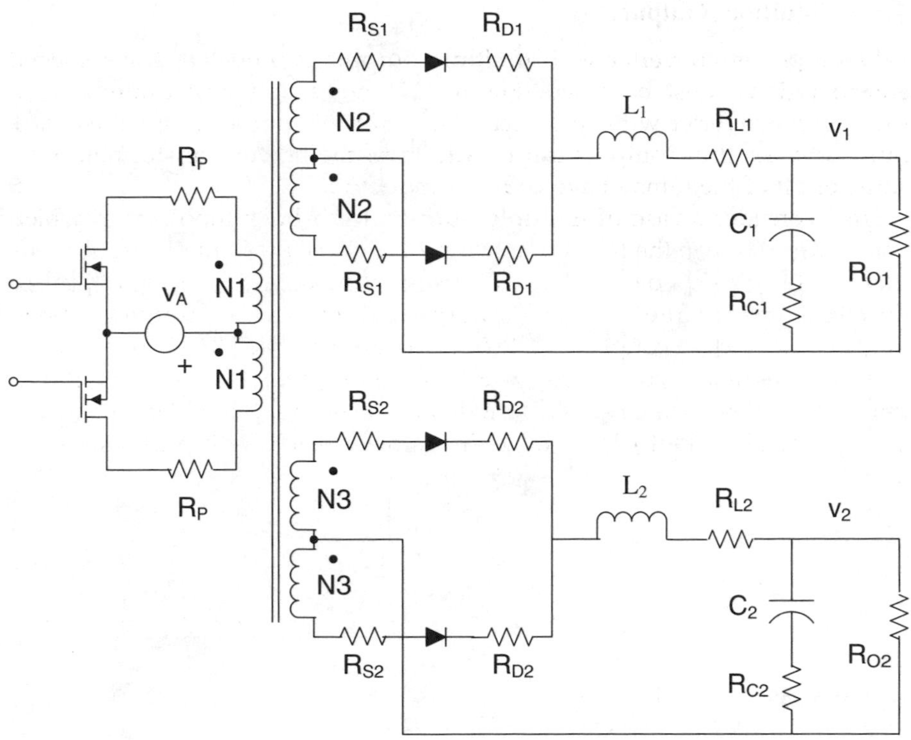

FIGURE 5.24
Multiple output push–pull buck converter with parasitic resistances.

determined by its individual conduction mode-detecting circuit as it loads the power transformer.

If the expected mode of all outputs is known at the outset, then the model can be simplified by eliminating the mode-detecting circuits and simply reflecting the output circuits or loads — whether continuous or discontinuous — to the transformer primary by the current sources F1 and F2. Incidentally, when synchronous rectifier outputs as described in Section 5.3.7 are used with conventional rectifiers on other outputs, this mixed continuous/discontinuous situation may still result.

Now look at this same model example with parasitic effects. Figure 5.24 shows the basic circuit and Figure 5.25 shows the circuit-averaged model for this circuit that may be used in a computer model. The model development is not shown here, but may be derived using the methods indicated in Polivka et al.[20] Another example showing a split secondary plus or minus output push–pull buck converter with parasitic resistances is shown in Figure 5.26. The accompanying circuit-averaged model is shown in Figure 5.27. Again, the derivation is not shown but may be derived using the methods in Polivka et al.[20]

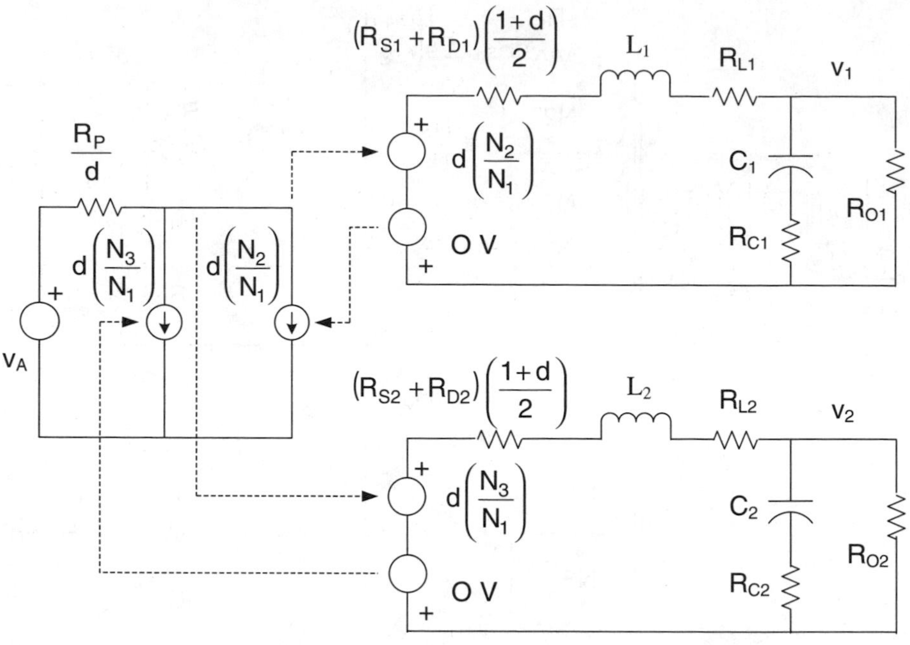

FIGURE 5.25
Circuit-averaged model of the converter in Figure 5.24.

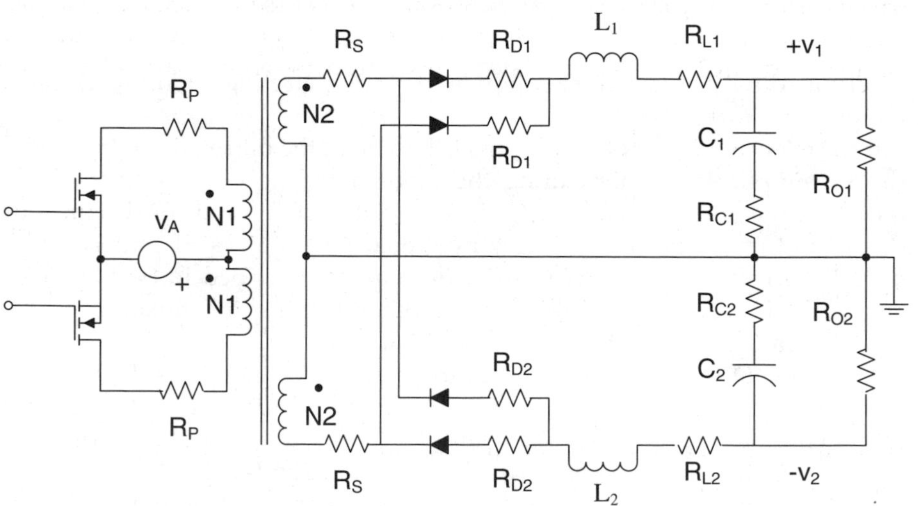

FIGURE 5.26
Plus or minus output push–pull buck converter with parasitic resistances.

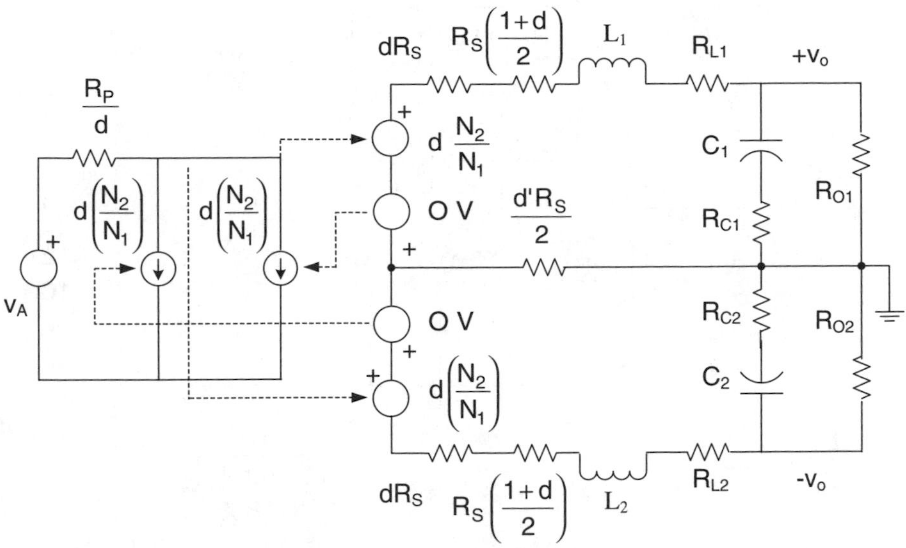

FIGURE 5.27
Circuit-averaged model of the converter in Figure 5.26.

5.3.6 Postregulated Output Circuits

In many instances, the outputs of an SMPS are not directly regulated or controlled by the circuitry that controls the PWM of the main power converter. This is always the case when multiple outputs (discussed in Section 5.3.5) are used. When this occurs and it is desired to have better regulation than is normally obtainable, a postregulator is inserted between the particular SMPS output and load.

In practice, postregulation can be provided in two ways. One may simply implement a dissipative linear voltage regulator; this may be acceptable for low-power outputs, but would ordinarily be prohibitive for higher powers. The more efficient option would be simply to insert a second post-SMPS. This SMPS may be implemented by a conventional converter design as described in Chapter 1; however, a more viable approach is to take advantage of the existing main converter switching action and use a magnetic amplifier scheme. Mammano and Mullett[25] provide information on the fundamentals of using mag-amps for this application. Magnetic amplifier design is a somewhat specialized endeavor and many subtleties are involved when the magnetic nonlinearities and reset requirements are addressed. Figure 5.28 shows a conceptual circuit of a two-output forward converter SMPS in which one output is directly regulated by the main PWM controller and the second output is controlled by a post-mag-amp regulator. Voltage waveforms are also shown.

When these types of mag-amp regulators are analyzed, it is important to recognize that they are essentially buck topology SMPSs in themselves. The model is created and generally analyzed in the same way as for any

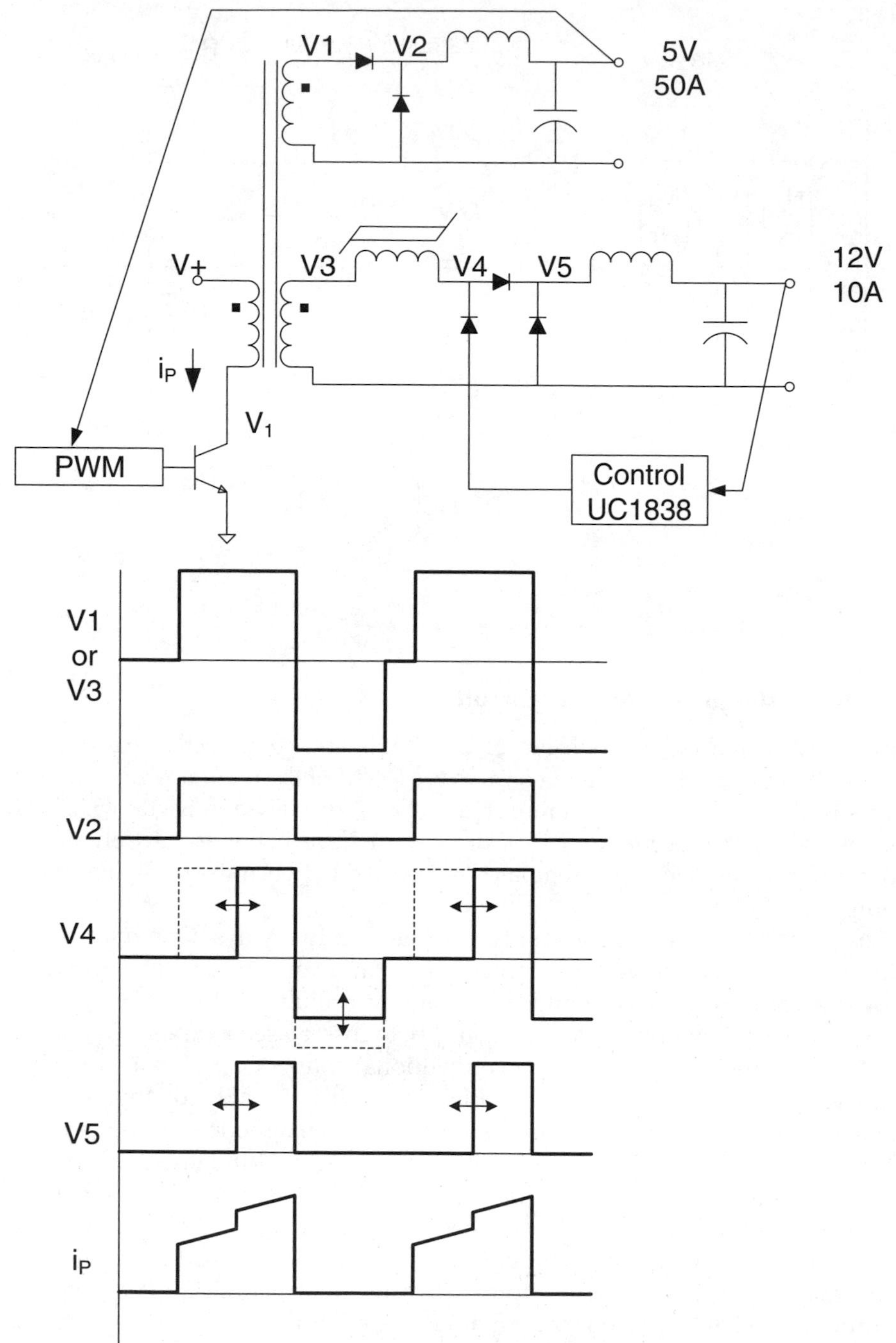

FIGURE 5.28
Conceptual forward converter SMPS with post-mag-amp regulator. (R.A. Mammano and C.E. Mullett, Unitrode Application Note U-109.)

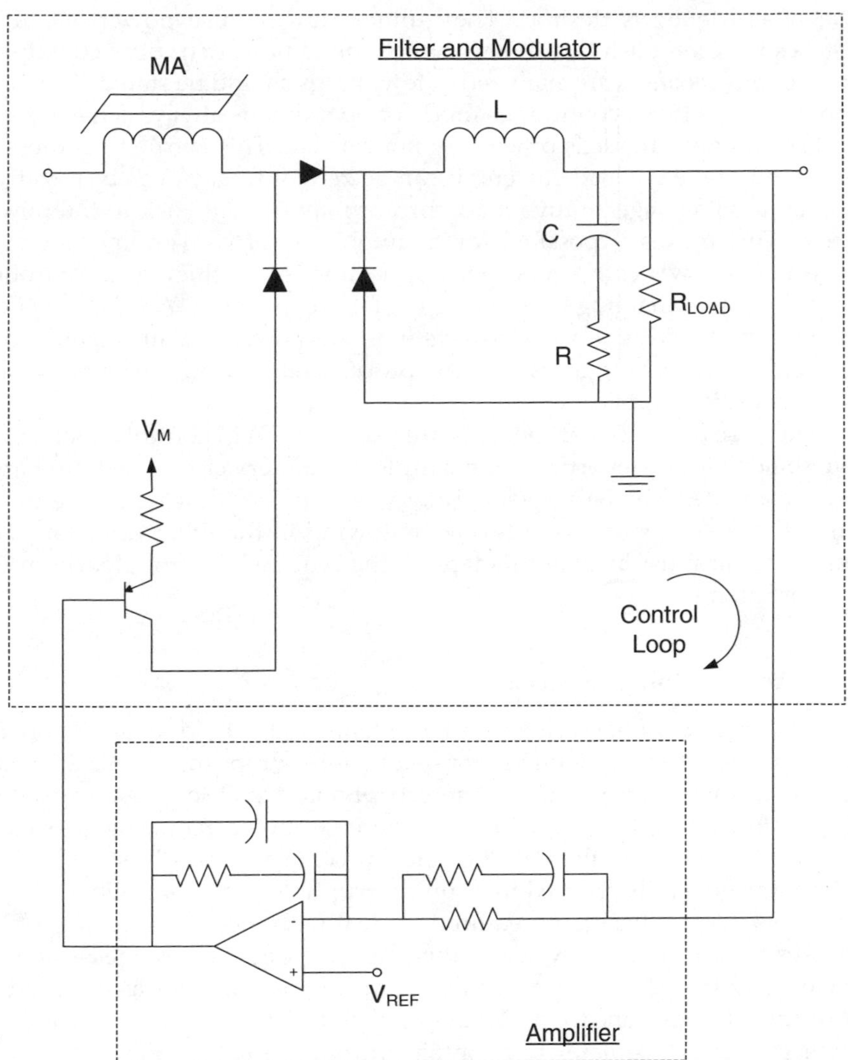

FIGURE 5.29
Typical mag-amp postregulator control loop. (R.A. Mammano and C.E. Mullett, Unitrode Application Note U-109.)

conventional SMPS, with one main exception: the gain and a pronounced phase delay associated with the magnetic modulator. Figure 5.29, excerpted from Mammano and Mullett,[25] shows a typical regulator control loop.

5.3.7 Synchronous Rectifier Circuits

Synchronous rectifiers are generally used in low output voltage, high current power converters to replace conventional diode rectifiers when the highest level of efficiency is desired. They almost always use power MOSFET switches to accomplish this. When switch mode power supplies containing synchronous rectifiers are analyzed, a few things should be stated. The most fundamental is that a continuous mode of operation is always achieved and the discontinuous mode is practically nonexistent. This is true regardless of the magnitude of the load current; it can be zero or it may even be negative. In some designs, regenerative load current may be sent back to the power source. This makes it possible for a unique set of converters known as bidirectional power converters. One application is for achieving a controlled charging and discharging for systems containing batteries. Another application may be when motor control systems are powered with regenerative capabilities. Power amplifiers with AC output voltages and currents are also a very important application.

When these bidirectional concepts are used, the PWM control circuits may have some unique properties. For example, the battery charge and discharge power converter may be a buck topology with power flowing in one direction and a boost topology with power flowing in the other direction. The control circuit must be able to adapt to the required control algorithm for both directions.

5.3.8 Energy Storage Systems

Energy storage peripheral circuits are implemented to hold up and maintain output power when the main input power source drops out. These dropouts may be relatively short transient interruptions, which can use capacitive energy storage, or lengthy or steady state dropouts requiring a backup battery or some other alternate power source. (These steady state backup systems are generally referred to as uninterruptible power supplies or UPS.) A control circuit is generally required to deal with these situations.

In almost all of these systems, the energy from the storage system is injected into the power supply at the input to the converter as opposed to the output. This is done for a number of practical reasons such as using the power converter/regulator to maintain output regulation and also single-point energy storage for multiple output converters. In most cases, more efficient energy storage and better performance are generally achieved when energy is injected at the input.

For very short transient dropouts, the normal input filter may provide enough energy storage and thus nothing extra added to the design except maybe a little more capacitance at the input. However, in cases in which this is insufficient, an additional capacitance storage bank and its accompanying control circuit are required. Figure 5.30 shows one approach to this problem. The capacitor bank is self-charged or "bootstrapped" from the main converter to a voltage higher than the normal DC input voltage in most cases by a simple resistor and rectifier circuit. When a line dropout occurs, a low DC input voltage is sensed and activates the discharge controller, allowing for a controlled energy transfer of the energy from the capacitor bank into

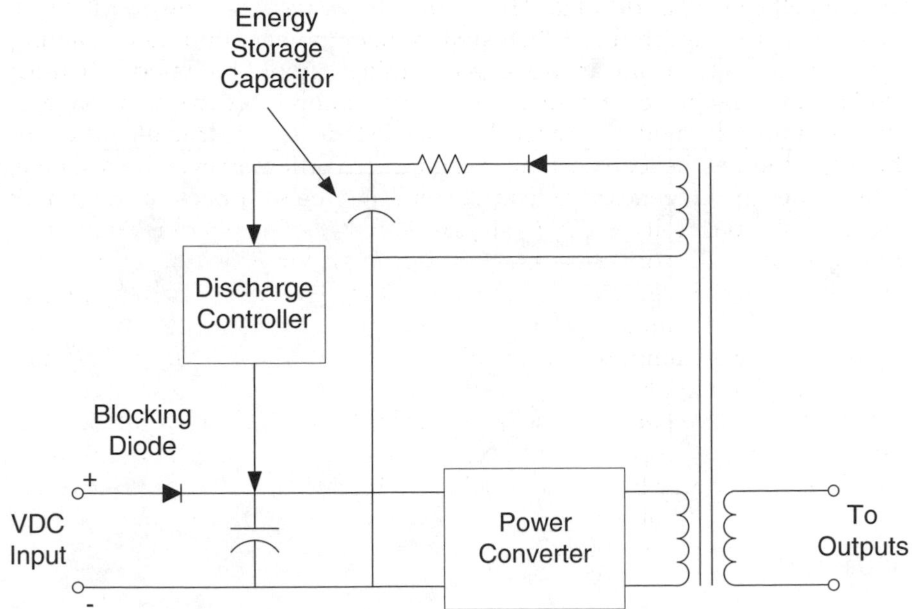

FIGURE 5.30
Capacitive energy storage system.

the main power converter input. In most cases this, is a simple voltage regulator attempting to regulate the DC input voltage. It may be a switching or a linear regulator, but will probably be the latter for simplicity and cost reasons.

When this approach is used with a DC power source, a blocking diode may be deemed necessary at the input to prevent loss of the stored energy back to the failed power source. This diode will lower the overall SMPS efficiency and must be taken into consideration. A controlled MOSFET switch may be used in some lower voltage and/or power applications in lieu of this blocking diode to increase the efficiency. Off-line applications do not have this problem because the input rectifiers provide the blocking function.

For outright total input power failure or extra long transient dropouts in which the capacitance approach is unfeasible, a battery backup system is required. The battery voltage is generally lower than the normally expected DC input voltage and is simply diode ORed into the input; it provides power when the source voltage drops out (see Figure 5.31). The battery is generally kept charged from the input power source by its own charger for proper battery conditioning. Battery chargers may also be linear or switching regulators and, in general, will be a constant current supply with some sort of trickle charge condition set up, possibly when a predetermined charge voltage is reached. Conversely, the charger may simply just go into a voltage

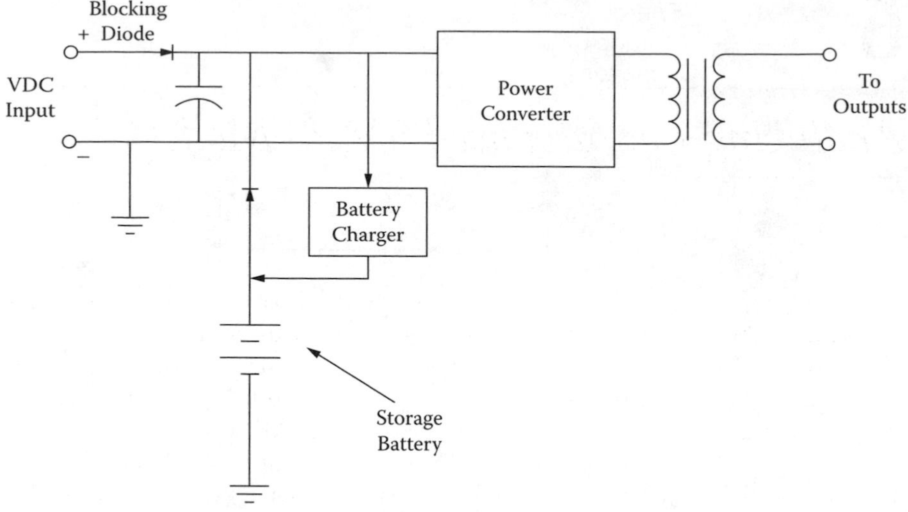

FIGURE 5.31

regulation mode at some peak charge voltage. When a switching power converter is used, it would most likely have a buck topology and use current mode control because this would be very desirable for constant current outputs.

5.4 Summary

This chapter has presented concepts and some development approaches to creating the real-world circuit component models necessary when a higher than fundamental level of SMPS analysis is desired. Also, a practical representation of most of the peripheral circuits generally associated with SMPSs has been presented along with some comments on modeling them for a circuit-averaged analysis. Of course, other peripheral circuits exist, but hopefully this chapter can guide the reader in a way that will make it possible for him to develop his concepts and circuit models.

6

Analyzing the Advanced SMPS Model

Chapter 4 considered most of the fundamental kinds of DC and AC analysis that one might encounter when analyzing the various properties of a very basic buck switching regulator. These were for the static power converter with no duty ratio control, the voltage mode controlled switch mode power supply (SMPS), and the current mode controlled SMPS. This chapter will continue with the general analysis theme of that chapter and consider the more advanced or comprehensive detailed analysis methods. These will include large signal transient analyses as well as the DC and AC analysis methods. To be on firm ground, the reader should consult some of the numerous references[14,24,26] regarding the fundamentals of switch mode power supplies and arm himself or herself with as much basic knowledge as is possible.

First, the use of the pulse width modulator controller macromodel developed in Appendix B will be demonstrated. Next, the analysis of a flyback (buck–boost) converter switch mode power supply taken from a widely published application note will be considered. This will encompass an advanced practical circuit topology. Following this, a low-output voltage buck converter along with its parasitic resistance effects will be presented. This model can be used for modeling resistances drops in series power paths as well as assisting in determining power converter efficiencies. Finally, a buck converter SMPS with some interesting large signal analysis results is shown.

6.1 Pulse Width Modulator (PWM) Controller Macromodel

When a computer analysis of an electronic circuit is performed, a practical approach may be to develop macromodel building blocks representing specific complex functions. When an SMPS that uses a commercially available integrated circuit controller is analyzed, it can be very useful to utilize a behavioral macromodel for this controller. These controllers all perform the function of PWM control and may sometimes be quite complex. Appendix B shows the general methodology for developing a *circuit-averaging* SMPS controller macromodel.

As a point of departure, the technique of using a controller macromodel to provide PWM control of a very basic power converter will be illustrated.

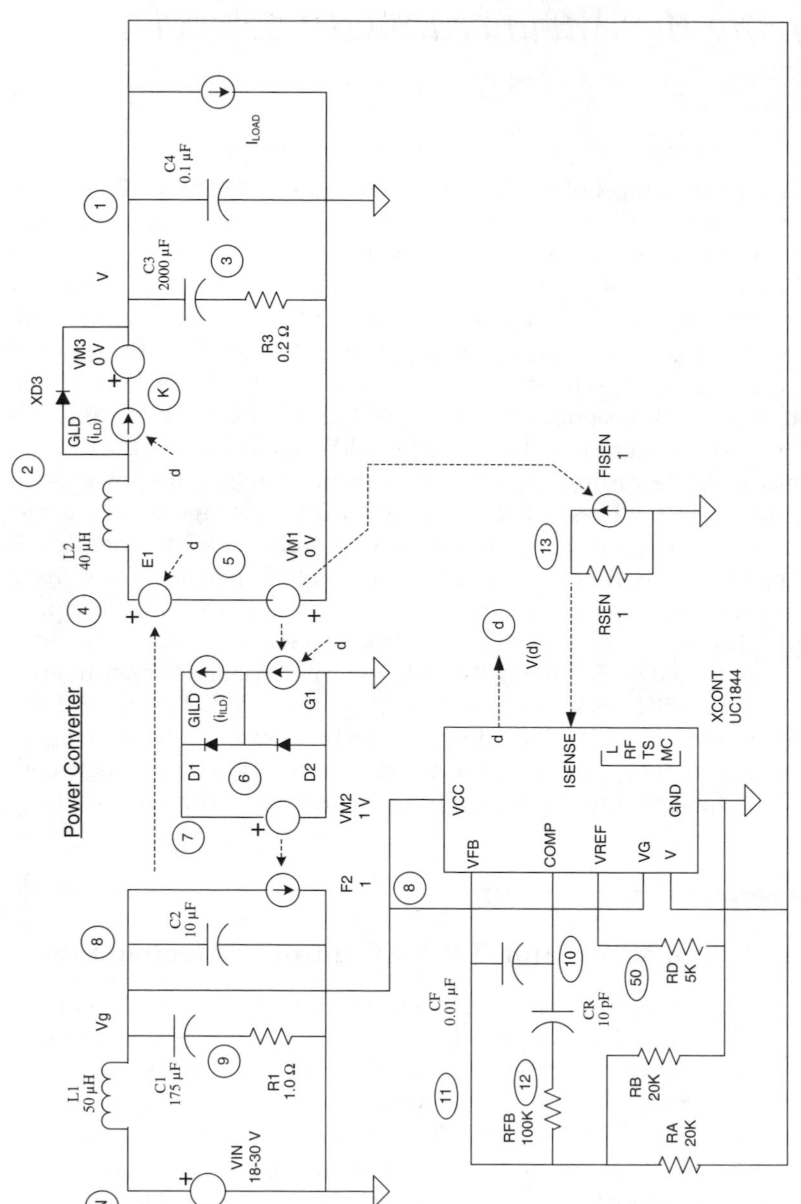

FIGURE 6.1
Basic current mode SMPS with UC1844 macromodel controller.

The basic current mode converter of Chapter 4 will be used, along with the UC1844 controller developed in Appendix B to regulate a DC output voltage of 10 V. Figure 6.1 shows the basic SMPS model to be analyzed. Two large signal transient analyses showing line and load regulation are demonstrated. A PSPICE netlist, labeled Netlist 6.1, is used for these simulations.

NETLIST 6.1

```
BUCK CONVERTER CURRENT MODE ANALYSIS (10.0 V0UT) NETLIST
FIGURE 6.1
*
.OPTIONS STEPGMIN ITL4=40
.NODESET V(13)=.1 V(1)=10 V(8)=30 V(D)=.33
*
********************************************************************************
*
** SOURCE-LOAD CONFIGURATION FOR LINE REGULATION ANALYSIS
** (FIGURE 6.2)
VIN N 0 PULSE(1U 30 .1 10 10 1M)
ILOADDC 1 0 DC 1
.TRAN 50M 40 0 5M
*
********************************************************************************
*
** SOURCE-LOAD CONFIGURATION FOR LOAD REGULATION ANALYSIS
** (FIGURE 6.3)
*VIN N 0 DC 30
*ILOAD 1 0 PULSE(0 4 1.5 1)
*.TRAN 1M 4 0 1M
*
********************************************************************************
*
**POWER CONVERTER MODEL
*
C4 1 0 .1U
C3 1 3 2000U
R3 3 0 .2
*
*L = 40U
*TS = 10U
*N = TRANSFORMER TURNS RATIO
*
**GLD 2 K VALUE = {V(d)**2*N*(V(8)/V(1))*(V(8)-V(1))*(TS/(2*L))}
GLD 2 K VALUE = {V(d)**2*1*(V(8)/V(1))*(V(8)-V(1))*(10U/(2*40U))}
VM3 K 1
XD3 2 1 DIDEAL
L2 4 2 40U
**E1 4 5 VALUE = {V(8)*V(d)*N}
E1 4 5 VALUE = {V(8)*V(d)*1}
VM1 0 5
GILD 6 7 VALUE = {I(VM3)*V(1)/V(8)}
**G1 0 6 VALUE = {I(VM1)*V(d)*N}
G1 0 6 VALUE = {I(VM1)*V(d)*1}
```

```
D1 6 7 D2
D2 0 6 D2
.MODEL D2 D
VM2 7 0 DC 1
F2 8 0 VM2 1
RVM2 7 0 1K
C2 8 0 10U
C1 8 9 175U
R1 9 0 1
L1 N 8 50U
*
*******************************************************
*
*CONTROL CIRCUIT MODEL
*
FISEN 0 13 VM1 1
RISEN 13 0 1
CF 10 12 .001U
RFB 12 11 100K
CR 10 11 10P
RA 11 1 60K
RB 11 0 20K
RD 50 0 5K
XCONT 8 50 11 10 13 1 8 d 0 UC1844
*
*******************************************************
*
**UC 1844 PWM CONTROLLER CIRCUIT AVERAGING MACRO
.SUBCKT UC1844 VCC 3      4     COMP 6        7  8    d GND
*                   VCC VREF VFB COMP ISENSE V VG d GND
**NOTE: ALL SIGNAL INPUTS MUST REFERENCED TO THE UC1844 GND
XUVLO VCC UVLO GND UVLO
XREF UVLO VCC 3 VREF/2 GND REF
XERRAMP 4 VREF/2 VCC COMP GND ERRAMP
XDRC COMP d 6 7A 8A UVLO GND DRC
**DEPENDENT GENERATORS FOR GROUND ISOLATION IF DESIRED
EVG 8A 0 8 GND 1
EV 7A 0 7 GND 1
.ENDS UC1844
*
.SUBCKT DRC 1    d 8     9 10 14  GND
*        COMP  d ISENSE V VG UVLO GND
VBK 1 1A DC 2
R1 1A C 200K
R2 C GND 100K
X1 C 3 DIDEAL
X2 GND C DIDEAL
VLIM 3 GND DC 1
RCONV1 6 0 1G
RCONV2 7 0 1G
RCONV3 4 0 1G
*
REL 10 9 1E8
GL 0 11 10 9 1
D1L 11 12 DX
```

```
D2L 0 11 DX
RLMIN 12 13 1
VLMIN 13 0 1U
*
DX1 4 0 DX
DX2 7 4 DX
DX3 6 7 DX
*
**IDRMAX VALUE = MAX LIMITED DUTY RATIO
IDRMAX 7 6 DC .48
DX4 0 6 DX
DX5 6 6A DX
VMD 6A 0
EMD d 0 VALUE = {I(VMD)*V(14)+1P}
RMD d 0 1
.MODEL DX D IS=1E-12
*
*L = 40U
*RF = 0.3
*TS = 10U
*MC = .3E6
*
**GCM 4 7 VALUE = {(V(C)-RF*V(8))/((TS*RF)/(2*L))*(2*MC*L+V(12)))}
GCM 4 7 VALUE = {LIMIT((V(C)-0.3*V(8))/((10U*0.3)/(2*40U)*(2*.3E6*40U+V(12))),0,1)}
*
**GDCM 0 4 VALUE = {V(C)/((TS*RF/L)*(MC*L+V(12))}
GDCM 0 4 VALUE = {LIMIT(V(C)/((10U*0.3/40U)*(.3E6*40U+V(12))),0,1)}
*
.ENDS DRC
*
.SUBCKT UVLO VCC UVLO   GND
*           VCC UVLO   GND
RIN VCC 2 1K
ISUP 2 GND 1M
XSUP GND 2 DIDEAL
G5 VCC GND VALUE = {.01*V(UVLO,GND)}
** G1 AND G2 SET UVLO START AND STOP LEVELS RESPECTIVELY
** START = 17
** STOP  = 10
G1 0 3 TABLE {V(VCC,GND)} = (0,0) (17,0) (18,20)
G2 3 0 TABLE {V(VCC,GND)} = (9,20) (10,0) (17,0)
CG2 3 0 20N
R1 3 0 1E3
G3 0 4 3 0 1
G3HYST 0 4 VALUE = {-100U+200U*V(UVLO,GND)}
CHOLD 4 0 20N
G4 GND UVLO 4 GND 1
**VOVRD TO BE INSERTED FOR UVLO OVERRIDE
*VOVRD UVLO GND DC 1
RG4 UVLO GND 1
CG4 UVLO GND 1N
X5 GND UVLO DIDEAL
X1 3 5 DIDEAL
X2 6 3 DIDEAL
X3 4 5 DIDEAL
```

```
X4 6 4 DIDEAL
VP 5 0 DC 1
VN 0 6 DC 1
.ENDS UVLO
*
*5.0 VOLT REFERENCE MACRO
.SUBCKT REF UVLO VCC   VREF   VREF/2      GND
*           UVLO VCC   VREF   VREF/2      GND
*
EIN 2 3 VCC GND .46E-3
RVT 3 4 .32 TC=0.2E-3
IVT GND 3 DC 3.125
VT GND 4 DC 1
V5 1 2 DC 4.99342
FUVLO 1 GND VMR 1
EUVLO VREF 9 VALUE = {V(1,GND)*V(UVLO,GND)}
VMR GND 9
EVREF/2 VREF/2 GND VREF GND .5
.ENDS REF
*
*ERROR AMP MACRO
.SUBCKT ERRAMP 6   7   8   1     GND
*              VIN VIP VCC OUTPUT GND
RIN 5 7 1MEG
EPSRR 6 5 VALUE = {(V(8)-15)*3E-4}
IBIAS GND 5 .3U
GA GND 2 7 6 1K
RG 2 GND 30
CG 2 GND 0.159M
VNC 4 GND DC 1
DN 4 2 D1
VPC 3 GND DC 5
DP 2 3 D1
RO 1 2 3K
RPS 8 0 1E8
.ENDS ERRAMP
.MODEL D1 D IS=1E-9
*
****************************************************
*
.SUBCKT DIDEAL 1 2
VAS 1 3 DC -1u
D1 3 2 D
D2 3 4 D
D3 4 2 D1
FAS 4 2 VAS 1
.MODEL D D IS=1E-6 EG=0 XTI=-4
.MODEL D1 D EG=0 XTI=0
CC 1 2 .1P
.ENDS DIDEAL
*
.PROBE
.END
```

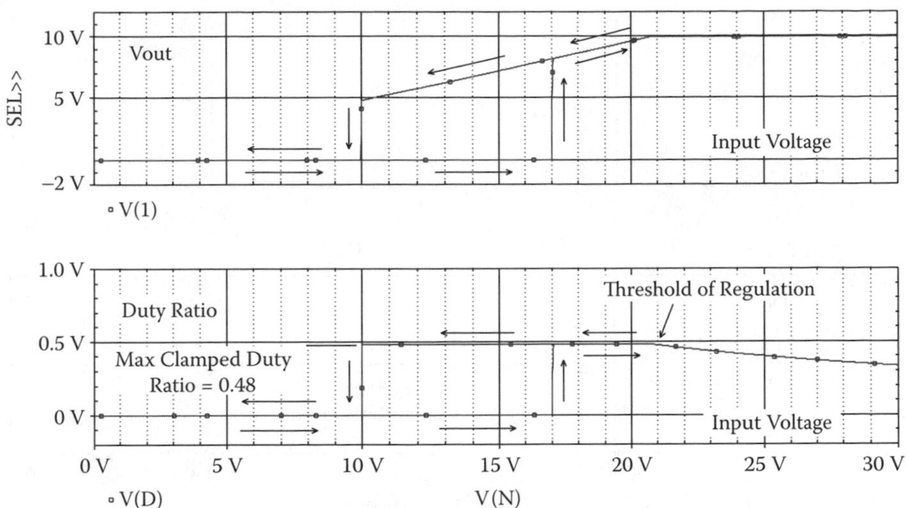

FIGURE 6.2
Line regulation characteristic of the basic 10-V output current mode SMPS.

Figure 6.2 shows the representative line voltage regulation curve of the basic 10-V output SMPS with a constant current 1.0-amp load. Note that the output voltage remains at 0 V until the input voltage rises to the undervoltage lockout (UVLO) level of 17 V. At this point, the converter is allowed to switch and the duty ratio immediately rises to its maximum clamped value of $d = 0.48$ with the output voltage rising to a level of around 8.2 V. As the input voltage is further increased, the output voltage rises accordingly until the threshold of regulation is reached at an input voltage of about 22 V. At this point, the feedback control loop takes over and regulates the output voltage to 10 V by controlling the duty ratio as the input voltage is further increased to 30 V. Now notice that, when the input voltage is reduced from the 30-V level downward and then drops below the 22-V level, the output voltage eventually decreases correspondingly. However, due to the hysteresis in the UVLO, the converter continues functioning down to an input voltage of 10 V before shutting down.

Figure 6.3 shows the representative load regulation characteristic with a constant 30.0-V input voltage. The overload characteristic with the output-regulated 10 V starts to drop at a load current of approximately 1.5 amps. This occurs when the control voltage, V_C, in the UC1844 saturates high at its maximum clamped value of approximately 1 V. (This is the current sense comparator input in the actual UC1844.) The power supply is therefore current limited at the load current corresponding to this clamped 1.0-V value of control voltage.

Some comments might be made here about Netlist 6.1. Note comment entries. Those at the beginning parts of the netlist indicate the correct stimuli and commands to be used for a *desired_analysis*. Here, only one analysis set is activated at a time by removing the asterisk at the beginning of the line entries.

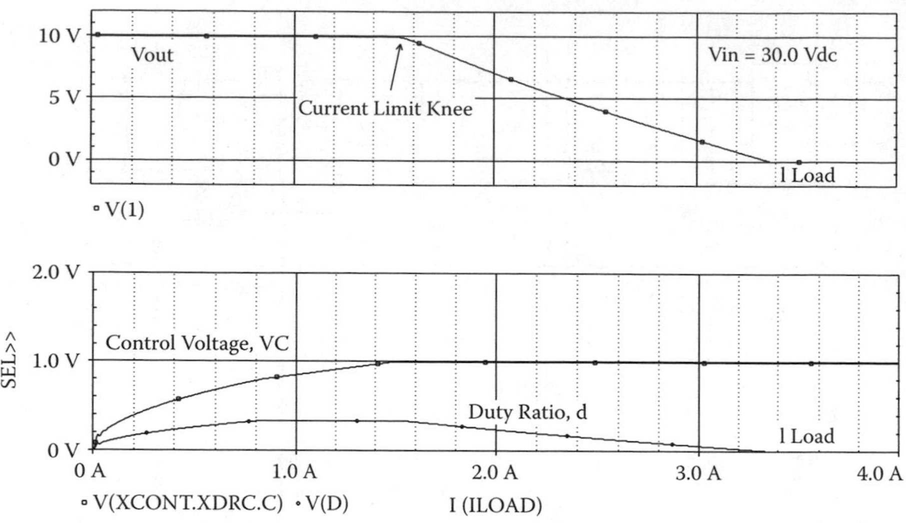

FIGURE 6.3
Load regulation characteristic of the basic 10-V output current mode SMPS.

Follow any instructions indicated shown for that particular analysis. Other comment lines shown further down and throughout the netlist indicate the elements or equation operations that must be set up for a *particular design*. These parameter inputs must be inserted for *both* the converter model and the UC1844 controller macromodel.

A word of caution from the outset regarding the difficulties that may be encountered when running these simulations: due to lack of convergence problems, they can sometimes require considerable experimentation. These models have regenerative latching circuits, as well as several nonlinear dependent equations with points of discontinuity, and can sometimes have convergence problems in DC and transient simulations. All simulations in this book were run on a personal computer using the relatively low-cost PSPICE™ software. More expensive workstation software simulators, such as SABER™, ACCUSIM™, and others, generally have less problems. One should familiarize himself with all the OPTIONS and time step control aspects of the software used in order to be able to overcome these convergence problems.

6.2 Practical Flyback SMPS Analysis

The schematic diagram of a practical flyback SMPS example taken from Unitrode Corporation Application Note No. 96A is shown in Figure 6.4. This example is selected because it is typical of many commercial SMPSs used in

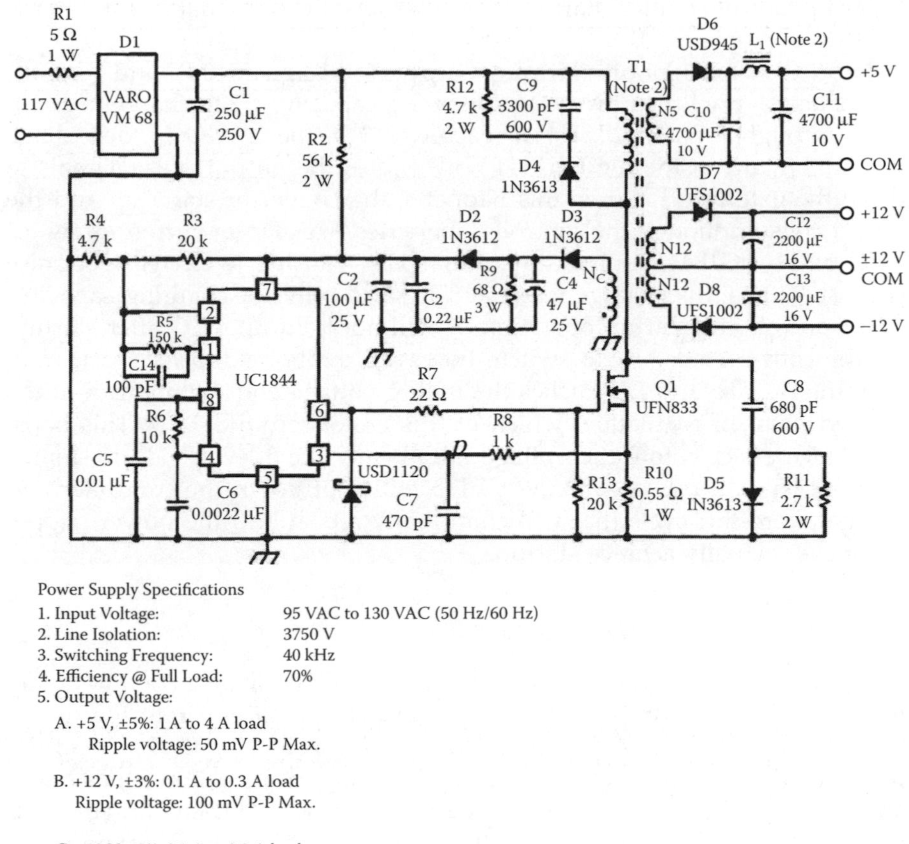

FIGURE 6.4
Practical flyback SMPS analysis example.

electronic equipment today. The power supply specifications are as indicated in Figure 6.4.

A brief description of the power supply is as follows: the flyback converter section of the power supply is composed of the power switch, *Q*1; transformer, *T*1; rectifier diodes, *D*6 through *D*8; and output filter capacitors, *C*10 through *C*13. The applied AC input power of 117 Vac is rectified by a full wave bridge, represented by the block, *D*1, and then filtered by capacitor *C*1. This gives a nominal DC voltage of about 165 V as input voltage to the converter section. The PWM controller, UC1844, obtains the peak inductor current information by sensing the voltage across resistor *R*10. In this case, the peak switch current is the same as the desired peak inductor current. A separate feedback winding labeled N_C provides not only the feedback voltage regulation signal, but also the power saving bootstrap voltage to power the

UC1844 controller after startup. (See the specifications datasheet for the UC1844 in Appendix B.)

Before the controller provides any output PWM signals to *R*7 and *Q*1-*Gate*, it consumes very little power, thus allowing capacitor *C*2 to charge to about 17 V through resistor *R*2. Upon reaching 17 V, the UVLO (undervoltage lockout circuit) inside the UC1844 now allows the output pulse signals to drive the gate of *Q*1. When this happens, the converter starts up and the power consumption of the controller increases. This larger current draw for VCC of the UC1844 results in the initial 17 V starting to decrease because resistor *R*2 has a very large value and is sized only for charging capacitor *C*2 prior to initial startup operation. It will not hold up VCC after startup. As the converter begins to switch, however, the bootstrap feedback from winding N_C, *D*2, and *D*3 catches the falling voltage and maintains it at the desired point of regulation, which in this case is around 13 V. This boost must take effect before the voltage falls below the 10-V UVLO hysteretic dropout level. If this occurs, the UVLO will shut down the controller and instigate a restart cycle that will continually repeat and the power supply will never actually achieve startup.

6.2.1 Flyback SMPS Model Setup

Figure 6.5 shows the equivalent circuit model, which will be used to obtain the desired analysis results. The steps undertaken to generate the model from the actual circuit of Figure 6.4 begin with considering the basic continuous and discontinuous conduction mode model from Section 3.6 as a separate macro identified as "flyback converter model." Then the circuit is expanded around this macro until the desired SMPS model is obtained. As a point of departure, consider the input circuitry by adding the power source and input filter. The power source, VIN, is the DC equivalent voltage of the bridge rectified 60-Hz input voltage and is simply the peak value of the specified AC 60-Hz input voltage. The approximate equivalent series resistance of the bridge rectifier, *R*1, along with the capacitor, *C*1, constitutes what might be considered the input filter.

Now consider the output circuitry. The outputs are +5 and ±12 Vdc. Also, the feedback voltage is connected as a load on the converter. All that is necessary is to reflect these loads to the output of the converter by the turns ratios of the transformer, *T*1, by using the ideal dependent generator transformer model. The rectifier diodes for the various outputs will here be modeled by simply inserting the specified (or equivalent) diode models as shown on the schematic of Figure 6.4. This is a simplification because, in actuality, the currents are chopped in the flyback converter and the correct modeling for these rectifier diodes should be as described in Section 5.1.3. This simplification will result in slightly higher output voltages than would be the actual case. Also, some series resistances effects of the power transformer are being neglected; again, this will tend to indicate a slightly higher output voltage than would be expected otherwise. Nevertheless, these simplifications will be used.

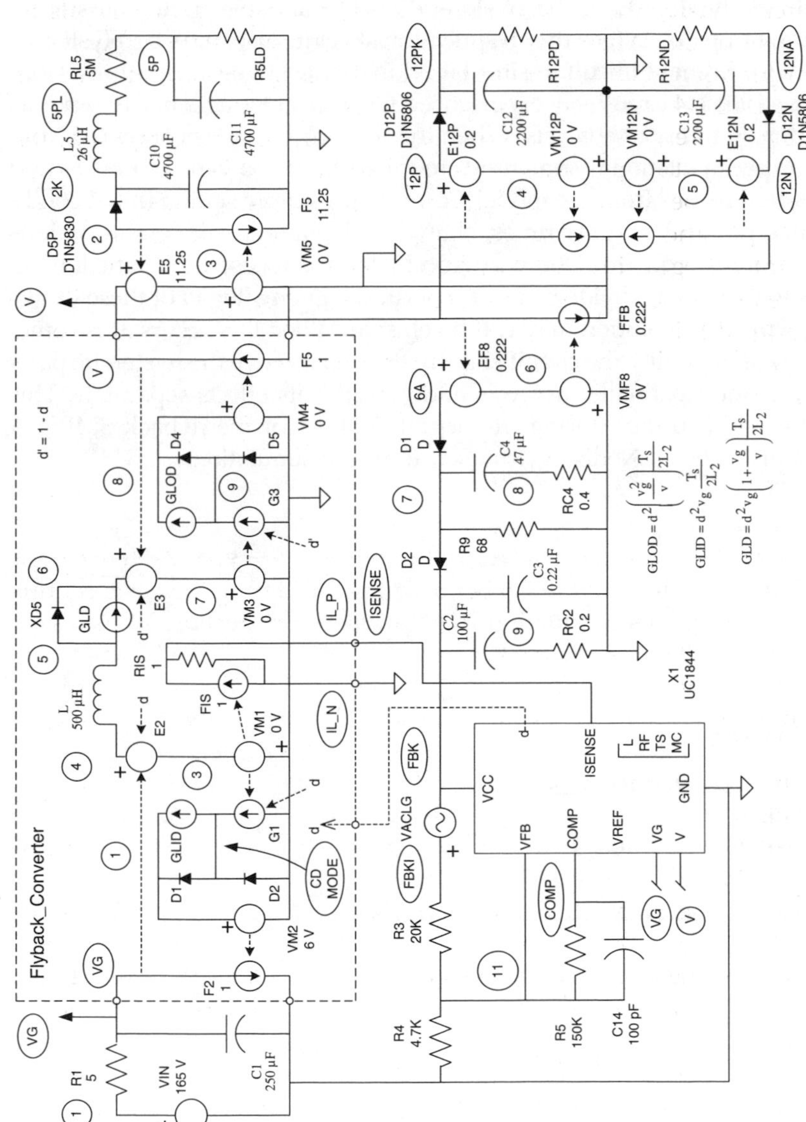

FIGURE 6.5
Flyback SMPS model.

Now all that remains is to insert the circuit-averaged macro of the controller (UC1844) along with its required external components. Then the signal and parametric inputs and outputs are added to the macro and the SMPS model is complete.

A word of caution should be made at this point. The controller macro has two hysteretic type circuits: the hysteretic UVLO and the hysteretic supply current draw from VCC. (See the specification sheet in Appendix B.) Hysteretic circuits are in general difficult to simulate due to their two-state natures. Convergence problems are routinely experienced when calculating initial bias points and also during transient analysis. When these problems are encountered, considerable experimentation is sometimes required to get past these hurdles and get the desired results from the model. Sometimes, simply setting initial conditions will suffice and, at other times, changing the simulation default options of the software is required. One very good action is to use limit functions in equations to prevent a "divide by zero" condition. Even after all of these things are done, a practical solution may still be elusive. When this occurs, it is sometimes practical to modify the circuit slightly or even remove the hysteretic parts of the circuit for a particular analysis and consider its effects separately. This will be discussed further during the actual analysis of the flyback SMPS. A PSPICE netlist, labeled Netlist 6.2, is used for these simulations.

NETLIST 6.2

```
FLYBACK SMPS ANALYSIS
FIGURE 6.5
*
** MINIMUM LOAD RESISTANCES
R5LD 5P 0 5
R12PD 12PK 0 120
R12ND 12NA 0 120
**
** MAXIMUM  LOAD RESISTANCES
*R5LD 5P 0 1.25
*R12PD 12PK 0 40
*R12ND 12NA 0 40
*
*********************************************************************************
*
** SOURCE CONFIGURATION & COMMANDS FOR TURN-ON/TURN-OFF TRANSIENT
**ANALYSIS
** (FIGURES 6.6 AND 6.7)
** (MAXIMUM AND MINIMUM LOAD RESISTANCE CASES)
*
VIN 1 0 PULSE(0 165 100M 1M 1M 1.7)
.TRAN 1M 3 0 500U
.OPTIONS STEPGMIN ITL4=40
*
*********************************************************************************
*
** SOURCE-LOAD CONFIGURATION AND COMMANDS FOR LOAD TRANSIENT
 ANALYSIS
```

```
** (FIGURE 6.8)
** (LOAD RESISTANCES SET TO MINIMUM)
*
*VIN 1 0 DC 165
*I5_LOAD 5P 0 PULSE(0 4 1.5 10U 10U 100M)
*I12POS_LOAD 12PK 0 PULSE(0 .2 1.7 10U 10U 100M)
*.TRAN 1M 2 0 500U
*.OPTIONS STEPGMIN ITL4=20
*.NODESET V(VG)=100 V(V)=30
*
*******************************************************************************
*
** SOURCE CONFIGURATION AND COMMANDS FOR LINE REGULATION ANALYSIS
** (FIGURE 6.9)
** (MINIMUM AND MAXIMUM LOAD RESISTANCE CASES)
*
*VIN 1 0 PWL (0,0) (100M,0) (101M,134) (2.5,133) (2.501,184) (2.700,184) (2.701,133) (3.0 134)
*.TRAN 1M 3 0 500U
*.OPTIONS STEPGMIN ITL4=40
*
*******************************************************************************
*
** SOURCE-LOAD CONFIGURATION AND COMMANDS FOR CURRENT LIMITING
** PERFORMANCE
** (FIGURE 6.10)
** (LOAD RESISTANCES SET TO MAXIMUM)
*
*VIN 1 0 PULSE(0 165 100M 1M)
*D_LOADSHORT 5P WW D1
*V_LOADSHORT WW 0 PULSE(7 0 1.8 1M)
*.TRAN 1M 4 0 200U
*.OPTIONS STEPGMIN ITL4=20
*
*******************************************************************************
*
** SOURCE CONFIGURATION AND COMMANDS FOR AC STABILITY ANALYSIS
** (FIGURE 6.11)
** (MAXIMUM AND MINIMUM LOAD RESISTANCE CASES)
** (ACTIVATE VOVRD IN UVLO SECTION OF THE UC1844 SUBCKT)
*
*VIN_MIN 1 0 DC 134
*VIN_MAX 1 0 DC 184
*VACLG FBKI FBK AC 1
VACLG FBKI FBK AC 0
*.AC DEC 50 10 .1MEG
*.OPTIONS STEPGMIN RELTOL=.004
*******************************************************************************
*
** SOURCE CONFIGURATION AND COMMANDS FOR AC LINE REJECTION ANALYSIS
** (FIGURE 6.12)
** (MAXIMUM AND MINIMUM LOAD RESISTANCE CASES)
** (ACTIVATE VOVRD IN UVLO SECTION OF THE UC1844 SUBCKT)
*
*VIN_MIN 1 0 DC 134 AC 1
*VIN_MAX 1 0 DC 184 AC 1
```

```
*.AC DEC 50 10 100K
*.OPTIONS STEPGMIN
*
********************************************************************************
*
**INPUT FILTER
C1 VG 0 2.5U
R1 1 VG 5
*
**STARTUP RESISTOR
R2 VG FBK 56K
*
*CONTROL CIRCUIT COMPONENTS
R3 11 FBKI 20K
R4 11 0 4.7K
R5 COMP 11 150K
C14 COMP 11 100P
EFB 6A 6 V 0 .222
VMFB 0 6
FFB V 0 VMFB .222
D1 6A 7 D
C4 7 8 47U
R9 7 0 68
RC4 8 0 .4
D2 7 FBK D
.MODEL D D
C3 FBK 0 .22U
C2 FBK 9 100U
RC2 9 0 .2
X1 FBK REF 11 COMP ISENSE 0 VG d 0 UC1844
*
*5 VOLT OUTPUT STAGE
VM5 3 0
E5 V 3 2 0 11.25
F5 0 2 VM5 11.25
D5P 2 2K D1N5830
C10 2K 0 4700U
L5 2K 5PL 26U
RL5 5PL 5P 5M
C11 5P 0 4700U
RCONV V 0 1E12
*
*+12 VOLT OUTPUT STAGE
E12P 12P 4 V 0 .2
F12P V 0 VM12P .2
VM12P 0 4
D12P 12P 12PK D1N5806/27C
C12 12PK 0 2200U
*
*-12 VOLT OUTPUT STAGE
E12N 5 12N V 0 .2
F12N V 0 VM12N .2
VM12N 5 0
D12N 12NA 12N D1N5806/27C
C13 12NA 0 2200U
```

```
*
********************************************************************************
*
**UC 1844 PWM CONTROLLER CIRCUIT AVERAGING MACRO
.SUBCKT UC1844 VCC 3    4   COMP 6      7 8  9 GND
*           VCC VREF VFB COMP ISENSE V VG d GND
**NOTE: ALL SIGNAL INPUTS MUST BE REFERENCED TO THE UC1844 GND
XUVLO VCC UVLO GND UVLO
XREF UVLO VCC 3 VREF/2 GND REF
XERRAMP 4 VREF/2 VCC COMP GND ERRAMP
XDRC COMP 9 6 7A 8A UVLO GND DRC
**DEPENDENT GENERATORS FOR GROUND ISOLATION IF DESIRED
EVG 8A 0 8 GND 1
EV 7A 0 7 GND 1
.ENDS UC1844
*
.SUBCKT DRC 1      d 8      9 10 14   GND
*         COMP   d ISENSE V VG UVLO GND
VBK 1 1A DC 2
R1 1A C 200K
R2 C GND 100K
X1 C 3 DIDEAL
X2 GND C DIDEAL
VLIM 3 GND DC 1
RCONV1 6 0 1G
RCONV2 7 0 1G
RCONV3 4 0 1G
*
REL 10 9 1E8
GL 0 11 10 9 1
D1L 11 12 DX
D2L 0 11 DX
RLMIN 12 13 1
VLMIN 13 0 1U
*
DX1 4 0 DX
DX2 7 4 DX
DX3 6 7 DX
*
**IDRMAX VALUE = MAX LIMITED DUTY RATIO
IDRMAX 7 6 DC .48
DX4 0 6 DX
DX5 6 6A DX
VMD 6A 0
EMD d 0 VALUE = {I(VMD)*V(14)+1P}
RMD d 0 1
.MODEL DX D IS=1E-12
*
*L = 500U
*RF = 0.55
*TS = 25U
*MC = 0
*
**GCM 4 7 VALUE = {(V(C)-RF*V(8))/((TS*RF)/(2*L))*(2*MC*L+V(12)))}
GCM 4 7 VALUE = {LIMIT((V(C)-0.55*V(8))/((25U*0.55)/(2*500U)*(2*0*500U+V(12))),0,1)}
```

```
*
**GDCM 0 4 VALUE = {V(C)/((TS*RF/L)*(MC*L+V(12))}
GDCM 0 4 VALUE = {LIMIT(V(C)/((25U*0.55/500U)*(0*500U+V(12))),0,1)}
*
.ENDS DRC
*
.SUBCKT UVLO VCC UVLO   GND
*         VCC UVLO   GND
RIN VCC 2 1K
ISUP 2 GND 1M
XSUP GND 2 DIDEAL
G5 VCC GND VALUE = {.01*V(UVLO,GND)}
** G1 AND G2 SET SET  UVLO START AND STOP LEVELS RESPECTIVELY
G1 0 3 TABLE {V(VCC,GND)} = (0,0) (17,0) (18,20)
G2 3 0 TABLE {V(VCC,GND)} = (9,20) (10,0) (17,0)
CG2 3 0 20N
R1 3 0 1E3
G3 0 4 3 0 1
G3HYST 0 4 VALUE = {-100U+200U*V(UVLO,GND)}
CHOLD 4 0 20N
G4 GND UVLO 4 GND 1
**VOVRD TO BE INSERTED FOR UVLO OVERRIDE
*VOVRD UVLO GND DC 1
RG4 UVLO GND 1
CG4 UVLO GND 1N
X5 GND UVLO DIDEAL
X1 3 5 DIDEAL
X2 6 3 DIDEAL
X3 4 5 DIDEAL
X4 6 4 DIDEAL
VP 5 0 DC 1
VN 0 6 DC 1
.ENDS UVLO
*
*5.0 VOLT REFERENCE MACRO
.SUBCKT REF UVLO VCC   VREF   VREF/2      GND
*          UVLO VCC   VREF   VREF/2      GND
*
EIN 2 3 VCC GND .46E-3
RVT 3 4 .32 TC=0.2E-3
IVT GND 3 DC 3.125
VT GND 4 DC 1
V5 1 2 DC 4.99342
FUVLO 1 GND VMR 1
EUVLO VREF 9 VALUE = {V(1,GND)*V(UVLO,GND)}
VMR GND 9
EVREF/2 VREF/2 GND VREF GND .5
.ENDS REF
*
*ERROR AMP MACRO
.SUBCKT ERRAMP 6  7  8  1     GND
*          VIN VIP VCC OUTPUT GND
RIN 5 7 1MEG
EPSRR 6 5 VALUE = {(V(8)-15)*3E-4}
IBIAS GND 5 .3U
```

```
GA GND 2 7 5 1K
RG 2 GND 30
CG 2 GND 0.159M
VNC 4 GND DC 1
DN 4 2 D1
VPC 3 GND DC 5
DP 2 3 D1
RO 1 2 3K
RPS 8 0 1E8
.ENDS ERRAMP
.MODEL D1 D IS=1E-9
*
*
X2 vg 0 d ISENSE 0 V 0 FLYBACK
******************************************************************************
*
*FLYBACK CONVERTER MODEL
.SUBCKT  FLYBACK vg    GNDP     d               IL_P     IL_N     v       GNDS
*          (VIN) (PRI GND) (DUTY RATIO) (INDUCTOR CURRENT) (VOUT) (SEC GND)
*TS = 25U
*L  = 500U
*
**GLOD 9 8 VALUE = {V(d)**2*(V(vg)**2/V(v))*(TS/(2*L))}
GLOD 9 8 VALUE = {V(d)**2*(V(vg)**2/V(v))*(25U/(2*500U))}
*
**GLID CDMODE 1 VALUE = {V(d)**2*V(vg)*(TS/(2*L))}
GLID CDMODE 1 VALUE = {V(d)**2*V(vg)*(25U/(2*500U))}
*
**GLD 5 6 VALUE = {V(d)**2*(V(vg)*(1+V(vg)/V(v))*(TS/(2*L))}
GLD 5 6 VALUE = {V(d)**2*V(vg)*(1+(V(vg)/V(v)))*(25U/(2*500U))}
*
F2 GNDP vg VM2 1
VM2 1 0 DC 6
D1 CDMODE 1 D2
D2 0 CDMODE D2
G1 0 CDMODE VALUE = {V(d)*I(VM1)}
E2 4 3 VALUE = {V(d)*V(vg)}
VM1 0 3
L 4 5 500U
XD5 5 6 DIDEAL
VM3 7 0
E3 6 7 VALUE = {(1-V(d))*V(v)}
G3 0 9 VALUE = {(1-V(d))*I(VM3)}
D4 9 8 D2
D5 0 9 D2
.MODEL D2 D IS=1E-7 CJO=1N
VM4 8 0 DC 6
F4 GNDS V VM4 1
*
RISEN IL_P IL_N 1
FISEN IL_N IL_P VM1 1
*
.ENDS FLYBACK
*
******************************************************************************
```

```
*
*
.SUBCKT DIDEALA 1 2
EID 3 1 TABLE {V(3,2)} = (-1,1U) (1u,1U) (1,1)
DIO 3 2 D
.MODEL D D IS=1E-12
CC  1 2 10p
.ENDS DIDEALA
*
.SUBCKT DIDEAL 1 2
VAS 1 3 DC -1u
D1 3 2 D
D2 3 4 D
D3 4 2 D1
FAS 4 2 VAS 1
.MODEL D D IS=1E-6 EG=0 XTI=-4
.MODEL D1 D EG=0 XTI=0
CC 1 2 .1P
.ENDS DIDEAL
*
.LIB DIODE.LIB
.PROBE
.END
```

6.2.2 Flyback SMPS Large Signal Power ON/OFF Analysis

Begin by conducting a large signal analysis of the SMPS. First, simply perform a turn-on and turn-off operation and investigate the various signals that occur. Set the loads to their minimum values of 5 Ω (1.0 amp) on the +5-Vdc output and 120 Ω (0.1 amp) on each of the ±12-Vdc outputs. Figure 6.6a through Figure 6.6c show some of the waveforms. A 0 to 165 V step is applied for turn-on and after that cycle settles out, the power supply is turned off by stepping the input voltage from 165 to 0 V. Figure 6.6a shows some results of this analysis.

After the 165-V input step voltage is applied, the VCC begins to rise. When the upper trip point of the UVLO is reached at VCC = 17 V, the UVLO signal goes from 0 to 1, thus allowing the controller *X*1 (UC1844) to come alive. At this point, the operating supply current of *X*1 loads VCC so that its voltage starts dropping, but eventually the rising bootstrap feedback from node V(7) catches it. The control loop then begins to regulate holding VCC at approximately 13.1 V. When the input voltage is suddenly reduced from 165 to 0 V for a normal turn-off, it is noted that the VCC begins to decrease with the converter duty ratio, *d*, rising to its maximum value of 0.48. The UVLO drops to 0 when VCC is reduced to its lower trip point of 10 V. The controller, *X*1, now shuts down by setting the converter duty ratio, *d*, to 0; all voltages continue to drop.

Figure 6.6b shows more waveforms indicating how the circuit is performing. Figure 6.6c provides a signal description of node CDMODE. This is a signal

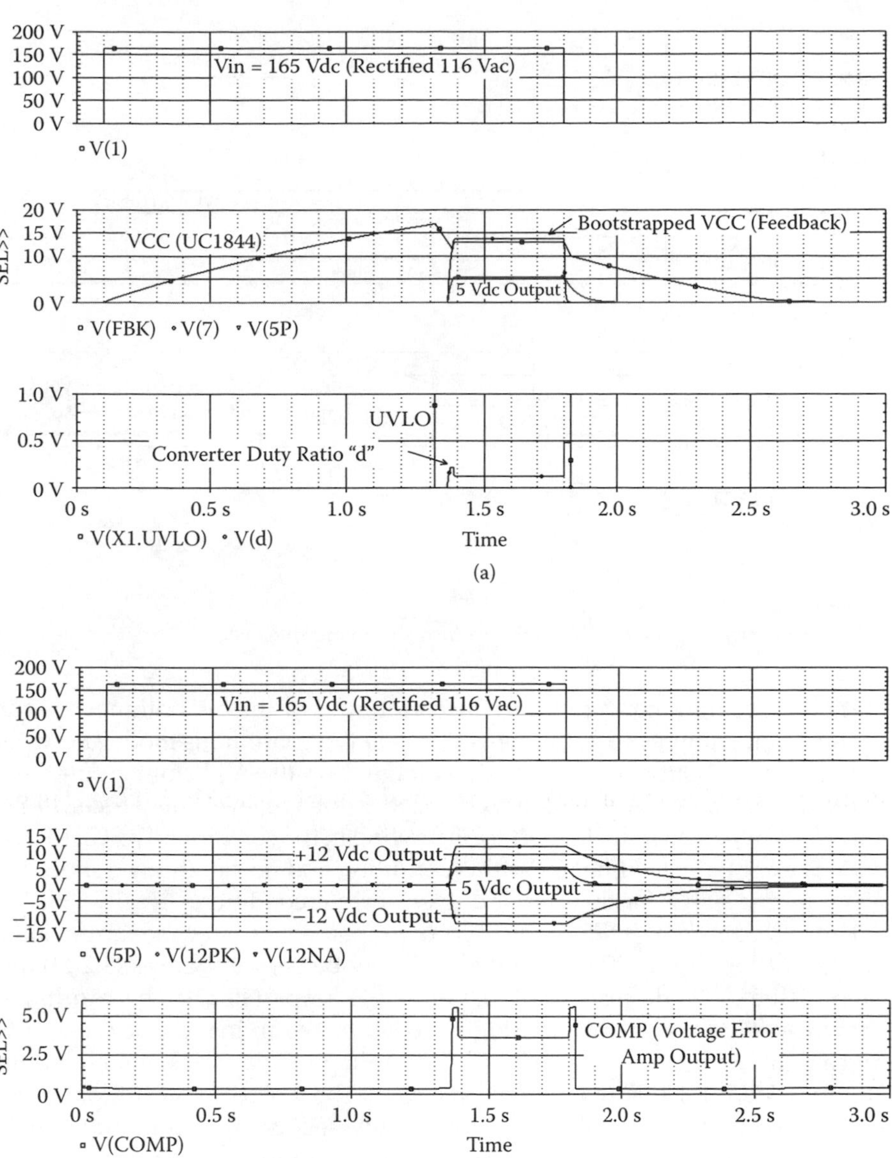

FIGURE 6.6
(a) Flyback SMPS turn ON/OFF waveforms (minimum loads). (b) More flyback SMPS turn ON/OFF waveforms (minimum loads).

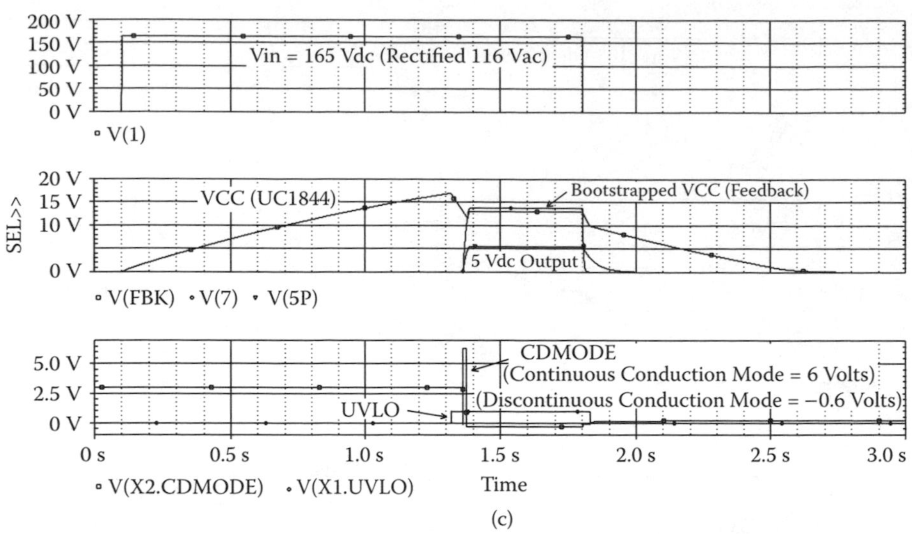

FIGURE 6.6c
Still more flyback SMPS turn ON/OFF waveforms (minimum loads).

that indicates if the converter is in the continuous or the discontinuous mode. When its magnitude is 6 V, the converter is in the continuous mode and when it is less than about a forward diode voltage drop (0.6 V) below ground, it is in the discontinuous conduction mode. Other voltage levels may occur in this waveform when no inductor current is flowing in the converter. The CDMODE node is essentially floating under this condition and no significance should be given to this signal for this zero inductor current condition.

The preceding ON/OFF analysis is now repeated with the maximum loads on the power supply. The 5-V load will now be set for 4 amps (1.2 Ω) and each of the ±12-Vdc loads will be set at 0.3 amps (40 Ω). The results are shown in Figure 6.7a through Figure 6.7c. An examination of the results reveals that, when powered, the COMP voltage is very near its maximum value of 5.0 V for this maximum load case. The converter is still operating in the discontinuous mode, however. It only seems to go continuous for a brief period at the thresholds of regulation. If the load is increased further, current limiting will take place. This will be shown in the next section when a load sweep analysis is performed.

6.2.3 Flyback SMPS Large Signal Load and Line Transient Analysis

Now some transient load and line tests are performed on the three models and the effects considered. As a point of departure, first add load resistors to the +5- and ±12-Vdc outputs that represent minimum loading on the supply. That would be 5 Ω (1.0 amp) for the +5-Vdc output and 120 Ω (0.1 amp) for each of the ±12-Vdc outputs. Then apply a pulse load to the +5 Vdc

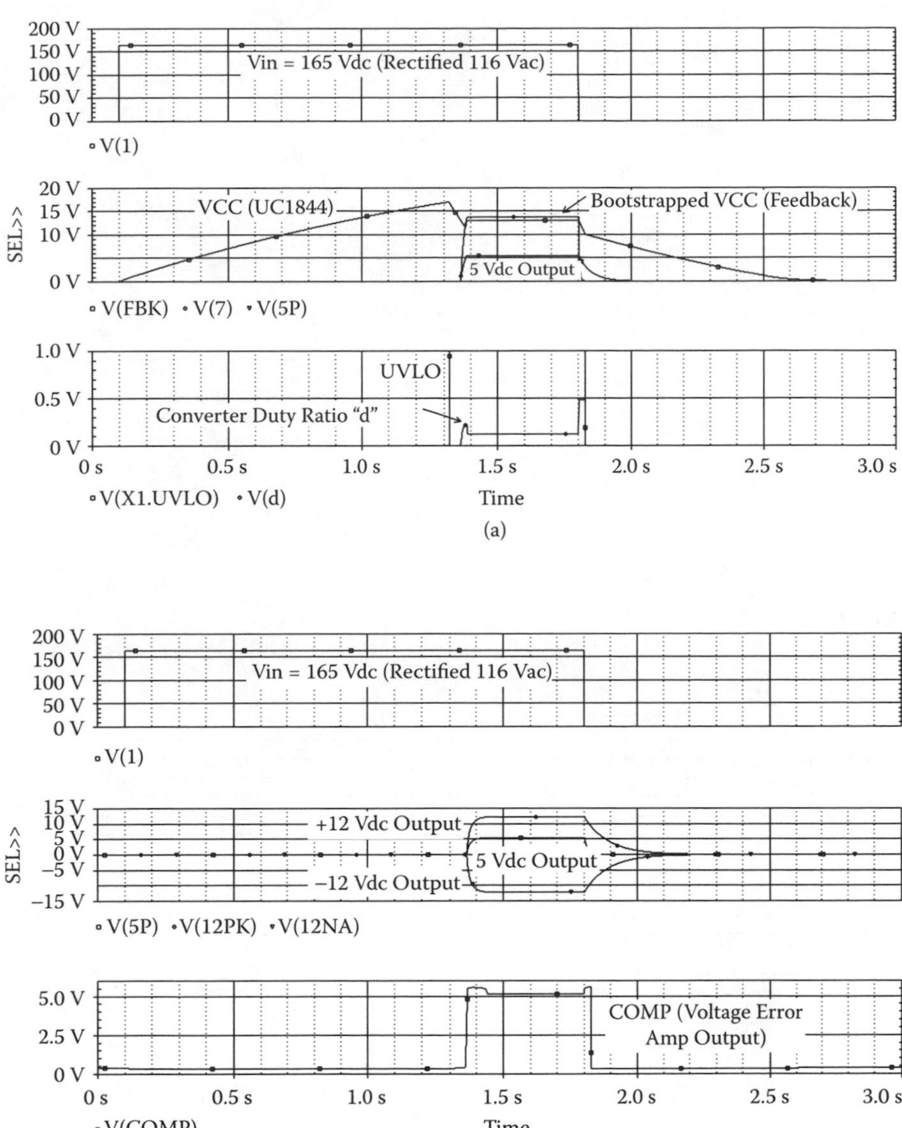

FIGURE 6.7
(a) Flyback SMPS turn ON/OFF waveforms (maximum loads). (b) More flyback SMPS turn ON/OFF waveforms (maximum loads).

of an additional 4.0 amps and subsequently apply a 0.2 amp pulse load to the +12-Vdc output; then look at the various circuit voltages. The input voltage is held constant at +165 Vdc (116 Vac). The results are shown in Figure 6.8a and Figure 6.8b. The results of Figure 6.8a show that the +5-V

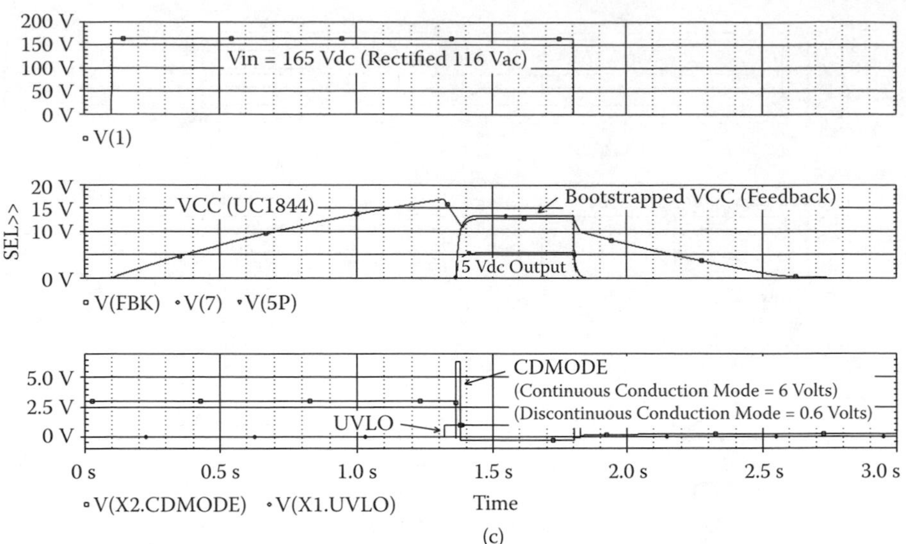

FIGURE 6.7c
Still more flyback SMPS TURN ON/OFF waveforms (maximum loads).

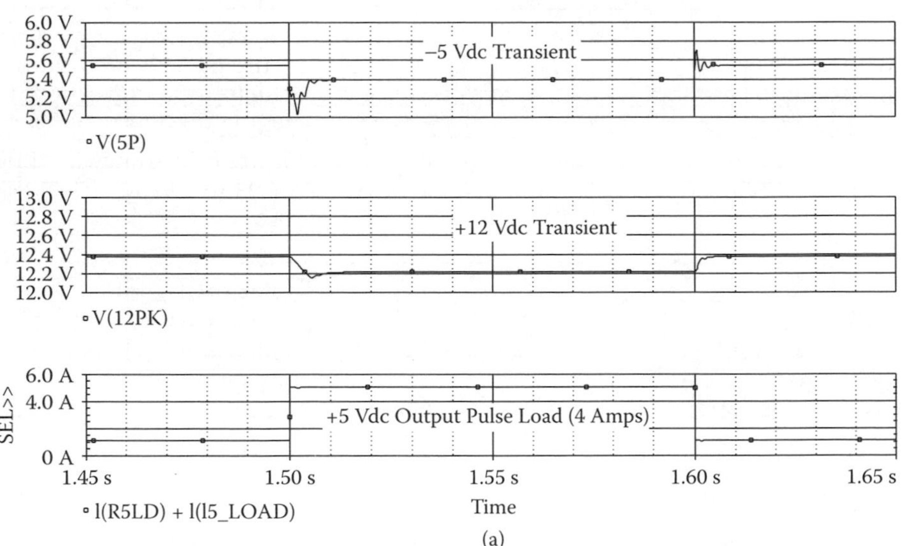

FIGURE 6.8a
Flyback SMPS load transient analysis (5-V load pulse).

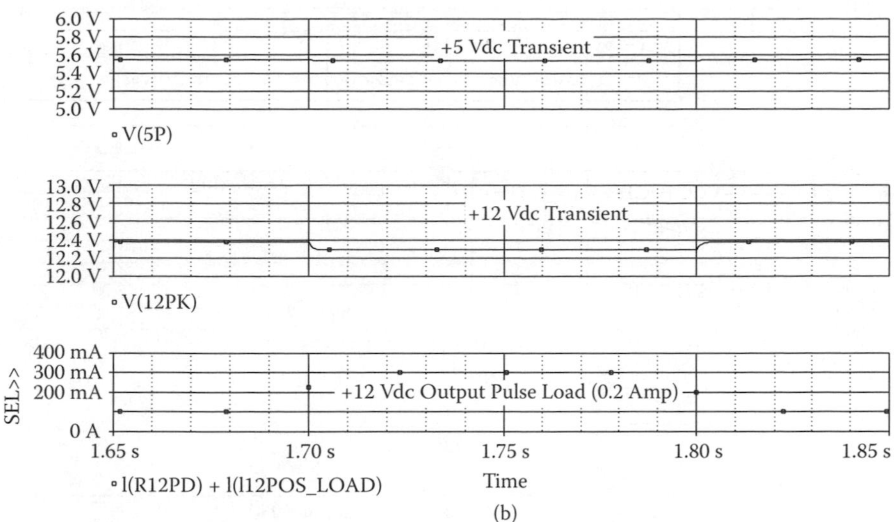

FIGURE 6.8b
Flyback SMPS load transient analysis (+12-V load pulse).

output initially dips by about 0.5 V and then the maximum load case settles to about 0.15 V less than the minimum load case. The +12-V output responds with about a 0.15 V sag. Examining the results of Figure 6.8b shows that the lighter transient load on the +12-V output has relatively little effect on any of the outputs, with only a 0.1 V sag on its own output.

Now look at the input pulsed line voltage case. Figure 6.9a shows the line voltage pulsed high and low over its entire range of 95 to 130 Vac with the minimum loads on the SMPS. (The converter DC input for this source configuration is 134 V with a superimposed 50-V pulse.) Note that output voltages dip very slightly with the increase in line voltage. Figure 6.9b shows this output transient again with the maximum loads on the SMPS. For this maximum load case, the output transients are negligible with the input line transient. As a sidelight, note that the power supply takes a considerably longer amount of time of over 2 sec to turn on after initial application of the lower input voltage of 134 Vdc (rectified 95 Vac) than it did from the 165-V input of the previous analysis.

6.2.4 Flyback SMPS Current-Limiting Performance

Provide an overload to the 5-V output and check the current-limiting performance of the power supply. This occurs for overloads in excess of about 6 amps. A load fault is simulated by suddenly applying a forward biased diode from the +5-Vdc output to ground after the power supply has just been powered up with all outputs at maximum loads. Figure 6.10 shows that, when the continuous overload is applied at a time of 1.8 sec, the +5-Vdc

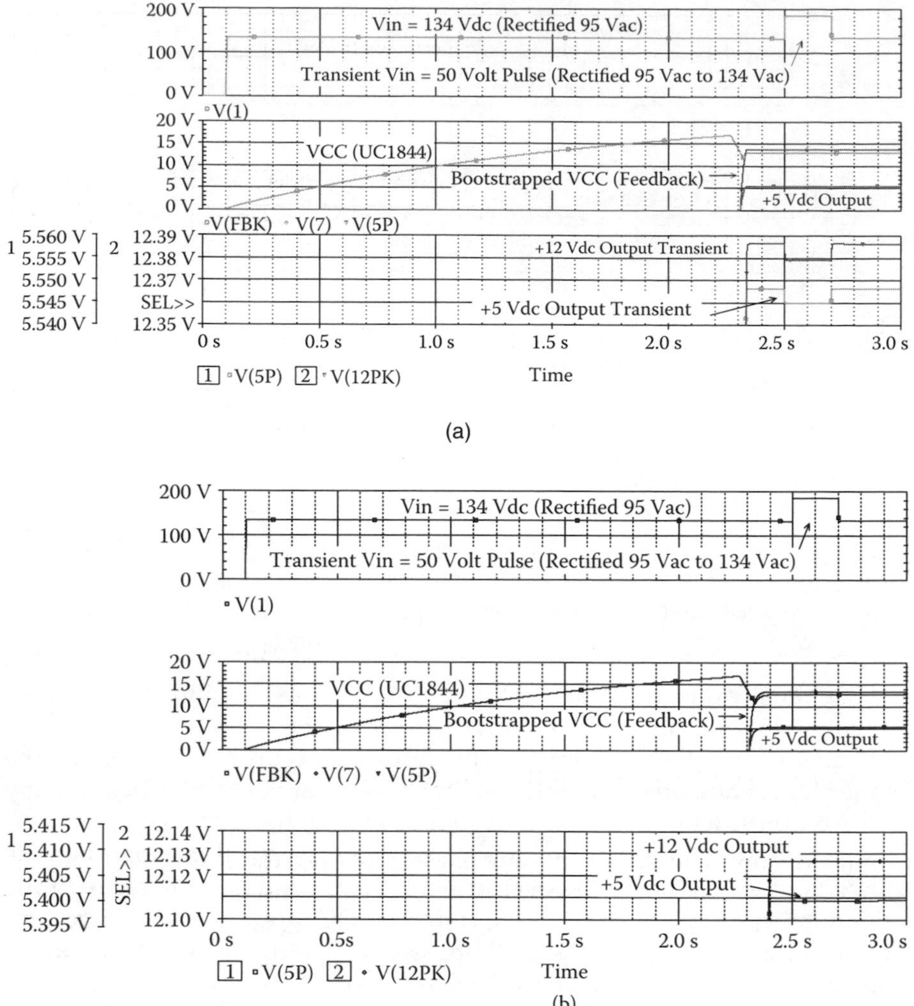

FIGURE 6.9
(a) Flyback SMPS line transient analysis (minimum load case). (b) Flyback SMPS line transient analysis (maximum load case).

output voltage immediately drops to about that of a forward biased diode (approximately 0.6 V). The bootstrap feedback voltage also drops but the +12-V output decays more slowly. The VCC voltage eventually drops to below its lower UVLO trip point of 10 V and the converter is switched off. The UC1844 controller supply current drops and the VCC voltage starts to rise again. When it reaches its upper UVLO trip level of 17 V, the converter attempts to start up again. With the overload still applied to the 5-V output,

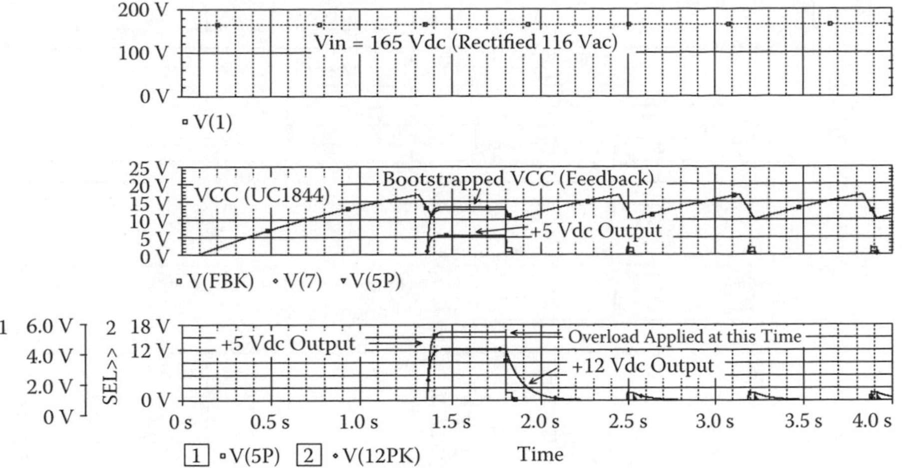

FIGURE 6.10
Flyback SMPS current limiting performance.

the previously described shutdown starts over again, thus starting a steady state limit cycle as shown in Figure 6.10.

6.2.5 Flyback SMPS AC Stability Analysis

The AC stability of the example flyback switching power supply will be examined now. The AC loop gain phase plots for the four conditions of maximum and minimum line and load will be discussed. Figure 6.11a through Figure 6.11d show the results. The stability margins are very good, with phase margins all in excess of 90°. The unity gain crossover is noted to be fairly independent of line voltage variations but does decrease from 300 Hz at minimum loads to 100 Hz at maximum loads.

6.2.6 Flyback SMPS AC Line Rejection Analysis

The AC line noise rejection characteristics of the flyback SMPS are now examined. This is an important part of any power supply analysis when power line noise is expected or when a stringent audio susceptibility specification is required to be met. An AC voltage is injected onto the input power lines and the frequency is swept over the range of 10 to 100 kHz. The results for the four extremes of maximum and minimum load current and line voltage are shown in Figure 6.12a through Figure 6.12d. The rejection is over 80 dB for all conditions and does not vary significantly over the ranges. This is characteristic of the discontinuous conduction mode in general.

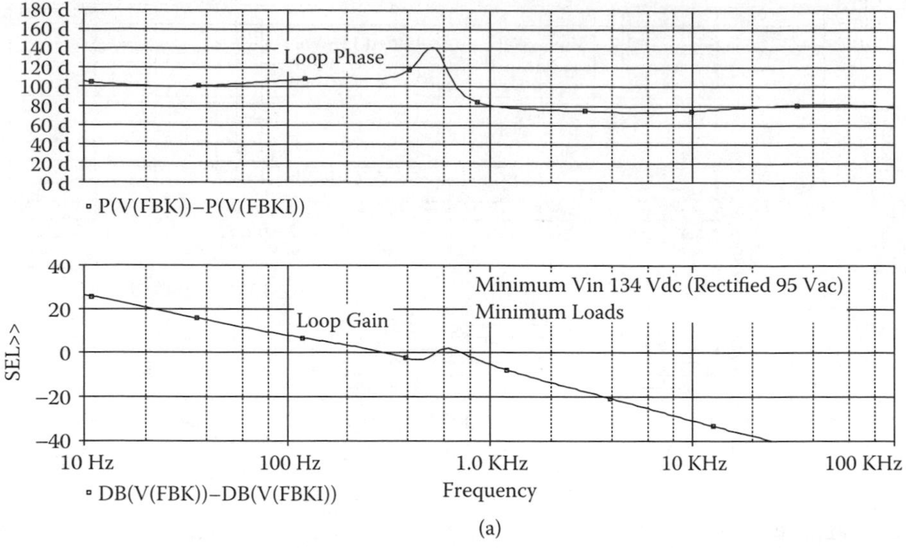

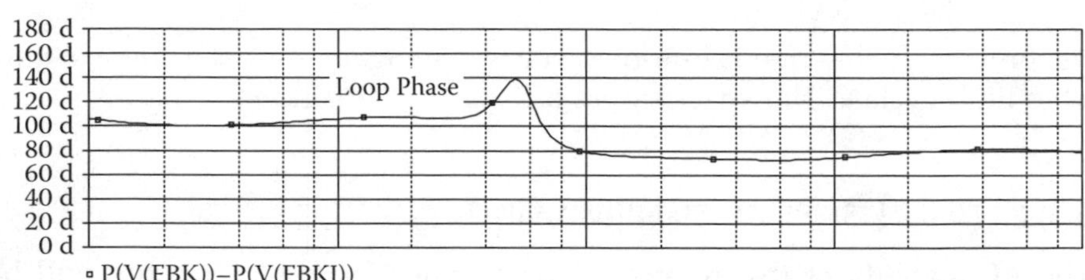

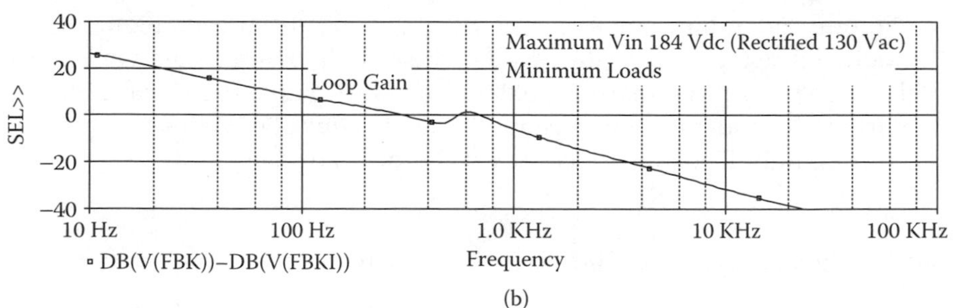

(b)

FIGURE 6.11a and 6.11b
(a) Flyback SMPS AC loop gain analysis (minimum line, minimum loads). (b) Flyback SMPS AC loop gain analysis (maximum line, minimum loads).

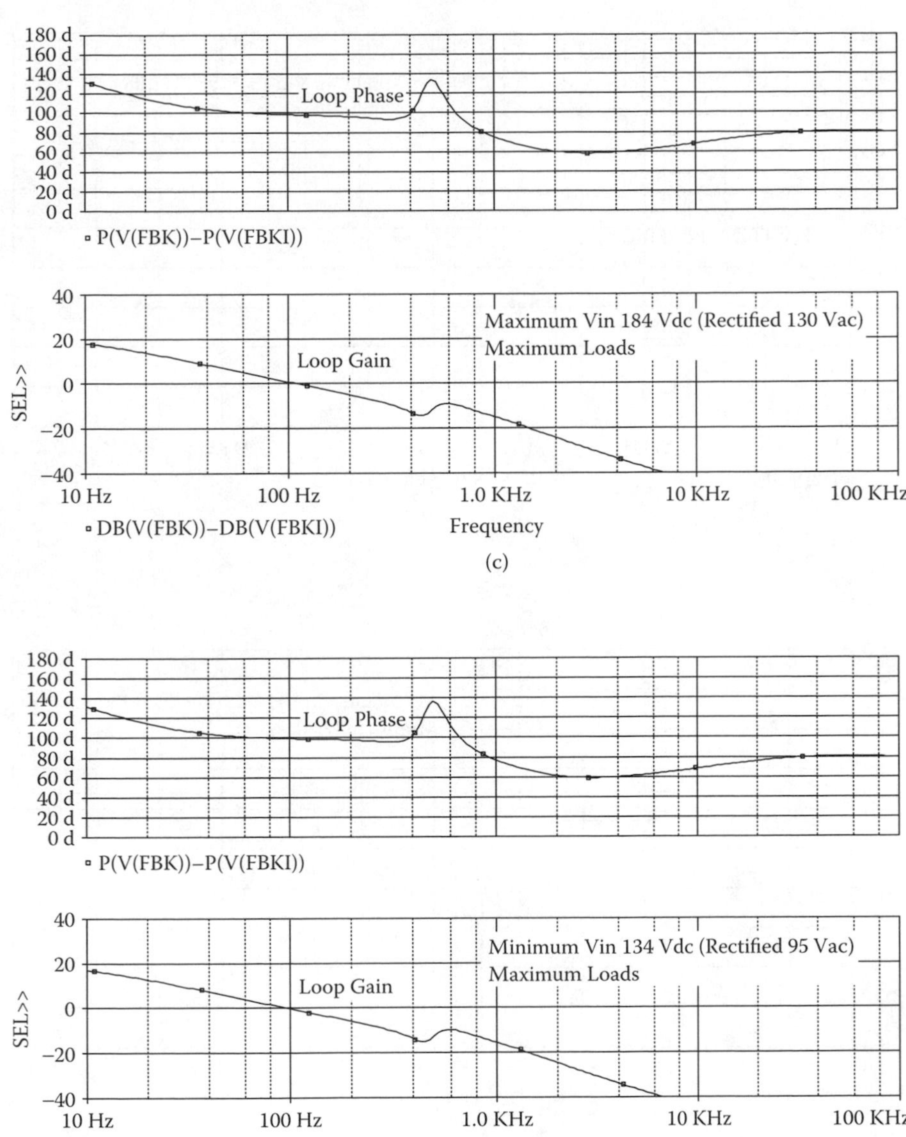

FIGURE 6.11c and 6.11d
(c) Flyback SMPS AC loop gain analysis (maximum line, maximum loads). (d) Flyback SMPS AC loop gain analysis (minimum line, maximum loads).

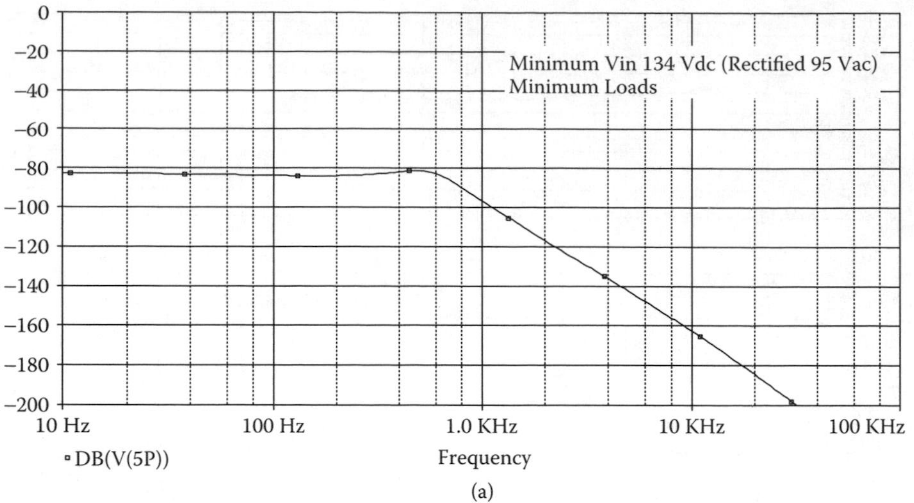

(a)

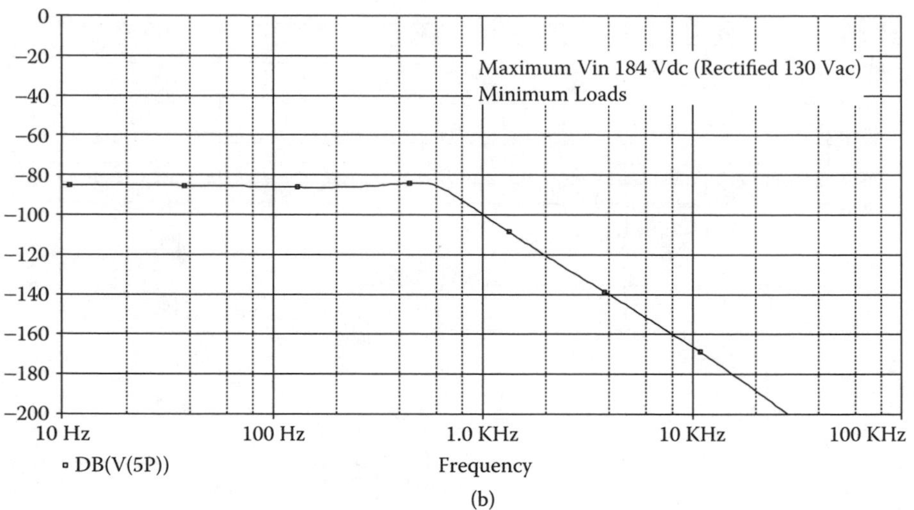

(b)

FIGURE 6.12a and 6.12b
(a) Flyback SMPS AC line rejection analysis (minimum line, minimum loads). (b) Flyback SMPS AC line rejection analysis (maximum line, minimum loads).

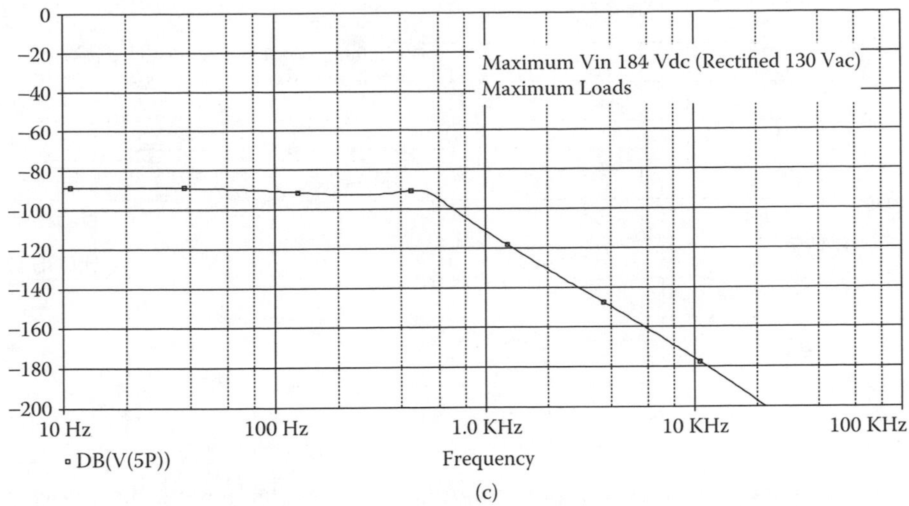

(c)

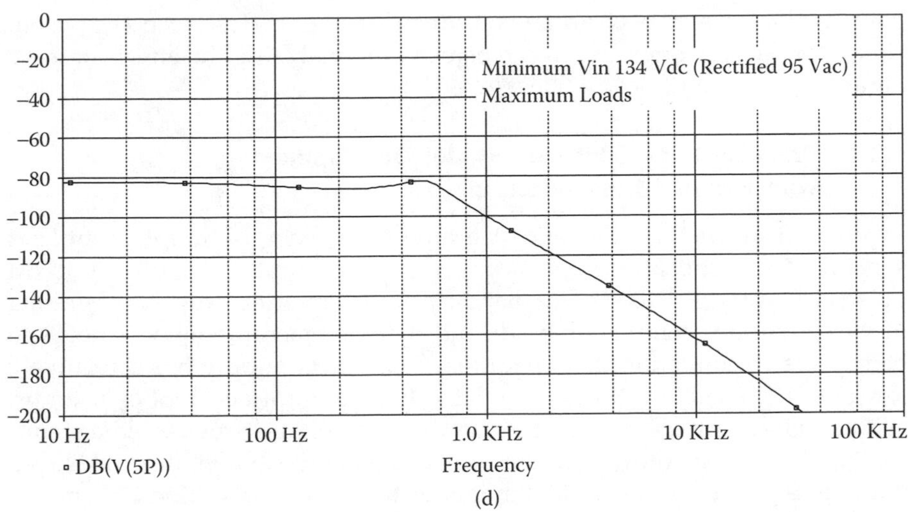

(d)

FIGURE 6.12c and 6.12d
(c) Flyback SMPS AC line rejection analysis (maximum line, maximum loads). (d) Flyback SMPS AC line rejection analysis (minimum line, maximum loads).

6.3 Practical Buck SMPS Analysis with Parasitic Resistances

Consider another example, which will show a buck topology switch mode converter when parasitic resistances are considered. Modern computer power supplies using low power supply output voltages and relatively large

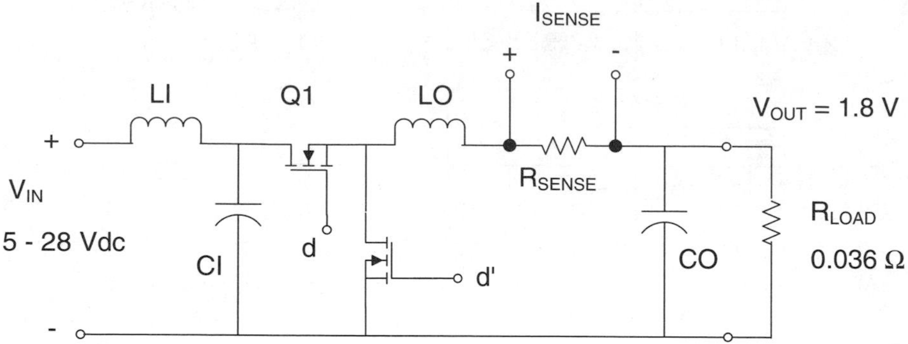

FIGURE 6.13
Buck converter with synchronous rectification.

load currents pose some unique design challenges in achieving high power conversion efficiencies. The example presented here illustrates how parasitic effects that may have previously been considered second and third order are now significant and need to be considered in the analysis. An equivalent circuit for the converter, which can be used for analyzing the effects of these parasitic effects, will now be developed.

6.3.1 Practical Buck Converter Model Development with Parasitic Resistances

Figure 6.13 shows the schematic diagram of the general example. It utilizes synchronous rectification when converting the input voltage to a PWM pulsed DC voltage. This implies that the power converter is always operating in the continuous conduction mode and that the macromodel only needs to be capable of emulating that condition. The power supply has a relatively low output voltage of 1.8 V and a relatively high output current of 50 amps. With an input voltage of 5 to 28 V, conduction voltage drops will be significant and a valid accounting of these is deemed necessary. Figure 6.14 shows the large signal circuit-averaged model of the converter section. The development of this model is explained next.

When Figure 6.14 is first examined, one of the very noticeable things is the considerable number (six) of series resistances elements in the model. All of these component resistances are undesirable because they decrease the efficiency of the power conversion process. Dr. Middlebrook's paper[20] provides the basis for determining the series resistance effects of elements that are functions of the converter duty ratio, *d*. These resistances can be quantified as follows:

- *RSENSE*. This is the only one of the series resistances that is not considered an inherent "parasitic" resistance because it was intentionally inserted for the purpose of measuring current. Unfortunately, however,

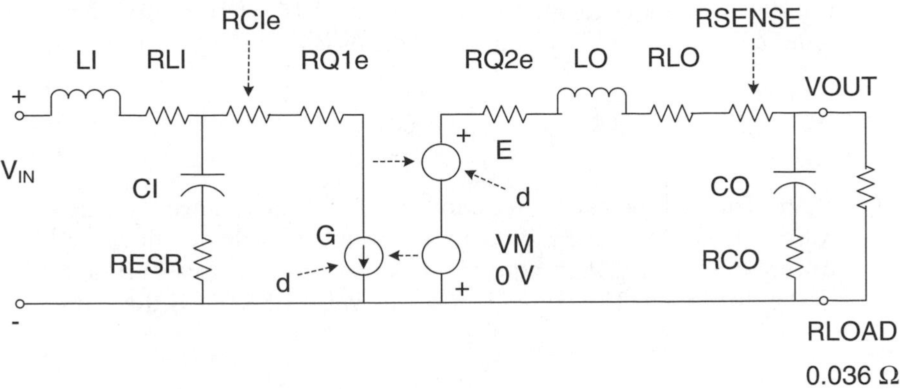

FIGURE 6.14
Buck converter model with parasitic resistances.

its effects on efficiency are still negative and should be considered. As the current in this resistance is equal to the inductor current and is not "chopped" by the power switches, its equivalent value is considered to be equal to its actual value for analysis purposes. On the other hand, if the inductor current has a significantly large ripple content, a duty ratio-dependent value of this resistance might be in order.

- *RLO*. This resistance is the series resistance of the averaging inductor, *LO*, and is in series with the sense resistor, *RSENSE*. Its circuit effects are the same as those of *RSENSE*. This resistance is in general a function of temperature and frequency. Once the temperature of the device has been determined, the winding resistance can be adjusted by the resistance temperature coefficient, *RTC*, of the coil winding material. (Most windings will be copper with an *RTC* of nominally +0.4%/°C.) Frequency dependency is determined by skin and proximity effects; however, for most power supply inductor designs, the AC ripple current is reduced by the inductor and these AC effects are not relevant. (When considering the effects of *transformer* windings, however, when the current is chopped and thus produces a considerable amount of AC current, the AC winding resistances now assume a whole new significance and generally must be considered.[16,17]) Equation 6.1 shows how this resistance varies with temperature.

$$RLO = R_{25°C} \times [(1 + .004(T - 25°C)] \tag{6.1}$$

- *RQ2e*. This resistance is a function of the ON resistance, $R_{DS(ON)}$, of MOSFET, *Q*2. The current through this device flows only during the interval, *d'*, so its equivalent series effects are a function of the duty ratio. For MOSFET switching devices, this resistance is composed of the silicon channel resistance and has the *RTC* of silicon, which is nominally +0.7%/°C. The equivalent series value of *RQ2e* is therefore:

$$RQ2e = R_{DS(ON)Q2} \times [(1 + .007(T - 25°C)] \times d' \tag{6.2}$$

- *RQ1e*. This resistance is a function of the ON resistance, $R_{DS(ON)}$, of MOSFET, *Q*1. Because the current through this device flows only during the interval, *d*, its equivalent series effects are also a function of the duty ratio. The equivalent series value of *RQ1e* is therefore:

$$RQ1e = \left(\frac{R_{DS(ON)Q1} \times [(1 + .007(T - 25°C)]}{d} \right) \tag{6.3}$$

- *RCIe*. This series resistance is a function of the equivalent series resistance, RESR, of the input filter capacitor, *CI*. *RESR* may sometimes be a complex function of temperature, *T*, and frequency, *F* (see Section 5.1.1). There are basically two components of AC currents flowing in this capacitor: the high-frequency converter switching frequency component and the low-frequency components associated with the dynamical behavior of the power supply. The high-frequency component of AC current in this capacitor is also a function of the converter duty ratio, *d*. Its series circuit resistance effect for the buck topology is as expressed by *RCIe* in Equation 6.4. The value of *RCI* is the value of *RESR* that exists at the switching frequency of the converter. (Although not proven here, this duty ratio modulated resistance, *RCIe*, is valid for power loss calculations.) The value of *RESR* is still maintained in series with capacitor, *CI*, when performing AC and transient analyses on the power supply.

$$RCIe = RCI \times \left(\frac{d'}{d} \right) \times f(T,F) \tag{6.4}$$

- *RLI*. This is the series resistance of the input filter inductor and is not considered to have any chopped current flowing in it. It is therefore simply represented by the actual value of the low-frequency DC resistance, *RLI*, and expressed by:

$$RLI = R_{25°C} \times [(1 + .004(T - 25°C)] \tag{6.5}$$

When these parasitic resistances are modeled, the wiring or conduction paths connecting these components have resistances and also must be inserted in the analysis model when deemed significant. The wiring resistances that are in series with the component resistances as indicated earlier are simply inserted in series with them and are treated with the same duty ratio modulation effects.

The converter of Figure 6.14 may be analyzed at the converter level or may simply be inserted into an appropriate control circuit to create the desired SMPS. Then, the complete switching regulator can be analyzed for changes of performance and efficiency by inserting, removing, or modifying these parasitic resistances as desired. These resistances may sometimes have a positive effect on the low-frequency performance of a power supply because the added resistances tend to increase circuit damping of the resonant LC filter circuits. This may result in less ringing and overshoot voltages during transient operating conditions. This damping may also help reduce high-frequency switching noise; however, in general, these resistances are not sufficient and additional damping circuits are required to deal with this.

6.4 Practical "Loop-Opening" Techniques for AC Analysis

When a feedback control loop for an AC loop gain analysis is "opened," a point of accessibility must be selected. Figure 6.15 shows the general case in which a unity gain voltage amplifier with output source impedance, Z_S, drives the input of voltage amplifier, A, whose input impedance is Z_I. For this general case, it is easy to see that the actual loop gain, T, is simply:

$$T = A\left(\frac{1}{1+\frac{Z_S}{Z_I}}\right) \tag{6.6a}$$

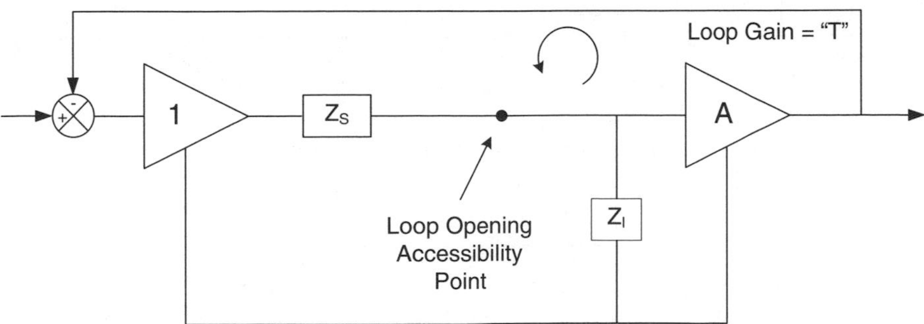

FIGURE 6.15
General feedback circuit showing loop opening accessibility for AC stability analysis.

Rearranging, T may also be expressed as:

$$T = \left(A \frac{Z_I}{Z_S} \right) \left(\frac{1}{1 + \frac{Z_I}{Z_S}} \right) \tag{6.6b}$$

As a point of departure, open the loop and insert a floating test voltage signal as shown in Figure 6.16a. It is important to note that the DC operating bias points of the circuit have been preserved because the AC characteristics could possibly be altered for different DC bias points in some circuits. Equation 6.7 shows the equation for an apparent loop voltage gain, T_V, expressed as a ratio of the voltages v_{OUT} and v_{IN}:

$$T_V = \frac{v_{OUT}}{v_{IN}} = A + \frac{Z_S}{Z_I} \tag{6.7}$$

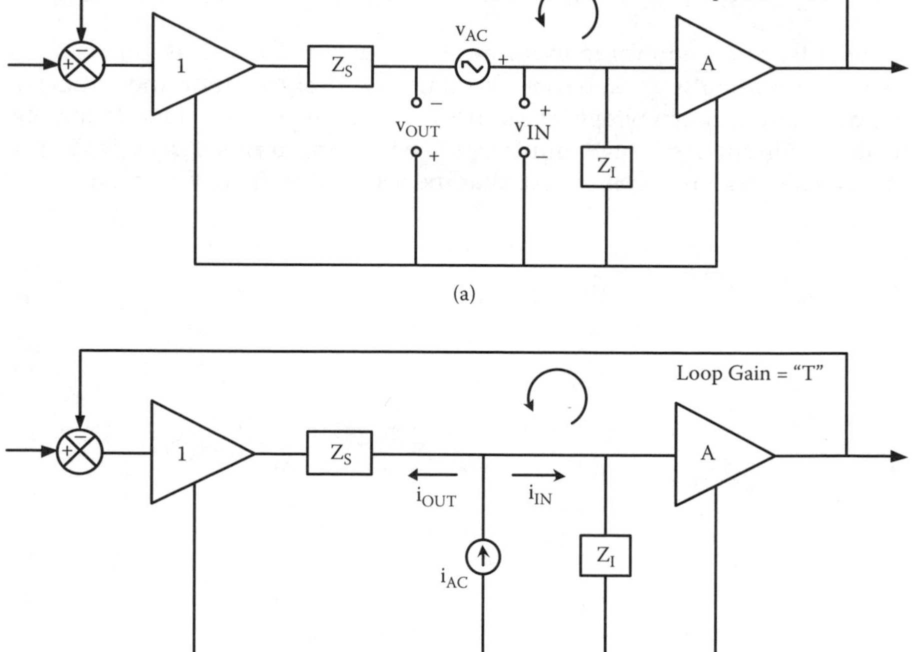

FIGURE 6.16
(a) General feedback circuit showing loop opening with floating voltage injected signal, v_{AC}.
(b) General feedback circuit showing loop opening with current-injected signal, i_{AC}.

Comparing Equation 6.6a and Equation 6.7, it is easily seen that, when Z_S is much less than Z_I, the loop gain reduces to A for both cases. Therefore, the easily obtained ratio, T_V, is an accurate indication of the actual loop gain, T, which is in this case simply A.

Now take the case in which the only point of accessibility is where Z_S is not much less than Z_I. It is now obvious that T_V is not the actual gain T. A very obvious thing to note is that, although the value of T now actually *decreases* with increasing values of Z_S (or decreasing values of Z_I), the value of T_V now *increases*, thereby giving a false impression that the gain is increasing when it is in reality decreasing. What is one to do now?

Apply the principal of circuit duality and examine the results obtained when a test current signal is injected as shown in Figure 6.16b. This is done in an attempt to obtain a current loop gain expression and see how it compares to the equation for the actual loop gain, T. Equation 6.8 shows the equation for this apparent current loop gain, T_I, expressed as a ratio of the currents i_{OUT} and i_{IN}:

$$T_I = \frac{i_{OUT}}{i_{IN}} = A\frac{Z_I}{Z_S} + \frac{Z_I}{Z_S} \tag{6.8}$$

By making the analogous comparison of Equation 6.6b and Equation 6.8, it is easily seen that when Z_I is now much less than Z_S, the loop gain reduces to $A\frac{Z_I}{Z_S}$ for both cases. Therefore, the ratio, T_I, is an accurate indication of the actual loop gain, T, which is in this case simply, $A\frac{Z_I}{Z_S}$. Unfortunately, in the case when Z_I is not much less than Z_S, the terms T and T_I do not compare favorably and it is obvious that T_I is not the actual gain T.

By combining Equation 6.6a, Equation 6.7, and Equation 6.8, one can now produce an equation that does yield the actual loop gain as expressed in Equation 6.9:

$$T = \frac{T_V T_I}{1 + T_V + T_I} \tag{6.9}$$

The results obtained from performing and combining the results of the *two* apparent gain tests — namely, T_V and T_I — now produce the true and actual loop gain T. Dr. Middlebrook has derived this equation in eloquent fashion.[29] The facilitation of this equation will be discussed later in this section.

As now might be surmised from Equation 6.6a and Equation 6.7 and the preceding discussion, it is in all probability most desirable to open the loop and inject a stimulus voltage test signal at a point at which a very low impedance source drives a very high impedance load. This might be where the low output impedance of an opamp circuit drives a high input impedance load as intimated in Figure 6.15. Z_S is much less than Z_I; thus, it is ensured that the measured feedback signal is produced solely by the loop gain amplified test signal and that the signal does not have any components to which any other factors attribute. All the examples presented in this book open the loop and inject the AC test signal at points at which this condition is considered met for the frequency ranges of interest.

Two loop opening methods have been used in the examples. One is with the use of the ultralow-pass filter method shown in Figure 4.9 and Figure 4.23. With the values of LOL and COL extremely large, it is easy to see that the AC feedback is opened and the DC feedback is maintained with DC bias points remaining unchanged. This ensures that the correct DC operating point is established for AC analysis. Also, the large value of LOL ensures a very high AC impedance to prevent any further possible loading of the measured AC feedback signal. This setup allows for a practical computer measurement of the actual loop gain, but obviously is impractical to implement in an actual physical test setup due to the extremely large values of LOL and COL. It is also easy to see that, when the impedance effects of Z_S and Z_I modify the actual gain, T, as stated earlier, this method will *not detect this effect*. It will only indicate the incorrect result and indicate a gain of A.

The second signal injection method is called the floating voltage injection signal method. This is the one illustrated in the general example of Figure 6.16a and yields the result of T_V. It is also used in the example shown in Figure 6.5. Voltage source, VACLG, is a floating injected signal connected between nodes FBKI and FBK. In the example, it is easy to see that the impedance seen looking back into node FBK is much lower than the impedance seen looking forward into node FBKI for the frequencies of interest.

To utilize Equation 6.9 in performing a practical computer AC loop gain analysis, one might set up two circuit models that will each provide a tabulation of the desired gain and phase vs. frequency computations of Equation 6.7 and Equation 6.8. The two sets of results are then combined in a spreadsheet program that uses Equation 6.9 to compute the actual values of T. The now computed values of gain and phase vs. frequency of T can be plotted to create a Bode plot providing the desired AC loop gain characteristics.

Another method that might be used to obtain the actual loop gain T more directly and more conveniently than the spreadsheet approach is shown in Figure 6.17. This approach obtains the true gain T directly from the simulation by setting up two identical circuits that are each stimulated with identical voltage amplitude and phase signals. A signal cancellation or "nulling" scheme is used in conjunction with a postsignal processor to determine the actual true loop gain, T, regardless of the impedance effects. The circuit in Figure 6.17a is simply an ultralow-pass filter opened loop and the measured signal, VOC, is simply the open circuit AC signal. With a unit input signal of VAC = 1, the value of VOC is simply:

$$VOC = A \tag{6.10}$$

Now if the signal VOC from circuit a) is inserted as E1 in circuit b), the AC equivalent circuit current into which i_1 flows is Z_I and has *zero* open circuit voltage source. This zero open circuit voltage is produced because E1 now exactly cancels out the amplified voltage produced by gain, A, in circuit b), which is also equal to VOC. Z_S and Z_I now form a simple current divider for

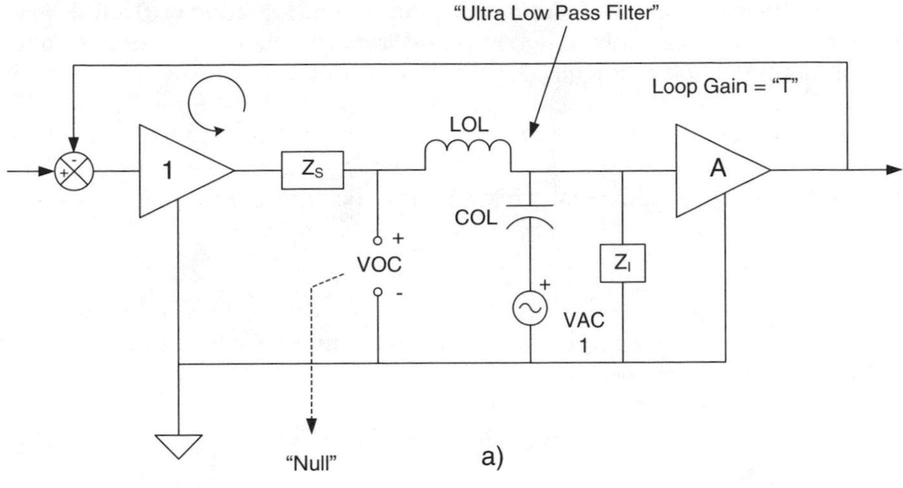

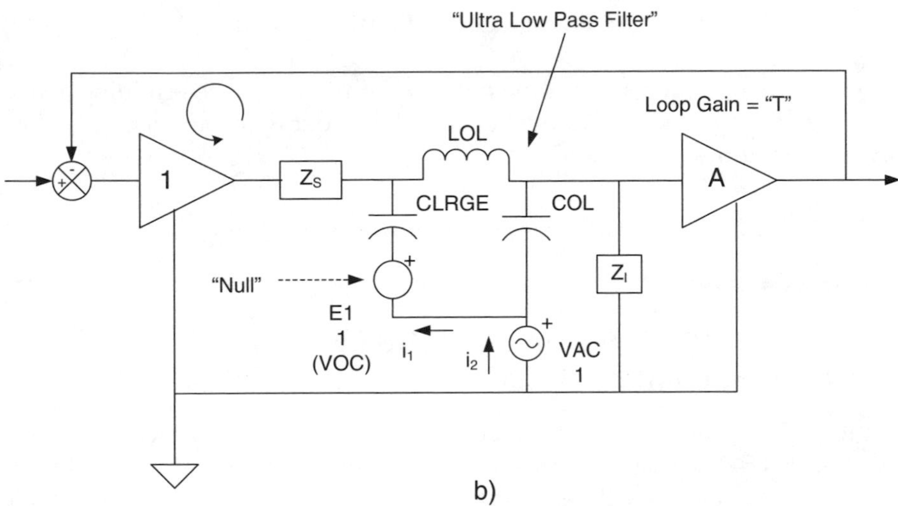

FIGURE 6.17
Analysis model for determining actual loop gain, *T*, using the null voltage method.

current, i_2, produced by the voltage source, VAC. (Keep in mind that capacitors COL and CLRGE are short circuits for AC signals, and LOL is an open circuit for AC signals.) From the divider relationship, it can be shown that

$$\frac{i_1}{i_2} = \left(\frac{1}{1 + \frac{Z_S}{Z_I}} \right) \tag{6.11}$$

Note that the easily obtained simulated quantities of Equation 6.10 and Equation 6.11 now have each of the two required terms needed in Equation 6.6a to determine the actual loop gain, T.

$$T = A\left(\frac{1}{1+\frac{Z_S}{Z_I}}\right) = (VOC)\left(\frac{i_1}{i_2}\right) \tag{6.12}$$

When Equation 6.10 and Equation 6.11 are routinely combined in a postsimulation processor equation, the desired gain in decibels and the phase of T can easily be obtained:

$$\mathrm{dB}(T) = \mathrm{dB}(VOC) + \mathrm{dB}(i_1) - \mathrm{dB}(i_2) \tag{6.13}$$

and

$$\mathrm{Phase}(T) = \mathrm{Phase}(VOC) + \mathrm{Phase}(i_1) - \mathrm{Phase}(i_2) \tag{6.14}$$

This method is probably not the only configuration for determining actual loop gain T directly from a simulation and the reader is encouraged to explore other possibilities.

6.5 Buck SMPS Soft Start and "Hiccup" Current Limit Analysis

Section 6.2 examined most of the pertinent performance characteristics of a flyback SMPS that uses a UC1844 PWM controller behavioral model. In this section, some analysis will be performed on a very simple continuous mode large signal buck converter model that uses the macromodel of the UC1825A PWM controller developed in Appendix B. The UC1825A controller is a more sophisticated device than the UC1844, with such features as soft start control and pulse-by-pulse current limiting. It also contains a latched overcurrent circuit, which facilitates a full soft restart of the power supply when an overcurrent condition is encountered. If the overcurrent persists, a hiccup on–off mode of operation is instituted. This section will demonstrate some of the performance characteristics that one might expect during turn ON and current-limiting situations when this controller is used.

Figure 6.18 shows a very basic SMPS using a simple continuous mode large signal buck converter model and the UC1825A PWM controller macromodel developed in Appendix B. Netlist 6.3 shows the circuit details for

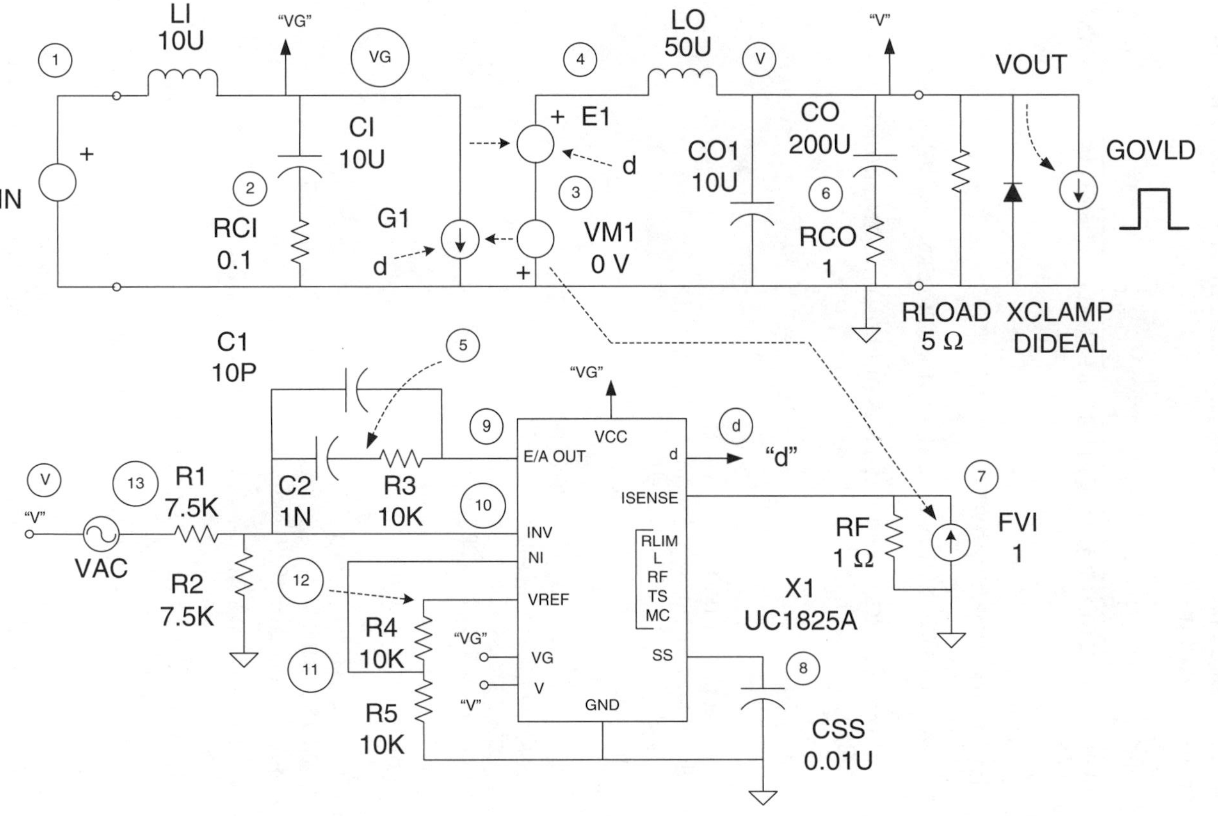

FIGURE 6.18
SMPS model using the UC1825A PWM controller circuit-averaged macromodel.

the analyses performed here. The model, by definition, is always in the continuous mode, which suggests that the converter would use synchronous rectification. Also, an AC loop gain and line voltage rejection analysis will be conducted to demonstrate how the UN1825A macromodel needs to be set up for this type of analysis.

NETLIST 6.3

```
BUCK CONVERTER SOFT START AND LOAD SHORT ANALYSIS USING UC1825A PWM
 CONTROLLER
FIGURE 6.18
*
**********************************************************************
*
** SOURCE/LOAD CONFIGURATION & COMMANDS FOR TURN-ON/TURN-OFF
** TRANSIENT ANALYSIS
** WITH SOFT START CAPACITOR (CSS) VALUES OF 0.01U, 0.1U, AND 0.001U
** (FIGURE 6.18)
*
** (Figure 6.19)
CSS SS 0 .01U
RLOAD V 0 5
VIN 1 0 PULSE(0 30 1M .1M 1M 10M)
.TRAN 10U 14M 0 10U SKIPBP
.OPTIONS RELTOL=.005 ITL4=60 PIVREL=1E-6
*
** (Figure 6.20)
*CSS SS 0 .1U
*RLOAD V 0 5
*VIN 1 0 PULSE(0 30 1M .1M 1M 80M)
*.TRAN 10U 100M 0 30U SKIPBP
*.OPTIONS RELTOL=.005 ITL4=80 PIVREL=1E-6
*
** (Figure 6.21)
*CSS SS 0 .001U
*RLOAD V 0 5
*VIN 1 0 PULSE(0 30 1M .1M 1M 2M)
*.TRAN 10U 6M 0 5U SKIPBP
*.OPTIONS RELTOL=.005 ITL4=60 PIVREL=1E-6
*
******************************************************************************************
*
** SOURCE/LOAD CONFIGURATION & COMMANDS FOR OUTPUT OVERLOAD & SHORT
** CIRCUIT TRANSIENT ANALYSIS
** WITH SOFT START CAPACITOR (CSS) VALUES OF 0.01U AND 0.1U
** (FIGURE 6.18)
*
*VIN 1 0 DC 20
*ROVLD P 0 1
*GOVLD V 0 VALUE = {LIMIT(V(V)*V(P),0,20)}
*.OPTIONS RELTOL=.01 ITL4=60 PIVREL=1E-6
*
*CSS SS 0 .01U
*RLOAD V 0 5
```

```
** (FIGURE 6.22) 2 OHM OVERLOAD
*IOVLD 0 P PULSE(0 .5 10m .01m .01m 18m)
** (FIGURE 6.23) 0.01 SHORT
*IOVLD 0 P PULSE(0 100 10m .01m .01m 18m)
*.TRAN 10U 40M 0 20U SKIPBP
*
*CSS SS 0 .1U
*RLOAD V 0 5
** (FIGURE 6.24) 2 OHM OVERLOAD
*IOVLD 0 P PULSE(0 .5 80m .01m .01m 50m)
** (FIGURE 6.25) 0.01 SHORT
*IOVLD 0 P PULSE(0 100 80m .01m .01m 50m)
*.TRAN 10U 200M 0 50U SKIPBP
*
****************************************************************************************
*
** SOURCE/LOAD CONFIGURATION & COMMANDS FOR LOAD SHORT AT TURN-ON/
** TURN-OFF TRANSIENT ANALYSIS
** WITH SOFT START CAPACITOR (CSS) VALUES OF 0.01U, 0.1U AND 0.001U
** (FIGURE 6.18)
*
** (Figure 6.26)
*CSS SS 0 .01U
*RLOAD V 0 .01
*VIN 1 0 PULSE(0 30 1M .1M 1M 10M)
*.TRAN 10U 14M 0 10U SKIPBP
*.OPTIONS RELTOL=.01 ITL4=80 PIVREL=1E-7
*
** (Figure 6.27)
*CSS SS 0 .1U
*RLOAD V 0 .01
*VIN 1 0 PULSE(0 30 1M .1M 1M 80M)
*.TRAN 10U 100M 0 50U SKIPBP
*.OPTIONS ITL4=60 PIVREL=1E-7
*
** (Figure 6.28)
*CSS SS 0 .001U
*RLOAD V 0 .01
*VIN 1 0 PULSE(0 30 1M .1M 1M 2M)
*.TRAN 10U 6M 0 5U SKIPBP
*.OPTIONS RELTOL=.005 ITL4=60 PIVREL=1E-6
*
****************************************************************************************
*
** SOURCE/LOAD CONFIGURATION & COMMANDS FOR AC LOOP GAIN ANALYSIS
** (FIGURE 6.29)
** NOTE: SUBCKT "XSOFTSTART" SHOULD BE REMOVED FROM NETLIST TO ACHIEVE
 BIAS SOLUTION
*
*RLOAD V 0 5
*CDUMMY SS 0 1
*VIN 1 0 DC 20
*VAC V 13 AC 1
VAC V 13 AC 0
*.AC DEC 20 100 1E6
```

```
*.OPTIONS ITL2=40
*
********************************************************************************************
*
** SOURCE/LOAD CONFIGURATION & COMMANDS FOR AC LINE REJECTION
** ANALYSIS
** (FIGURE 6.29)
** NOTE: SUBCKT "XSOFTSTART" SHOULD BE REMOVED FROM NETLIST TO ACHIEVE
 BIAS SOLUTION
*
*RLOAD V 0 5
*CDUMMY SS 0 1
*VIN 1 0 DC 20 AC 1
*.AC DEC 20 100 1E6
*.OPTIONS ITL2=40
*
********************************************************************************************
*
** BUCK CONVERTER
*
LI 1 VG 10U
CI VG 2 100U
RCI 2 0 1
G1 VG 0 VALUE = {I(VM1)*V(D)}
E1 4 3 VALUE = {V(VG)*V(D)}
VM1 0 3
LO 4 V 50U
CO1 V 0 10U
CO V 6 200U
RCO 6 0 1
FVI 0 7 VM1 1
RF 7 0 1
R1 13 10 7.5K
R2 10 0 7.5K
R3 9 5 10K
C2 5 10 .001U
C1 9 10 10P
R4 12 11 10K
R5 11 0 10K
X1 VG 12 10 11 9 7 SS V VG D 0 UC1825A
*
**UC1825A PWM CONTROLLER CIRCUIT AVERAGING MACRO
.SUBCKT UC1825A VCC 3    4   5  COMP    6    SS 7 8  D GND
*             VCC VREF INV NI E/A_OUT RAMP SS V VG d GND
**NOTE: ALL SIGNAL INPUTS MUST BE REFERENCED TO THE UC1825A GND
XUVLO VCC UVLO GND UVLO
XREF UVLO VCC 3 GND REF
XERRAMP 4 5 VCC COMP GND ERRAMP
XDRC COMP d 6 7A 8A UVLO SD CL GND DRC
** NOTE: SUBCKT "XSOFTSTART" SHOULD BE REMOVED FOR BIAS SOLUTIONS
XSOFTSTART 6 7A 8A 3 UVLO d COMP SS SD CL GND SOFTSTART
RSSCKT SS GND 1E9
**DEPENDENT GENERATORS FOR GROUND ISOLATION IF DESIRED
EVG 8A 0 8 GND 1
EV 7A 0 7 GND 1
```

```
.ENDS UC1825A
*
.SUBCKT DRC C      d 1      9 10 14    SD CL GND
*          COMP   d ISENSE V VG UVLO  SD CL GND
RCONV1 6 0 1G
RCONV2 7 0 1G
RCONV3 4 0 1G
R2 C GND 1E6
VCS 8 1 DC 1.25
*
REL 10 9 1E8
GL 0 11 10 9 1
D1L 11 12 DX
D2L 0 11 DX
RLMIN 12 13 1
VLMIN 13 0 1U
*
DX1 4 0 DX
DX2 7 4 DX
DX3 6 7 DX
*
**IDRMAX VALUE = MAX LIMITED DUTY RATIO
IDRMAX 7 6 DC .48
DX4 0 6 DX
DX5 6 6A DX
.MODEL DX D IS=1E-12
VMD 6A 0
EMD d 0 VALUE = {LIMIT(I(VMD)*(1-V(CL))*V(14)*(1-V(SD)),1U,.99)}
RMD d 0 1
*
**PULL DOWN THAT IS REQUIRED WHEN XSOFTSTART IS REMOVED FROM CIRCUIT
RCL CL 0 1E3
IRCL 0 CL DC 1P
RSD SD 0 1E3
IRSD 0 SD DC 1P
*
*L = 50U
*RF = .5
*TS = 10U
*MC = 0
*
**GCM 4 7 VALUE = {(V(C)-RF*V(8))/((TS*RF)/(2*L))*(2*MC*L+V(12)))}
GCM 4 7 VALUE = {LIMIT((V(C)-.5*V(8))/((10U*.5)/(2*50U)*(2*0*50U+V(12))),0,1)}
*
**GDCM 0 4 VALUE = {V(C)/((TS*RF/L)*(MC*L+V(12))}
GDCM 0 4 VALUE = {LIMIT(V(C)/((10U*.5/50U)*(0*50U+V(12))),0,1)}
*
.ENDS DRC
*
.SUBCKT UVLO VCC UVLO   GND
*          VCC UVLO   GND
RIN VCC 2 1K
ISUP 2 GND .1M
XSUP GND 2 DIDEAL
G5 VCC GND VALUE = {.028*V(UVLO,GND)}
```

```
**G1 AND G2 SET UVLO START AND STOP LEVELS RESPECTIVELY
**START = 9.2
**STOP  = 8.4
G1 0 3 TABLE {V(VCC,GND)} = (0,0) (9.2,0) (10.2,30)
G2 3 0 TABLE {V(VCC,GND)} = (7.4,30) (8.4,0) (9.2,0)
CG2 3 0 20N
R1 3 0 1E3
G3 0 4 3 0 1
G3HYST 0 4 VALUE = {-100U+200U*V(UVLO,GND)}
CHOLD 4 0 20N
G4 GND UVLO 4 GND 1
**VOVRD TO BE INSERTED FOR UVLO OVERRIDE
*VOVRD UVLO GND DC 1
RG4 UVLO GND 1
CG4 UVLO GND 1N
X5 GND UVLO DIDEAL
X1 3 5 DIDEAL
X2 6 3 DIDEAL
X3 4 5 DIDEAL
X4 6 4 DIDEAL
VP 5 0 DC 1
VN 0 6 DC 1
.ENDS UVLO
*
*5.0 VOLT REFERENCE MACRO
.SUBCKT REF UVLO VCC VREF  GND
*          UVLO VCC VREF  GND
*
EIN 2 3 VCC GND .25E-3
RVT 3 4 .56 TC=0.2E-3
IVT GND 3 DC 1.786
VT GND 4 DC 1
V5 1 2 DC 5.097555
FUVLO 1 GND VMR 1
EUVLO VREF 9 VALUE = {V(1,GND)*V(UVLO,GND)}
VMR GND 9
.ENDS REF
*
*ERROR AMP MACRO
.SUBCKT ERRAMP  6   7   8   1     GND
*               VIN VIP VCC OUTPUT GND
RIN 6 7 10MEG
EPSRR 6 5 VALUE = {(V(8)-12)*3E-5}
IBIAS 5 GND .6U
GA GND 2 7 6 1K
RG 2 GND 56
CG 2 GND 13U
VNC 4 GND DC 1.1
DN 4 2 D1
VPC 3 GND DC 4.1
DP 2 3 D1
RO 1 2 10
RPS 8 0 1E8
.ENDS ERRAMP
.MODEL D1 D IS=1E-9
```

```
*
.SUBCKT SOFTSTART ISENSE V VG VREF UVLO d C SS 11 8  GND
*              ISENSE V VG VREF UVLO d C SS SD CL GND
*
** L = 50U
** RF = .5
** TS = 10U
*
**PEAK CURRENT CALCULATOR
**EPKDCM 15 0 VALUE = {V(d)*(V(VG)-V(V))*(TS/L)*RF}
**NOTE: V(V) IS ZERO FOR ALL TOPOLOGIES EXCEPT THE BUCK
**      THE BUCK EQUATION IS SHOWN HERE FOR THE GENERAL CASE
EPKDCM 15 0 VALUE = {V(d)*(V(VG)-V(V))*(10U/50U)*.5}
XPKDCM 15 IPK DIDEAL
XPKCCM 16 IPK DIDEAL
E4 16 17 15 0 .5
EPKCCM 17 0 ISENSE 0 1
RPK IPK 0 1E3
*
**OVER CURRENT SENSE COMPARATOR 1.2
** RLIM = 0.5
**GIPK 0 8 VALUE = {V(IPK)*RLIM}
GIPK 0 8 VALUE = {V(IPK)*.5}
XP1 0 8 DIDEAL
IPK 8 0 DC 1.2
XP2 8 6 DIDEAL
V12 6 0 DC 1
EUV 6 19 UVLO GND 1
*
**OR GATE
** NOTE: G1 MAY BE REMOVED HERE TO DISABLE "HICCUP" CURRENT LIMIT MODE.
**      AN EQUIVALENT OF PULSE-BY-PULSE CURRENT LIMITING WILL RESULT.
**      THE VALUE OF IPK ABOVE MAY BE CHANGED FROM 1.2 TO 1.O TO THE
**      ACTUAL CURRENT LIMIT COMPARATOR THRESHOLD IF DESIRED.
G1 0 9 8 0 2
G2 0 9 19 0 2
X2 0 9 DIDEAL
IOR 9 0 DC 1
XR 9 7 DIDEAL
VOR 7 0 DC 1
*
**AND/OR LOGIC
G3 0 4 11 0 .75
G4 0 4 2 0 .75
G5 0 4 19 GND 1.5
X5 0 4 DIDEAL
IOL 4 0 DC 1
X6 4 5 DIDEAL
V6 5 0 DC 1
*
**SS RESTART COMPARATOR
GSS 10 0 SS GND 1
XR1 0 10 DIDEAL
IRS2 0 10 DC .2
XR2 10 1 DIDEAL
```

```
VP2 1 0 DC 1
*
*SS COMPLETE COMPARATOR
G5P 0 2 SS 0 1
XS1 0 2 DIDEAL
I5P 2 0 DC 5
XS2 2 3 DIDEAL
VP3 3 0 DC 1
*
*SS/COMP
XCP C 18 DIDEAL
ECOMP 18 GND SS GND 1
ITS 0 SS DC 9U
GLS SS GND VALUE = {LIMIT(V(13)*250U,1U,260U)}
XCLPH SS VREF DIDEAL
XCLPL GND SS DIDEAL
.MODEL D D
*
XRS1 10 9 11 12 GND RS-LATCH
XRS2 12 4 13 14 0 RS-LATCH
*
.ENDS SOFTSTART
*
.SUBCKT RS-LATCH R S Q QBAR GND
GR 3 GND R GND 1
GS GND 3 S GND 1
C3 3 GND 20N
G4 GND Q 3 GND 1
CG4 Q GND 20N
GHYS GND 3 VALUE = {-100U+200U*V(Q,GND)}
GRS GND Q VALUE = {.75*(V(R,GND)+V(S,GND))}
X1 GND Q DIDEAL
X2 Q 5 DIDEAL
X3 6 3 DIDEAL
X4 3 5 DIDEAL
EINV 5 QBAR VALUE = {V(Q)*(1-V(R)*V(S))}
VN GND 6 DC 1
VP 5 GND DC 1
.ENDS RS-LATCH
*
.SUBCKT DIDEAL 1 2
VAS 1 3 DC -1u
D1 3 2 D
D2 3 4 D
D3 4 2 D1
FAS 4 2 VAS 1
.MODEL D D IS=1E-6 EG=0 XTI=-4
.MODEL D1 D EG=0 XTI=0
CC 1 2 .1P
.ENDS DIDEAL
*
.LIB DIODE.LIB
.PROBE
.END
```

First, perform a normal power supply turn ON and OFF cycle and observe the soft start operation for this normal operating condition. Soft start capacitor values of 0.01, 0.1, and 0.001 μF are used in this analysis to show the kinds of performance that might occur for these different values. Figure 6.19 shows some of the expected transient voltages waveforms for the case when a 0.01-μF soft start capacitor is used. Note that the soft start voltage and comp voltage initially rise together; however, at the threshold of regulation of the 5-Vdc output, the comp voltage levels off at about 1.3 V and the soft start continues to rise to the saturated high value of about 5.1 V.

Also, the voltage at the overcurrent sense comparator rises to a level of 1.1 V due to the turn-on current surge in the output filter capacitors of the converter. This is very near the 1.2-V threshold of the overcurrent sense comparator and suggests that the value of the soft start capacitor is very close to being too small. (In the actual circuit, a pulse-by-pulse current limit condition would probably result with this signal exceeding the current limit comparator threshold of 1.0 V. This threshold has no effect here, however, because this comparator is not used explicitly in the circuit-averaged simulation.) In any case, a turn-on surge current signal this close to the thresholds of 1.0 and 1.2 V would probably be considered unacceptable. As power-down continues at the point shown as the 30-Vdc input power ramp down, the comp voltage begins to rise, but the UVLO signal shuts down everything at the UVLO stop level of about 8.4 V. The 5-Vdc output voltage has incidentally dropped to about 4.7 at this point.

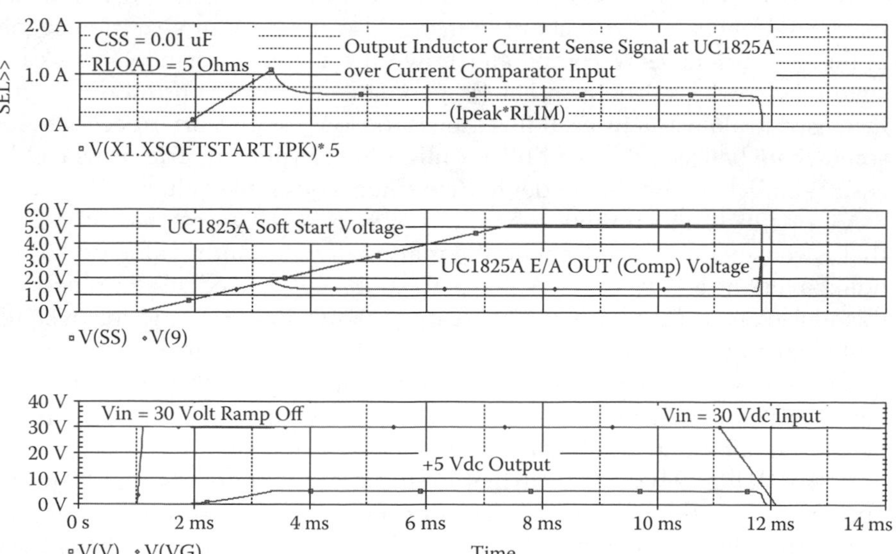

FIGURE 6.19
Normal turn ON and OFF of buck SMPS with UC1825A controller (CSS = 0.01 μF).

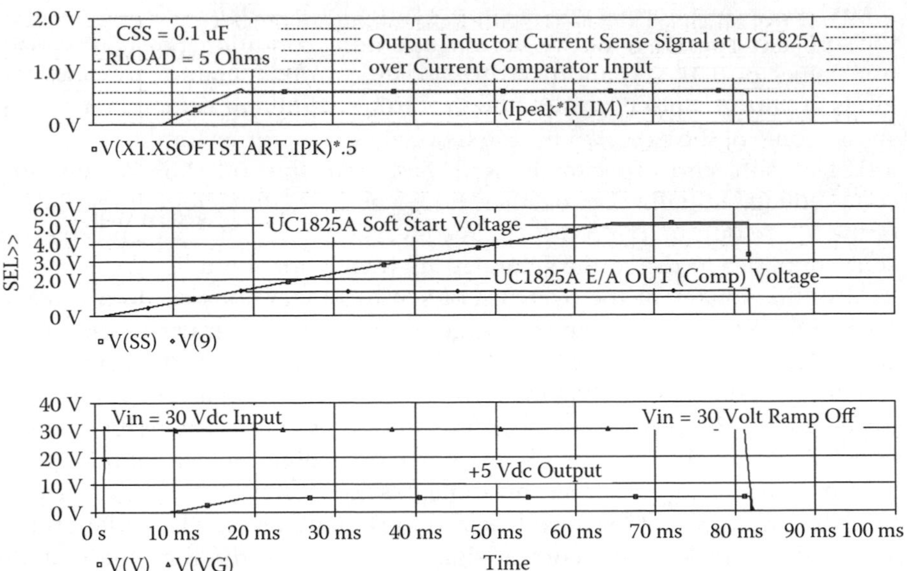

FIGURE 6.20
Normal turn ON and OFF of buck SMPS with UC1825A controller (CSS = 0.1 μF).

Figure 6.20 shows what happens when the value of the soft start capacitor is increased from 0.01 to 0.1 μF. It is first noted that the turn-on rise time has increased from about 2.5 to about 18 msec. Also, the peak surge current signal at the overcurrent sense comparator now only slightly overshoots its steady state value of 0.6 V. The turn here is very smooth, but the turn on time has increased to almost 20 msec. If this time is too long, a soft start capacitor value somewhere between 0.01 and 0.1 μF might be considered. Trial runs can be easily simulated with the model to determine a desirable value.

As a point of illustration, see what happens when a soft start capacitor that is considerably smaller than the marginally minimum value of 0.01 μF noted in Figure 6.19 is used. A value of 0.001 μF will be used; in all probability, this will allow a surge current large enough in magnitude to trip the overcurrent comparator at its level of 1.2 V. The results are shown in Figure 6.21 and this is exactly what happens. In fact, the hiccup current limit mode that will subsequently be considered is seen to result. From this, one should obviously be careful not to use a soft start capacitor that is too small.

Now consider the normal current limit modes of operation that result by applying an additional "soft" 2-Ω load on the supply and thus creating an overcurrent limit situation. After this, a "hard" 0.01-Ω short will be applied on the output of the power supply. The performance for each of the soft start capacitor values of 0.01 and 0.1 μF will be examined.

Figure 6.22 shows the case of the 2-Ω overload with the soft start capacitor value of 0.01 μF. The overload is switched on at the time of 10 msec — well

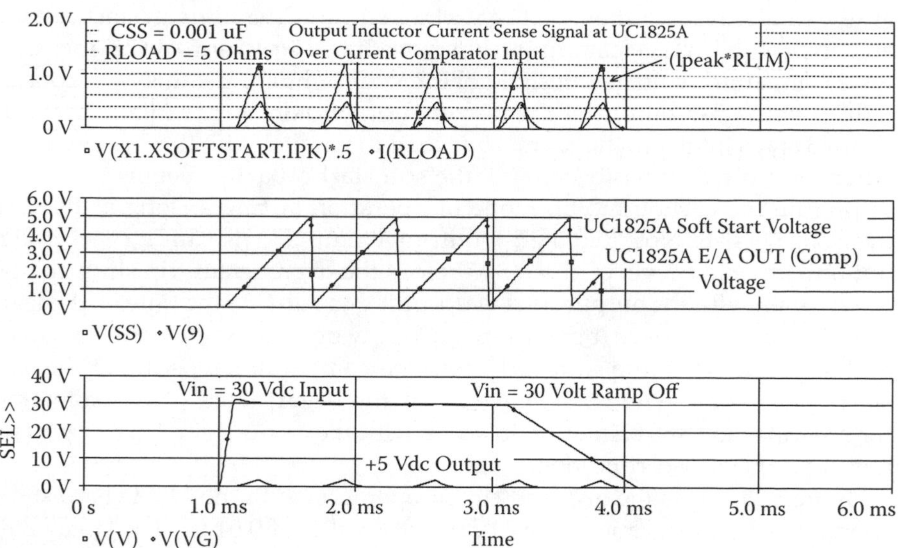

FIGURE 6.21
Normal turn ON and OFF of buck SMPS with UC1825A controller (CSS = 0.001 μF). Note that CSS is too small. The soft start voltage rises so fast that the turn-on surge current trips the overcurrent comparator setting up a hiccup current limit mode.

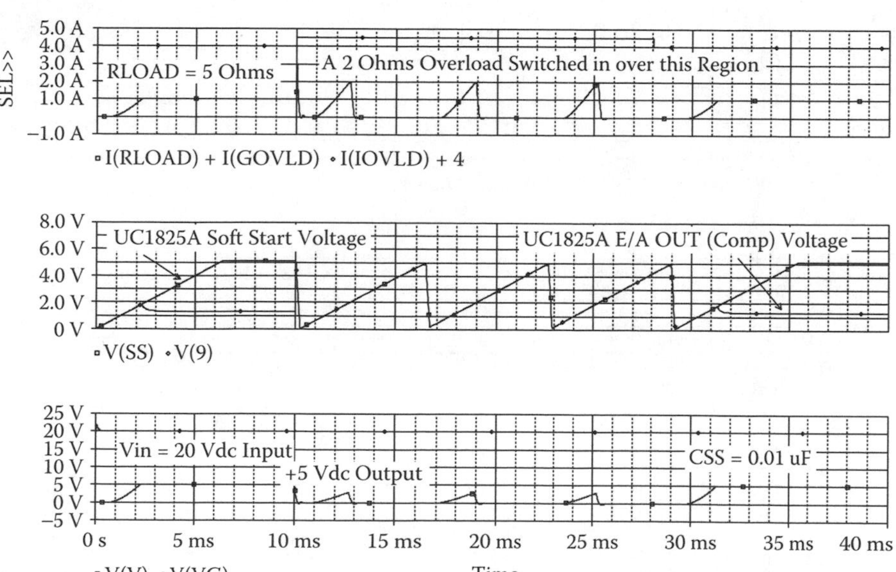

FIGURE 6.22
2.0-Ω overload hiccup mode of buck SMPS with UC1825A controller (CSS = 0.01 μF).

after steady state operation has been achieved. Note that the soft start voltage and the comp voltages are immediately reduced to less than the soft start reset comparator voltage threshold of 0.2 V. Accompanying this is, of course, a drop in the 5-V output of the power supply and a load surge current of about 3 amps total into the normal 5-Ω load in parallel with the 2-Ω overload. After the fault latch has been reset, the soft start capacitor begins to charge, setting up the cycling hiccup mode of operation lasting as long as the 2-Ω overload persists. Note that the 5-V output attempts to rise during each cycle, but only rises to a level of about 3 V before the overcurrent trip shuts things down again with the output current dropping to zero. At the time of 28 msec, the overload is removed, but the power supply does not immediately recover at that point. When the soft start complete comparator senses that the soft start voltage has again risen above 5.0 V, the restart latch is reset allowing for a normal soft start situation to occur with the 5-V output returning to its normal preoverload condition.

Figure 6.23 shows the overload circuit operation with the 0.01-Ω hard short condition using the same soft start capacitor value of 0.01 μF. The circuit still functions in the same manner except that the 5-V output is now continuously clamped to near 0 V. The initial load surge current from the output capacitors is now obviously much larger. The output current does not return to zero during each hiccup cycle as before, but ripples along at about a 1.5-amp average. Figure 6.24 and Figure 6.25 show the overload test results of the soft 2-Ω and the hard 0.01-Ω short overload condition using the larger 0.1-μF soft

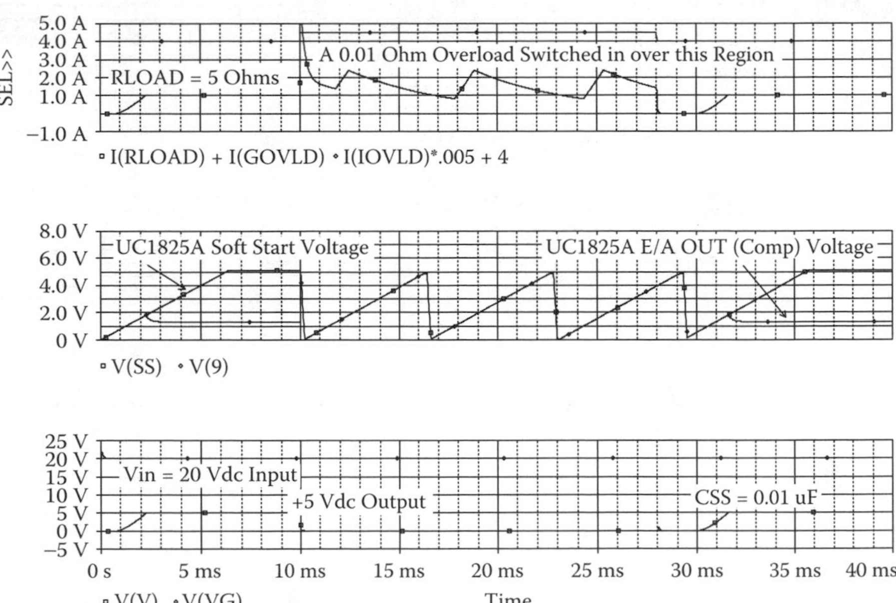

FIGURE 6.23
Short circuit overload hiccup mode of buck SMPS with UC1825A controller (CSS = 0.01 μF).

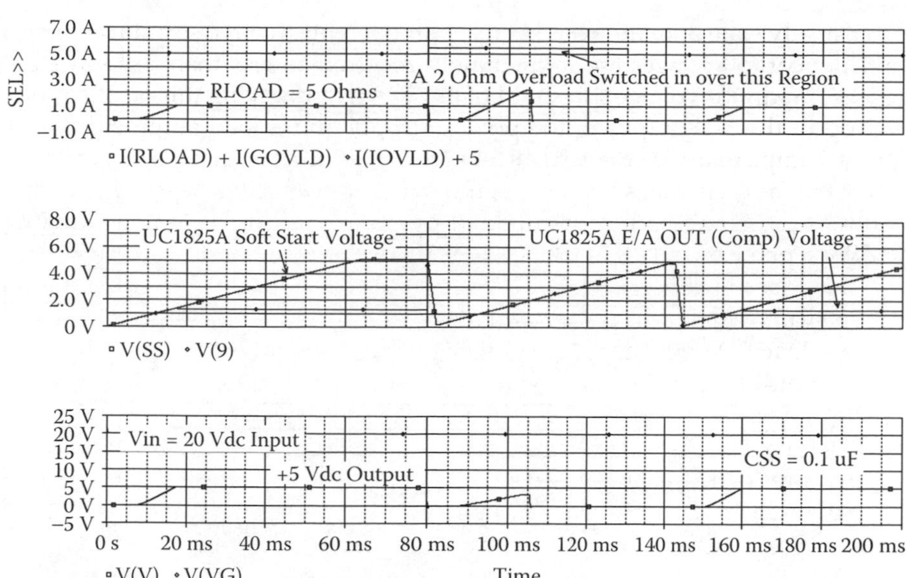

FIGURE 6.24
2.0-Ω overload hiccup mode of buck SMPS with UC1825A controller (CSS = 0.1 μF).

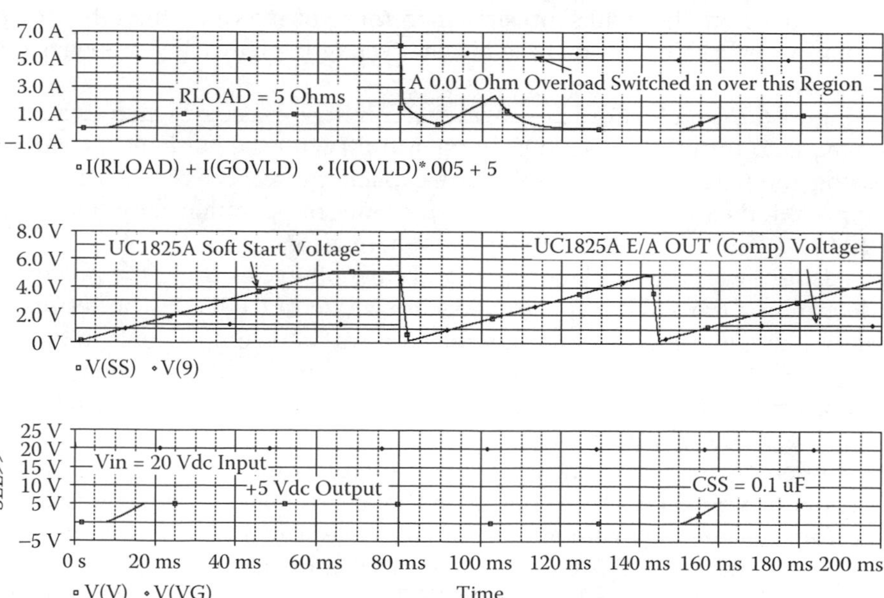

FIGURE 6.25
Short circuit overload hiccup mode of buck SMPS with UC1825A controller (CSS = 0.1 μF).

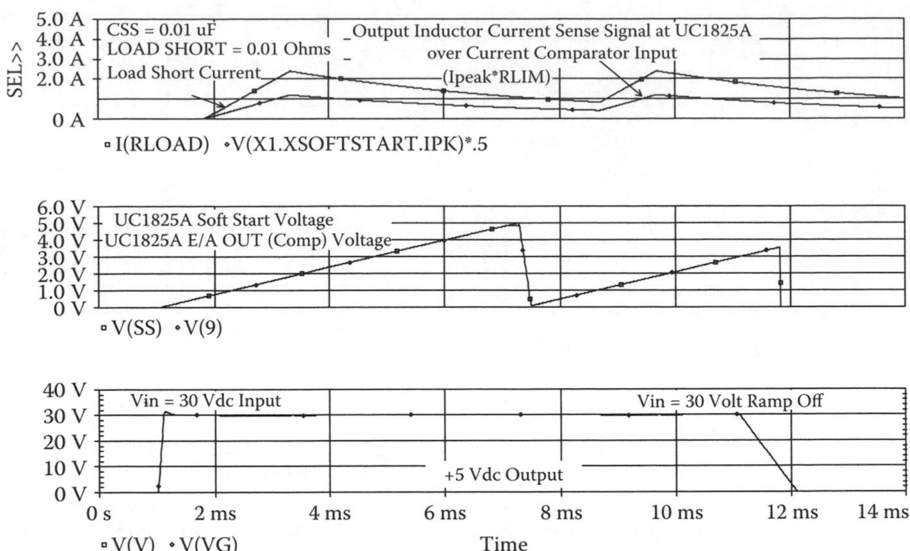

FIGURE 6.26
Load short turn ON and OFF of buck SMPS with UC1825A controller (CSS = 0.01 μF).

start capacitor. The results are similar to those of the case when the 0.01-μF capacitor is used except that the hiccup cycling is predictably of longer time duration.

Figure 6.26 and Figure 6.27 show what might be expected when an attempt is made to turn the power supply on and off with the hard short condition using the 0.01- and 0.1-μF soft start capacitors, respectively. As might be expected, the same overcurrent hiccup modes of operation shown in Figure 6.24 and Figure 6.25, respectively, exist here also.

Also shown here are the turn-on and turn-off cycles using the previously determined too small soft start capacitor value of 0.001 μF. The results are shown in Figure 6.28. Comparing the soft start and comp signals to those in Figure 6.21, it is seen that when soft start capacitor used is too small, it may be difficult to determine whether an actual overload condition exists when turning on the power supply.

To demonstrate how an AC analysis might be performed on the power supply using the UC1825A controller, the macromodel is now configured to perform an AC loop gain and an AC line rejection analysis. It is necessary to go into the UC1825A macromodel and comment out — in other words, remove the soft start subcircuit (XSOFTSTART) from the netlist. This prevents the hysteretic latches in the subcircuit from causing convergence problems when seeking the bias solution required for the AC analysis. In some

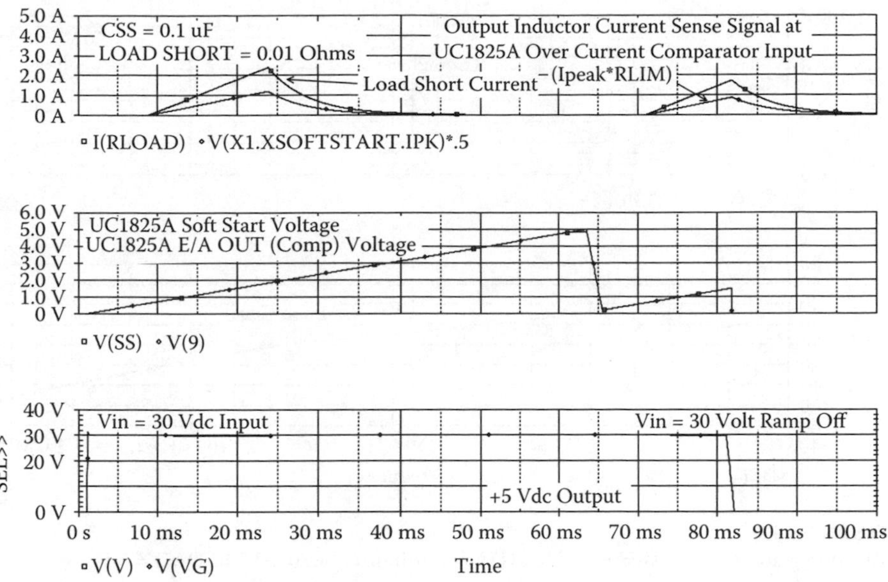

FIGURE 6.27
Load short turn ON and OFF of buck SMPS with UC1825A controller (CSS = 0.1 μF).

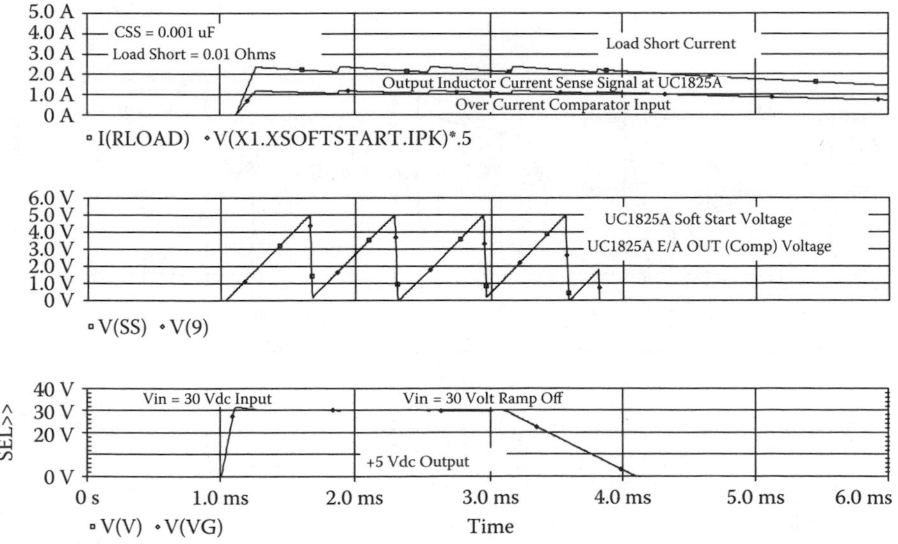

FIGURE 6.28
Load short turn ON and OFF of buck SMPS with UC1825A controller (CSS = 0.001 μF).

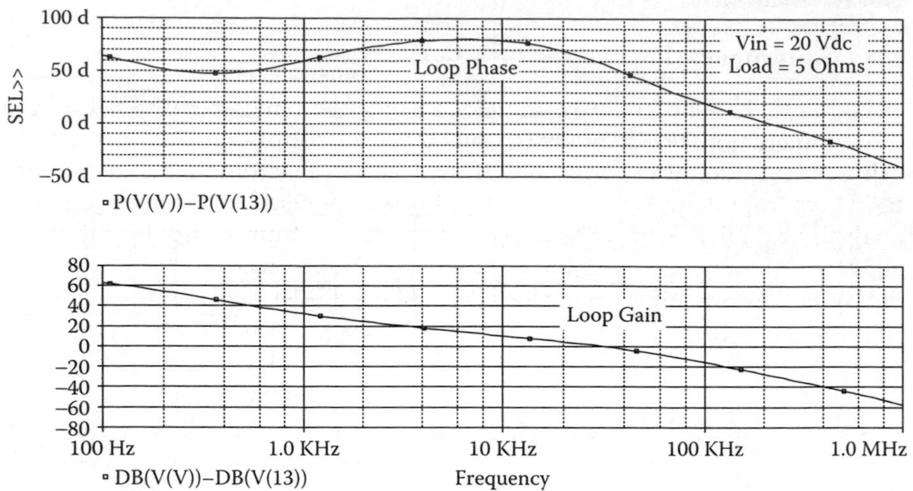

FIGURE 6.29
AC loop gain of buck SMPS with UC1825A controller. V_{in} = 20 Vdc; RLOAD = 5 Ω.

instances, the UVLO subcircuit (which contains a latch) may need to be overridden by inserting the VOVRD voltage source; however, that was not necessary here. The results for the loop gain are shown in Figure 6.29 and the results for the AC line rejection are shown in Figure 6.30.

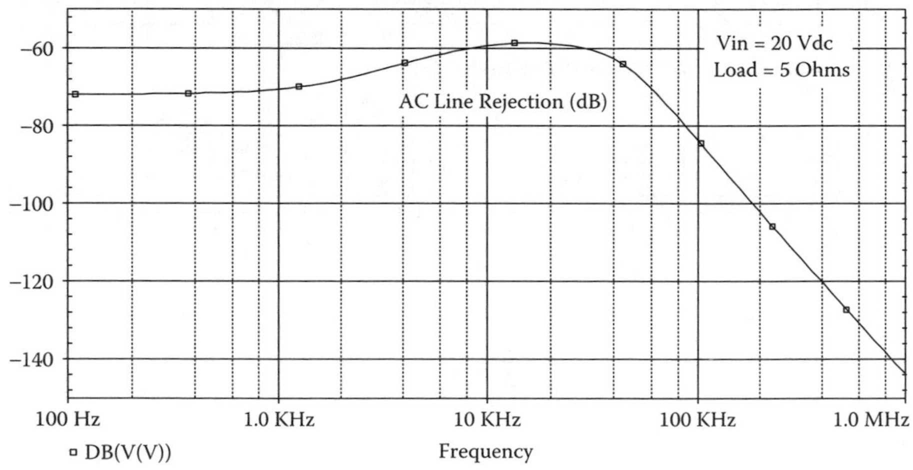

FIGURE 6.30
AC line rejection of buck SMPS with UC1825A controller. V_{in} = 20 Vdc; RLOAD = 5 Ω.

6.6 Summary

This chapter has shown an example of a method that may be used to analyze a fundamental buck PWM switch mode power supply quickly and practically. An additional example has shown how a macromodel of a modern PWM control integrated circuit (UC1844) may be used to obtain a practical analysis of a flyback power supply example taken from a widely published application note. This macromodeling approach is considered to be in keeping with the modern approaches to modeling and simulating electronic circuits in general. Another example shows how parasitic resistances may be inserted into a continuous conduction mode buck converter to analyze their effects on it. Next some possible methods of opening feedback control loops for determining AC stability were shown. Finally, how the UC1825A circuit-averaged macromodel functions under soft start and output current-limiting conditions was demonstrated.

As a footnote to this chapter, the reader should be cautioned about the difficulties that can often arise due to nonconvergence problems when circuits with the level of complexity presented here are run. All the simulations done here were run using PSPICE. Lower cost PC version software will generally present more problems than the more expensive workstation simulation software. However, it is felt that, with the experience and persistence that all computer circuit analysts must sometimes exhibit, the desired results can be obtained without too much difficulty. In the netlists in this chapter, take note of the various OPTIONS commands that were used in the examples. These commands may need to be experimentally altered when making circuit alterations.

Appendix A

Design Fundamentals of SMPS Input Filters

A.1 General Requirements

One of the most constraining aspects of any switch mode power supply (SMPS) design is the input filter. This circuit has to satisfy several unique and distinct requirements and not produce any unacceptable negative side effects on the SMPS performance simultaneously. Section 5.3.1 has provided a list of the more general requirements, but some of these general requirements will be reiterated here from a design perspective.

- *Conducted emissions*. The input filter is required to attenuate AC ripple currents emanating from the SMPS back to the power source to an acceptable specified level. This requirement is generally referred to as the "conducted emissions specification." This ripple is in general produced by the internal switching action of the SMPS, but it may also be produced by the AC load ripple current reflected back through the supply from the output to the input.
- *Audio susceptibility*. The input filter may be required, if only in part, to reduce the effects of specified undesirable AC input ripple voltages that are superimposed on the input power lines. They must be suppressed to levels at which SMPS performance requirements are satisfied. This requirement is generally referred to as the "audio susceptibility specification." The frequencies in this requirement are generally not limited to those in the audio range, but because those in this low-frequency band are usually the most difficult with which to deal, the adjective "audio" is customarily used in describing this requirement.
- *Output impedance*. The input filter output impedance must not contain any high Q resonant peaks that, when combined with the negative incremental input impedance of the power converter, could produce a system AC instability. Also, even if stability is verified, any allowed impedance peaking must be controlled to an acceptable magnitude so that reflected load current transients do not produce unacceptable voltage transients on the filter output.

- *Input impedance*. The AC input impedance must not contain any high Q resonant dips, which could excessively load the input power source over specified frequency bands.
- *Input transients' susceptibility*. Input power transients may exist in several forms, such as narrow, high-voltage spikes. They may also exist as sustained input overvoltage power surges or may exist as undervoltage power sagging or even a complete power dropout for some specified period of time. In the absence of surge protectors or extra energy storage components, the input filter may be required to protect the SMPS from these input anomalies. Specific solutions to these concerns are not addressed here, but any approach must be based on energy absorption and/or storage capability of the input filter while still maintaining performance and protection.

A.2 Fundamental Single-Stage LC Section Examples

All SMPS input filters are basically low-pass filters. They may exist as a single-section LC filter for simple applications, or they may be composed of multiple sections of possibly more complex topologies for the more sophisticated designs. As a point of departure, first consider the functioning of a very fundamental single LC-section design.

Figure A.1 shows the basic circuit. From the outset, basically three design parameters must be considered: voltage transfer function, H; output impedance, Z_O; and input impedance, Z_I. These parameters must be determined from examination of the requirements listed previously. For example, the audio susceptibility and the conducted emissions specification are key in determining the voltage transfer function, H. (The reverse current transfer ratio is numerically equal to H and its requirement is dictated by the conducted emissions limit.) The maximum value of the parameter Z_O is determined initially from examining the necessary requirements for power converter stability.[2,13] The limits for Z_I are sometimes specified by the power source provider; however, in many cases, component stresses occurring during audio susceptibility testing can be the determining factor in setting the limits.

From the outset, the high Q resonant peaks and dip are the worst case points to consider. They must be limited by lowering the Q of the filter. To accomplish this with the initial circuit topology, it would be necessary to increase the values of resistances R_L and/or R_C. Unfortunately increasing these values generally reduces filter performance and/or increases power losses and voltage regulation to unacceptable levels. None of these options is considered a desirable or practical thing to do. It now becomes necessary to reduce the Q of the filter by adding other damping components in such a way that the desired performance of the initial LC selection is not significantly compromised.

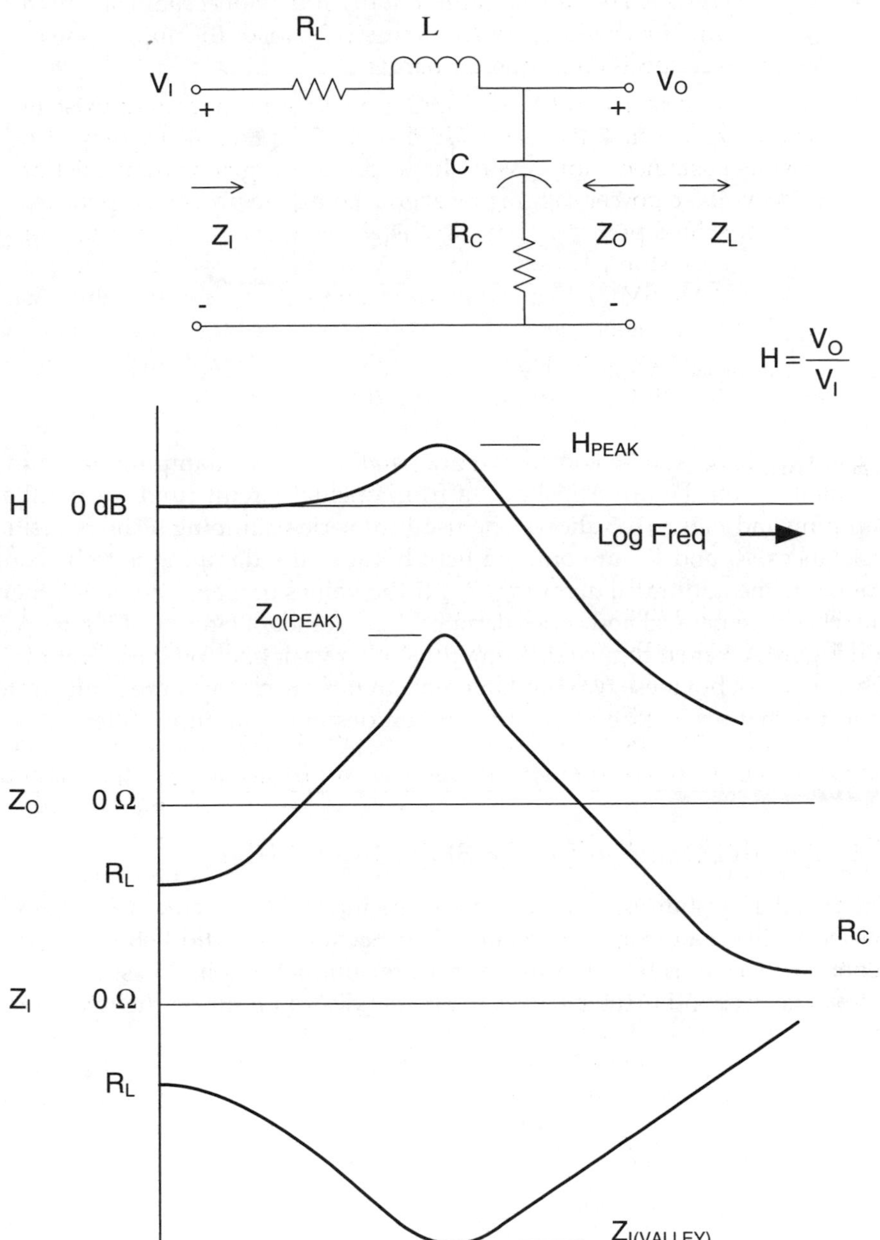

FIGURE A.1
Simple LC section input filter.

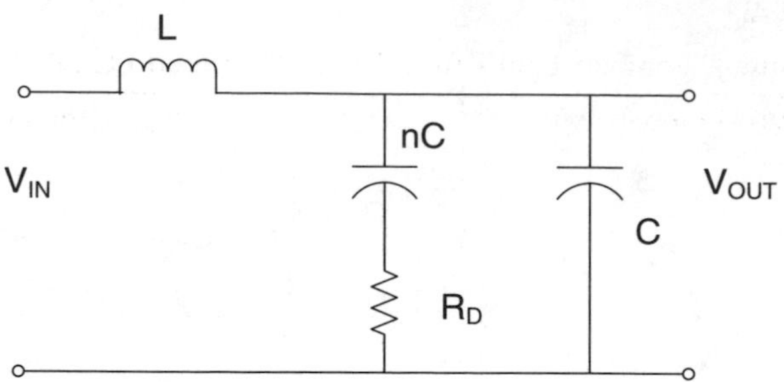

FIGURE A.2
Fundamental parallel damped LC filter.

The two options to be considered are *parallel* or *series* damping of the LC resonant circuit. Figure A.2 shows a fundamental circuit used for parallel damping and Figure A.3 shows one used for series damping. (The parasitic resistances, R_L and R_C, are omitted here because the damping is to be controlled by the additional damping, R_d.) If the values of n are very large, then the respective parallel and series damped RLC resonant circuits of Figure A.2 and Figure A.3 are recognized. Numerous other variations on these damping schemes exist, but these two fundamental circuits are chosen here to illustrate concepts that need to be understood when designing an input filter.

A.3 Parallel Damped Single-Stage Input Filter

The parallel, or shunt, damped single-stage input filter circuit of Figure A.2 has been given an excellent treatment in Section 4 of Middlebrook.[13] The general conclusions stated in this paper are summarized in this section. For any single-section LC filter with a realizable desired limit on the maximum

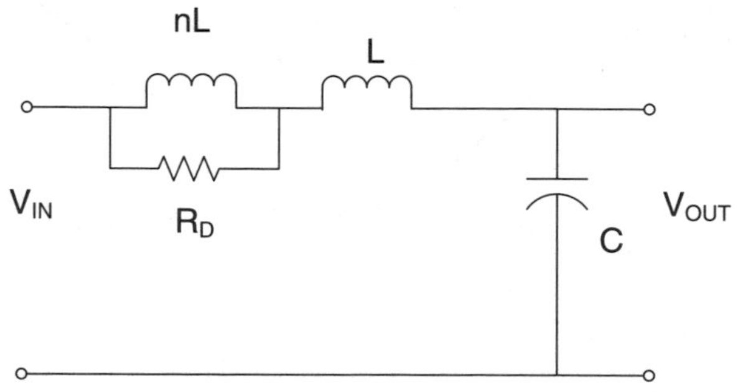

FIGURE A.3
Fundamental series damped LC filter.

TABLE A.1

Optimum Parameter Points for a Parallel Damped LC Filter

Parameter (normalized)	Q_{opt}	ω_{mm}/ω_o (normalized)
$H_{mm} = \dfrac{2+n}{n}$	$\sqrt{\dfrac{(1+n)(2+n)}{2n^2}}$	$\sqrt{\dfrac{2}{2+n}}$
$\dfrac{Z_{Omm}}{Z_C} = \sqrt{\dfrac{2(2+n)}{n^2}}$	$\sqrt{\dfrac{(4+3n)(2+n)}{2n^2(4+n)}}$	$\sqrt{\dfrac{2}{2+n}}$
$\dfrac{Z_{Imm}}{Z_C} = \sqrt{\dfrac{n^2}{2(1+n)(2+n)}}$	$\sqrt{\dfrac{2(1+n)^3(4+n)}{n^2(2+n)(4+3n)}}$	$\sqrt{\dfrac{2+n}{2(1+n)}}$

peak values of H and Z_O and minimum valley value of Z_I, an optimum value of n exists to help ensure that each of these requirements is met. For a given design, limits are specified for H_{MAX}, $Z_{O(MAX)}$, and $Z_{I(MIN)}$ for various frequencies.

The equations shown in Table A.1 and derived in Middlebrook[13] show the conditions in which the optimum normalized values of H_{PK}, $Z_{O(PK)}$, and $Z_{I(VAL)}$ occur. (For the remainder of this section, the parameters H_{PK}, $Z_{O(PK)}$, and $Z_{I(VAL)}$ are recognized as normalized parameters.) The *optimum* normalized values are designated as H_{mm}, Z_{Omm}/Z_C, and Z_{Imm}/Z_C. The "mm" designation denotes the minimum of the maximum (resonant peak) or the maximum of the minimum (resonant valley). These optimum values occur at an optimum Q (identified as Q_{opt}) and frequency ω_{mm}. The impedance values are normalized to the characteristic impedance, Z_C, and the frequency values are normalized to the corner frequency, ω_O, as defined in the following equations:

$$Z_C = \sqrt{\frac{L}{C}} \tag{A.1}$$

and

$$\omega_O = \sqrt{\frac{1}{LC}} \tag{A.2}$$

For purposes here, the *parallel* damping, Q, is defined as:

$$Q = \frac{R_d}{Z_C} \tag{A.3}$$

The table shows that the optimum condition for H_{mm} and Z_{Omm} occurs at the same frequency, but the frequency at which Z_{Imm} occurs is at a different

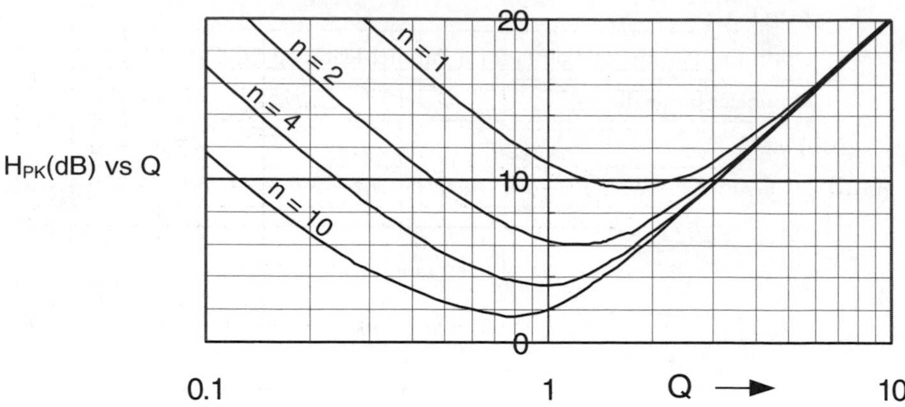

FIGURE A.4.1
Parallel damped. H_{PK} (dB) vs. Q.

frequency. Also, note that the optimum Q, Q_{opt}, for each of the three different parameters is different in each one. The conclusion from this is that the optimum design cannot be achieved simultaneously for all parameters and that some or perhaps all three will be chosen to operate at points that are shifted away from their optimum points.

The curves in Figure A.4.1, Figure A.4.2, and Figure A.4.3 show the normalized values of H_{PK}, $Z_{O(PK)}$, and $Z_{I(VAL)}$ as Q varies away from its optimum value. Families of curves for various values of n are depicted. (Note that the valleys

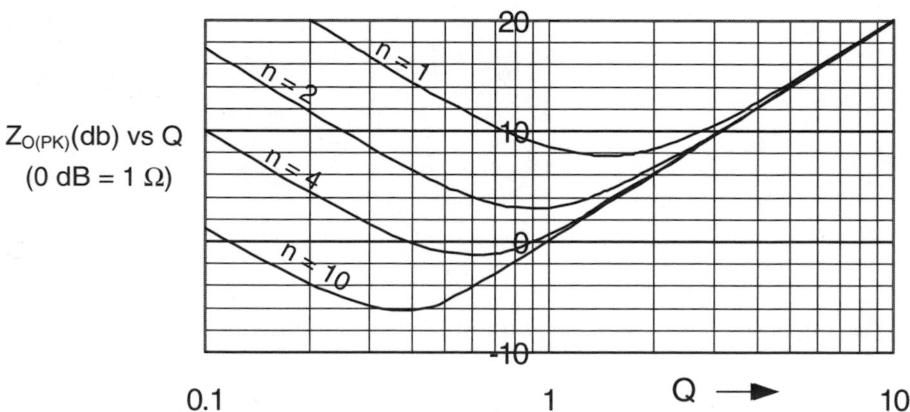

FIGURE A.4.2
Parallel damped. $Z_{O(PK)}$ (dB) vs. Q (0 db = 1 Ω).

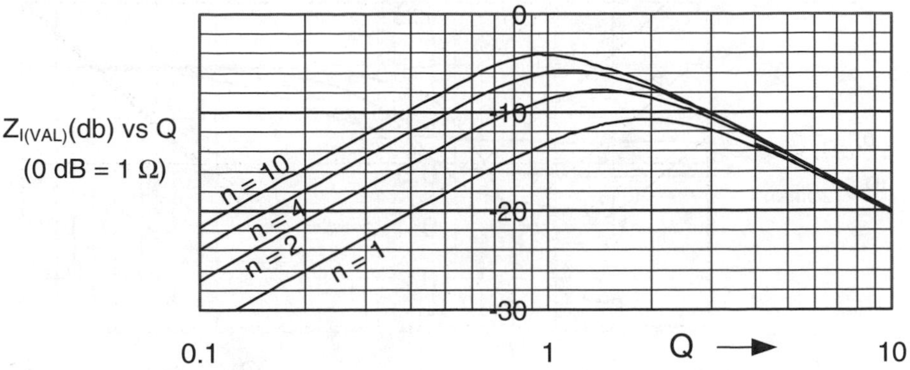

FIGURE A.4.3
Parallel damped. $Z_{I(VAL)}$ (dB) vs Q (0 db = 1 Ω).

and peaks of these plots occur at the optimum Q for the corresponding values of n.) These curves will assist in designing adequate damping and in determining when the design requirements are met. Figure A.4.4, Figure A.4.5, and Figure A.4.6 show the normalized frequencies, ω/ω_o, vs. Q that correspond to the normalized values of H_{PK}, $Z_{O(PK)}$, and $Z_{I(VAL)}$. A design example illustrating the design method presented here will now be examined.

Consider an SMPS that includes the following among its general specifications:

P_{OUT} = 30 W max
V_{IN} = 22 to 36 Vdc

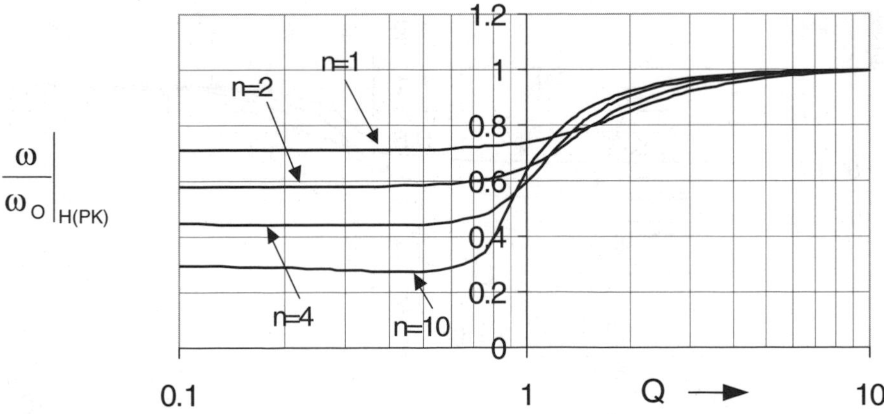

FIGURE A.4.4
Parallel damped.

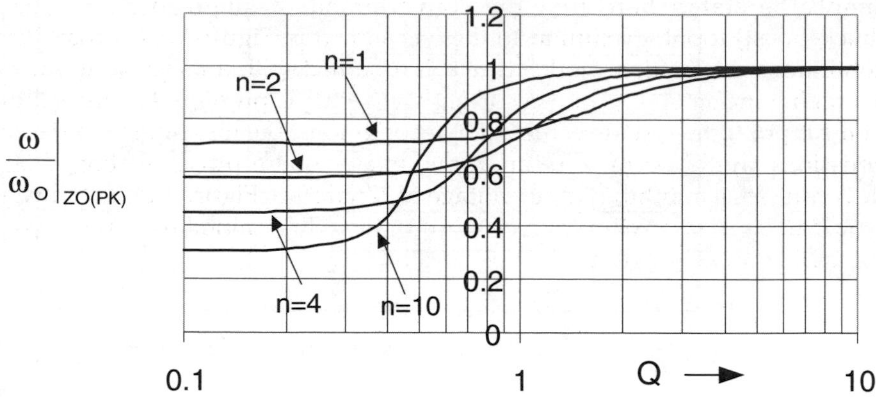

FIGURE A.4.5
Parallel damped.

V_{OUT} = 5 Vdc ± 1%

Efficiency = 85% minimum at max load

Converter switching frequency = 100 kHz (10-µS period)

Electromagnetic compatibility requirements per MIL-STD-461E

First, a value of the filter basic LC values must be determined. Several criteria may be used to make an initial estimate as to what the values of LC will be. As a point of departure, the reverse transfer function, H, will be used to ensure that the conducted emissions specifications are satisfied for this case. Without showing any extensive details of the power converter, it will

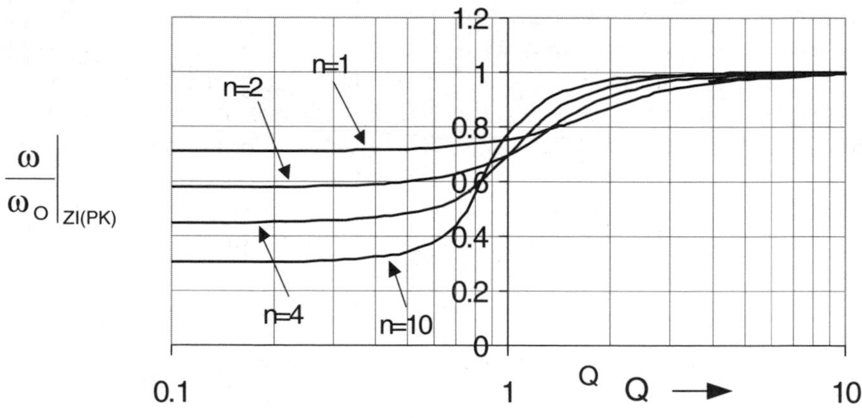

FIGURE A.4.6
Parallel damped.

simply be stated here that the converter has a single-switch flyback (buck–boost) topology similar to the one shown in Figure 1.6. At max load, continuous conduction mode current is assumed with a large value for the averaging inductor, *L*. A pulse current waveform is presented to the output of the input filter and, under the stated conditions, the maximum conducted emissions would occur with an approximate square pulse current of 50% duty ratio for a minimum input voltage of 22 Vdc (see Figure 1.6). The current would have an approximate peak amplitude as determined from the following basic power equations.

$$P_{IN} = \frac{P_{OUT}}{\eta} = \frac{30W}{0.85} = 35.3W \tag{A.4}$$

$$I_{PK} = \frac{I_{IN}}{D} = \frac{\left(\frac{P_{IN}}{V_{IN}}\right)}{D} = \frac{\left(\frac{35.3W}{22V}\right)}{0.5} = 3.21A \tag{A.5}$$

When only the AC components of a Fourier series expansion of this waveform are considered, the DC component is neglected and an actual peak AC value of the square wave will be one half of the value of I_{PK} or 1.605 A. The fundamental or first harmonic, component of this square pulse will have an RMS value of:

$$I_{RMS} = \left(\frac{I_{PK}}{\sqrt{2}}\right)\left(\frac{4}{\pi}\right) = \left(\frac{1.605}{\sqrt{2}}\right)\left(\frac{4}{\pi}\right) = 1.45A \tag{A.6}$$

This occurs, of course, at the switching frequency of 100 kHz. Figure A.5 and Figure A.6, excerpted from MIL-STD-461E, show the allowable specified limits for this power supply. At a frequency of 100 kHz, Figure A.6 indicates the limit to be approximately 74 dBμA or 5 mA. (This is interpreted as a current that has a magnitude 74 dB or 5000 times greater than a 1-μA reference.) Comparing this to the generated current of 1.45 A reveals that the input filter must provide a 100 kHz attenuation of at least

$$\frac{1}{1+\frac{X_L}{X_C}} \le \frac{5mA}{1.45A} = 0.0035 \text{ or } -49 \text{ dB} \tag{A.7}$$

This yields an impedance ratio of

$$\frac{X_L}{X_C} = 285 = \omega^2 LC \text{ @ } 100 \text{ kHz} \tag{A.8}$$

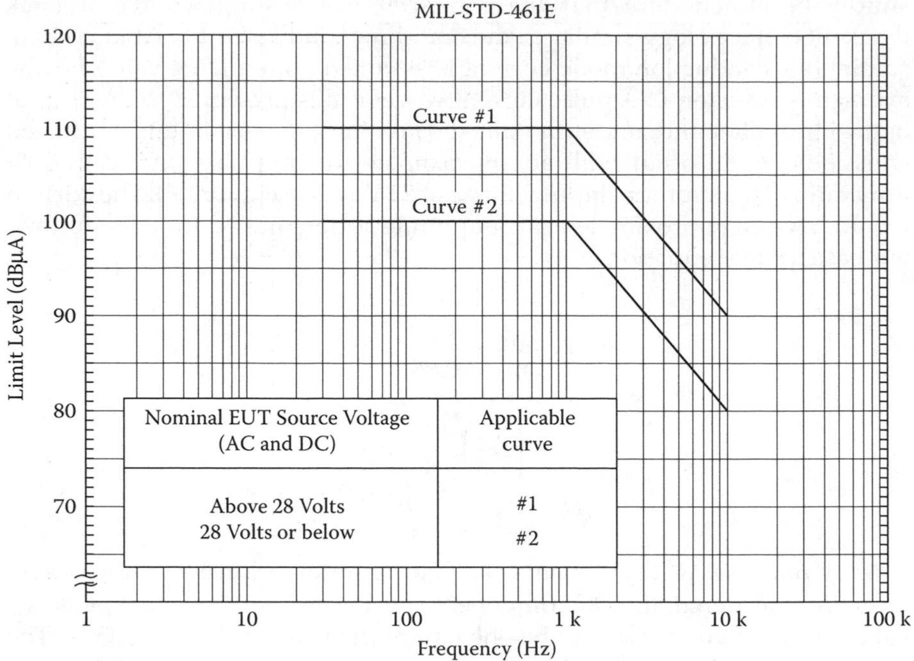

FIGURE A.5
MIL-STD-461E low-frequency conducted emissions limit.

or an LC product of

$$LC = \frac{285}{(2\pi)^2(100kHz)^2} = 722 \times 10^{-12} \tag{A.9}$$

The corner frequency, or LC value, for the required attenuation at 100 kHz has now been determined. To determine a value of L or C, it is necessary to look for other circuit limitations. From previous experience, assume a desire to limit the 100 kHz peak-to-peak voltage ripple, ΔV, presented to the converter to less than 1.0 Vp-p. (This may help in preventing possible erratic switching operation of the power converter stage or some other noise-related problems.) To accomplish, the filter capacitor value must be

$$C \geq \frac{(I_{PK})(\Delta T)}{\Delta V} = \frac{(3.21A)(5\mu S)}{1.0V} = 16\mu F \tag{A.10}$$

The term ΔT is 5 μS or one half the switching period of 10 μS at the 50% duty ratio. From Equation A.9, a corresponding value of L = 45 μH can be calculated. At this point, some design margin is added and these actual

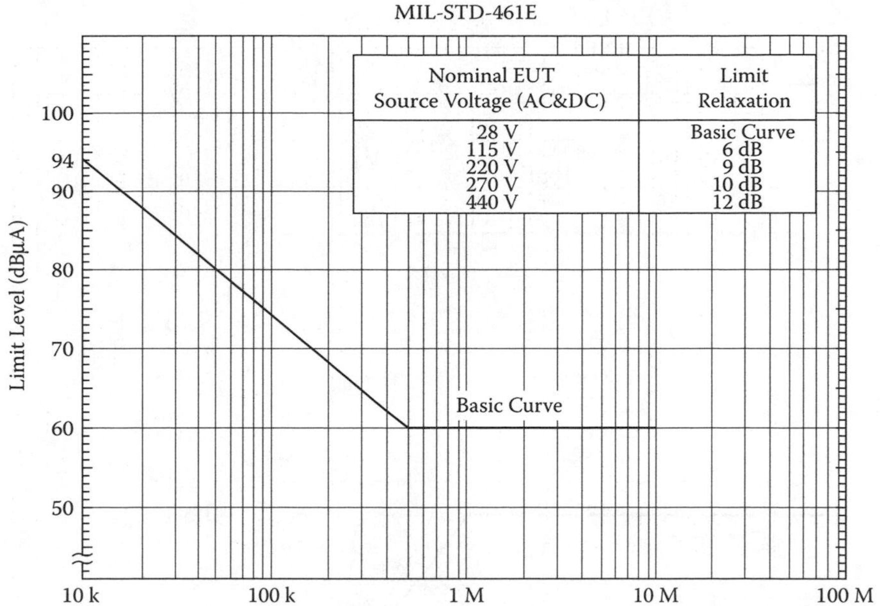

FIGURE A.6
MIL-STD-461E high-frequency conducted emissions limit.

values are increased to $C = 25\ \mu F$ and $L = 75\ \mu H$. Reasons for this will become evident later; now, when the filter is loaded with the negative resistance of the power converter, it will tend to decrease the effects of the positive resistance damping that will be added.

The basic low-pass LC filter of Figure A.7 shows the values of L and C providing the required attenuation. (As a practical side light, some capacitor types that might be selected may have an equivalent series resistance, ESR, or even equivalent series inductance, ESL, that could dominate characteristics at 100 kHz; this must be considered when calculating the required attenuation and might be a determining factor in going to a multiple stage filter topology as discussed later. In the case here, however, only the ideal capacitor case is considered.) The values to be used are, as stated earlier, $L = 75\ \mu H$ and $C = 25\ \mu F$. The LC filter attenuates the higher harmonics at a faster rate (40 dB/decade) than the specified limit curve in Figure A.5 (20 dB/decade) and the higher frequency harmonics are also lower in amplitude; thus, it can be concluded that when the 100-kHz requirement is met, the narrowband conducted emissions requirement is also met at all the other higher harmonic frequencies.

With these values selected, the conducted emissions requirement and a voltage switching ripple requirement have been met; however, an undamped resonant filter has been created that will produce all the undesirable side effects mentioned earlier. Now the filter will be damped out with the proposed parallel damping circuit. At this point, note that the converter utilizes

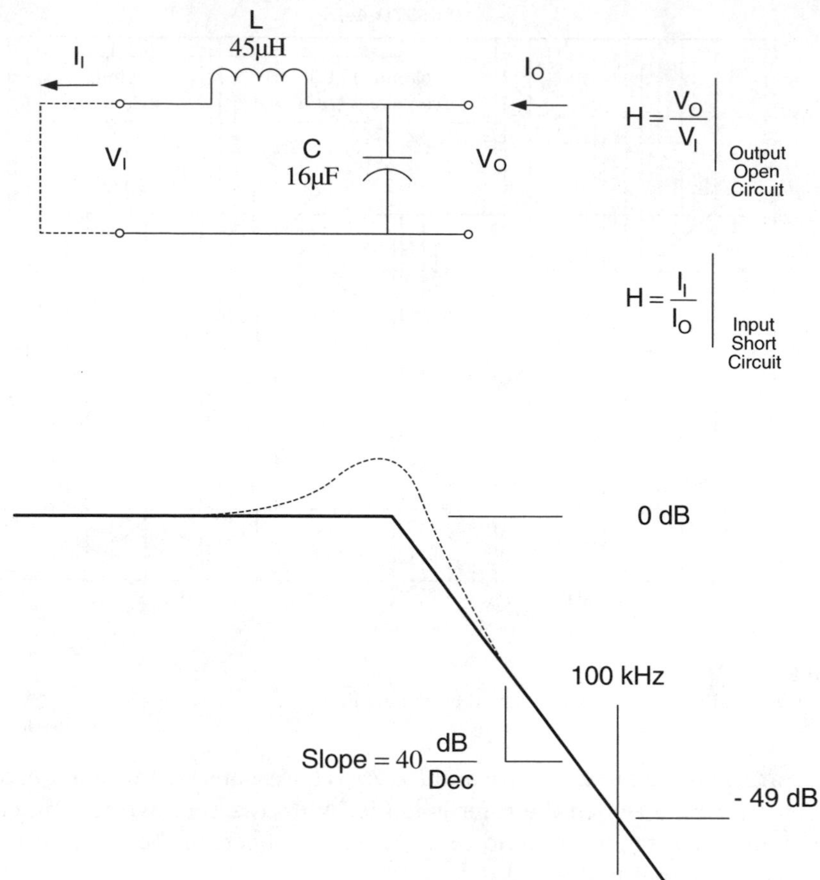

FIGURE A.7
Conducted emissions attenuation of LC filter.

current mode control and the input to this converter looks like a negative incremental resistance at frequencies from DC to approximately one-sixth of the switching frequency or 16 kHz in the stated case. This maximum negative incremental resistance loading (minimum resistance) occurs at the minimum input voltage and maximum load power:

$$R_O \geq \frac{V_{IN(MIN)}^2}{P_{MAX}} = \frac{(22V)^2}{35.3W} = -13.7\Omega \tag{A.11}$$

To ensure adequate stability margin and prevent an input filter instability (Section 2.2.2), it is desired to keep the peak magnitude of the input filter

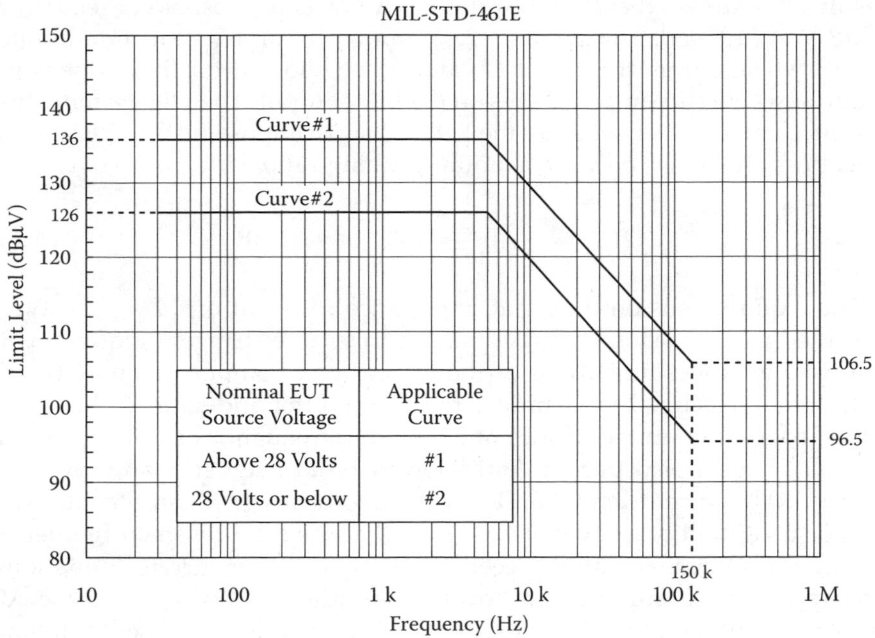

FIGURE A.8
MIL-STD-461E conducted susceptibility limits.

output impedance to around three times less than the minimum R_O or, in this case, one-third of 13.7 Ω or

$$Z_{O(MAX)} \leq \frac{R_O}{3} = \frac{13.7\Omega}{3} = 4.57\Omega \tag{A.12}$$

This limitation is to exist over the frequency range of dc to 16 kHz (see Figure 2.7). From Equation A.2, the resonant corner frequency is

$$f_O = \frac{\omega_O}{2\pi} = \frac{1}{2\pi}\sqrt{\frac{1}{LC}} = \frac{1}{2\pi}\sqrt{\frac{1}{(75\mu H)(25\mu F)}} \approx 3.68kHz \tag{A.13}$$

This will cause a resonant peak in Z_O to exist within the concerned frequency range of dc to 16 kHz.

Now consider another requirement forcing limitation of the peaking of the transfer function, H, at the resonant frequency, f_O. Figure A.8, another excerpt from MIL-STD-461E, shows an audio susceptibility specification of 126 dBμV (or 2.0 Vrms) to be imposed on the input power lines in the frequency range of 30 to 5000 Hz with no degradation of performance (regulation) occurring over the complete input voltage range of 22 to 36 Vdc. Possibly from lab test results, it is known that the converter has a low end threshold of regulation of 18 V at full load. This allows a 22-V minus 18- or 4-V reduction in filter output voltage from the 22-V DC value to be allowed for these sine wave audio susceptibility-imposed dips on the filter output. The voltage reduction produced at the filter input by the 2.0 V_{rms} superimposed AC is 2.828 V_{peak}. This possibly allows only a maximum peaking ratio, H_{MAX}, of

$$H_{MAX} = \frac{4V}{2.818V} = 1.42 \quad \text{or } +3.1 \text{ dB} \tag{A.14}$$

The requirements on the input impedance resonant dip, $Z_{I(MIN)}$, may be specified by the customer, but in many cases more stringent requirements imposed by stress limitations of the actual filter components may be the determining factor. For example, the input filter inductor, L, has been specified to have an inductance of 75 μH at a frequency of 100 kHz, but it will also have a peak current limitation to avoid magnetic saturation and a peak RMS current limitation to avoid overheating. When the 2.0-Vrms audio susceptibility test voltage is applied at the filter resonant frequency, the increased current could exceed one or both these current limitations. The resonant AC current also flows in the filter capacitor, C, so the AC ripple current stress might be excessive here as well. All these limitations must be assessed.

From Equation A.5, the peak input current at the minimum input DC voltage is 3.21 amps. Assume that the peak saturation current rating for the inductor is 8.0 amps and that this is the primary limiting concern. To design the filter to stay below this 8.0-amp value, the allowable peak AC current would be

$$I_{PK(AC)} \leq I_{L(MAX)} - I_{PK(DC)} = 8.0A - 3.21A = 4.79A \tag{A.15}$$

or an AC current of

$$I_{AC} = \frac{I_{PK(AC)}}{\sqrt{2}} = \frac{4.79}{\sqrt{2}} = 3.4Arms \tag{A.16}$$

With the 2.0-V_{rms} audio susceptibility test signal, the AC input impedance will now be limited to

$$Z_{I(MIN)} \geq \frac{2.0Vrms}{3.4Arms} = 0.59\Omega \tag{A.17}$$

Using somewhat unstructured but valid methods, the required limitations for the three parameters of H_{MAX}, $Z_{O(MAX)}$, and $Z_{I(MIN)}$ have been discerned. The design generally requires an iterative approach and the method being used makes these iterations quick and easy.

To reiterate, the parameters are listed before the design of the input filter parallel damping circuit is begun:

$H_{MAX} \leq 1.42$ or +3.1 dB
$Z_{O(MAX)} \leq 4.57\ \Omega$
$Z_{I(MIN)} \geq 0.59\ \Omega$

The normalized curves of Figure A.4.1, Figure A.4.2, and Figure A.4.3 will be used to determine the value of n and Q from which R_d can subsequently be determined from Equation A.3 and Equation A.1. First, however, calculate the characteristic impedance, Z_C, from Equation A.1 to obtain the normalized values:

$$Z_C = \sqrt{\frac{L}{C}} = \sqrt{\frac{75\mu H}{25\mu F}} = 1.732\Omega \tag{A.18}$$

Calculate the normalized limit values of $Z_{O(PK)}$ and $Z_{I(VAL)}$:

$$Z_{O(PK)} = \frac{Z_{O(MAX)}}{Z_C} = \frac{4.57\Omega}{1.732\Omega} = 2.64 \text{ or } +8.43 \text{ dB} \tag{A.19}$$

and

$$Z_{I(VAL)} = \frac{Z_{I(MIN)}}{Z_C} = \frac{0.59\Omega}{1.732\Omega} = 0.34\Omega \text{ or } -9.37 \text{ dB} \tag{A.20}$$

Now list these normalized limits, recognizing that the normalized H_{PK} is here equivalent to H_{MAX}:

$H_{PK} = H_{MAX} = +3.1$ dB
$Z_{O(PK)} = +8.43$ dB
$Z_{I(VAL)} = -9.37$ dB

Draw these normalized limits on the derived curves of Figure A.4.1, Figure A.4.2, and Figure A.4.3 and show the resultant intercepts in Figure A.9. Some important observations about what must be done can now be made. First, to ensure that H is always less than +3.1 dB, all values of n must be greater than 4, as the +3.1 dB line is practically tangential to the $n = 4$ plot at the lowest, or Q_{opt}, point. If the marginal value of $n = 4$ were selected, Q would have to be stabilized at this value of 1 because any significant shift in either direction would allow H_{PK} to exceed +3.1 dB.

Next, examine the restrictions for keeping $Z_{O(PK)}$ less than +8.43 dB. A lot of margin is present when selecting n and Q for this requirement.

Even with a small value of $n = 2$, Q could range from 0.3 to 2.6 and $Z_{O(PK)}$ would be less than the limit of +8.43 dB. Now look at the plot for input impedance Z_I. Note that for values of $n = 2$ and Q ranging between 0.9 and 2.2, input impedance requirements of $Z_{I(VAL)}$ being greater than –9.37 dB are satisfied.

With these observations, what should one now do? It appears that, if the requirement for H_{PK} can be satisfied at a value of Q near 1.0, then all the other requirements for $Z_{O(PK)}$ and $Z_{I(VAL)}$ will be satisfied. This unloaded minimum value is, as stated previously, $n = 4$. Because the eventual effects of the incremental negative resistance of the power converter have not yet been considered, add a conservative margin and select an n of 8. With H the primary concern, design with a Q equal to the H_{PK} Q_{opt} value of 0.845. This value may be calculated from the equation in Table A.1 or may be taken from the curve in Figure A.4.1. (Incidentally, it is not necessary to select integral values of n, but it just seems a logical thing to do because capacitors of the same type may be connected in parallel.)

Interpolating logarithmically from the H_{PK} curve in Figure A.9 and an $n = 8$ plot, an observation indicates that Q might be allowed to range from about 0.5 to 1.1 and still limit H_{PK} to less than the allowable +3.1 dB in the unloaded case. This allowable Q ranging might be a consideration when the significant and possibly highly variable equivalent series resistance of some capacitor types is utilized for damping instead of actually inserting a separate resistance for R_d.

Now, with these assessments, design the parallel damping network for the filter and then analyze the results to ensure that the requirements are met. For $Q = 0.845$, R_d can be determined from Equation A.3 and Equation A.18.

$$R_d = (Q)(Z_C) = (0.845)(1.732\Omega) = 1.464\Omega \qquad \text{(A.21)}$$

The circuit in Figure A.10.1 is the final parallel damped filter circuit. A computer analysis was run to see how this compares with requirements in the unloaded case. The results are shown in Figure A.11; thus far, filter damping design goals have been met in a practical manner without implementing excessive design margins in any of the three parameters. The results are summarized:

$$H_{MAX} = +1.934\text{ dB} \le +3.1\text{ dB}$$

$$Z_{O(MAX)} = 1.50\ \Omega \le 4.57\ \Omega$$

$$Z_{I(MIN)} = 1.003\ \Omega \ge 0.59\ \Omega$$

For the important worst-case tests, load the filter with the worst-case negative resistance loading of the power converter and verify that requirements are still met. The circuit of Figure A.10.2 shows the circuit with the negative resistance load of –13.7 Ω. This analysis will hopefully verify that allowed design margins are adequate. The applicable results of the computer analysis for H_{MAX} and $Z_{I(MIN)}$ in this loaded case are shown in Figure A.12;

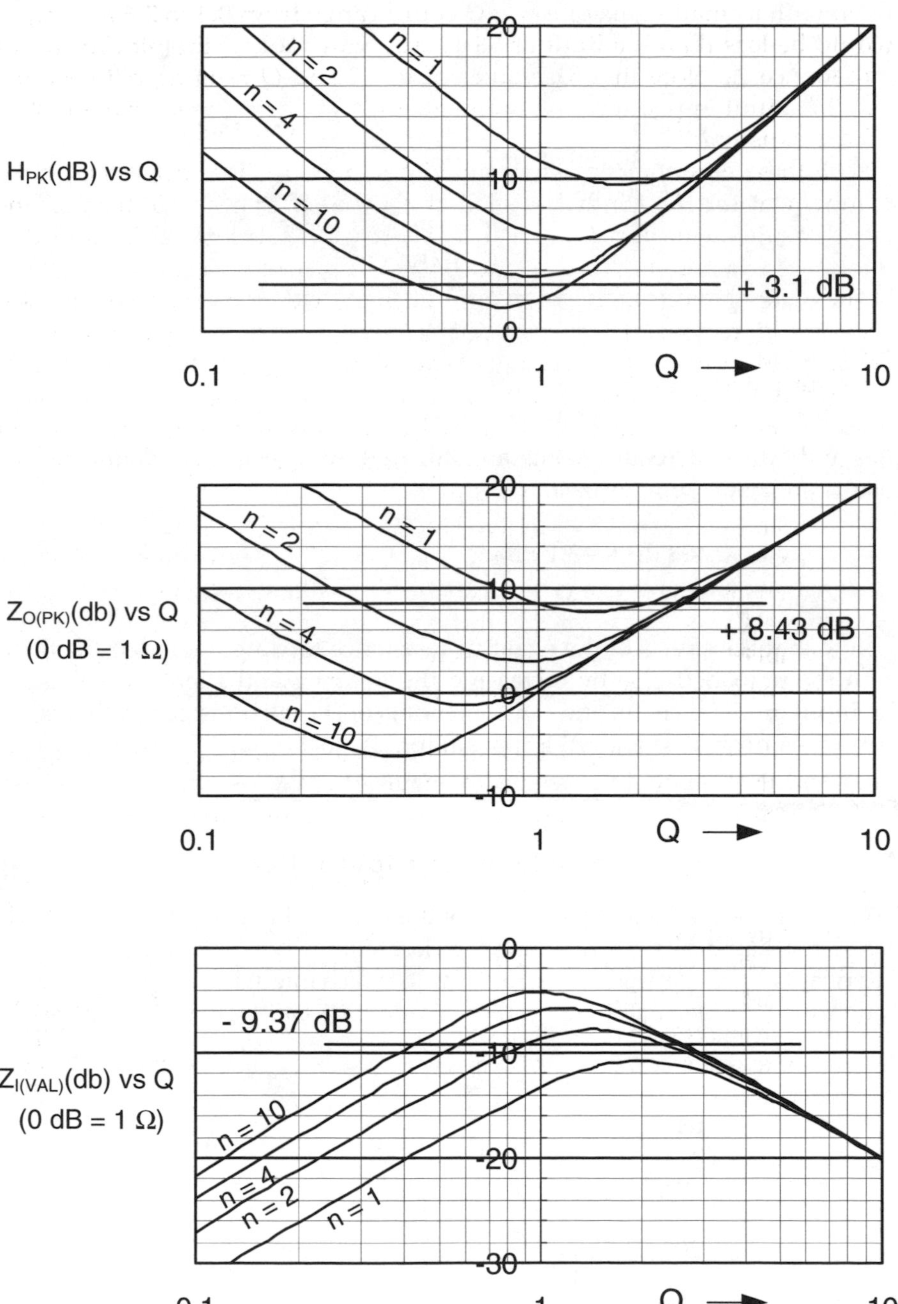

FIGURE A.9
Parallel damped filter design example.

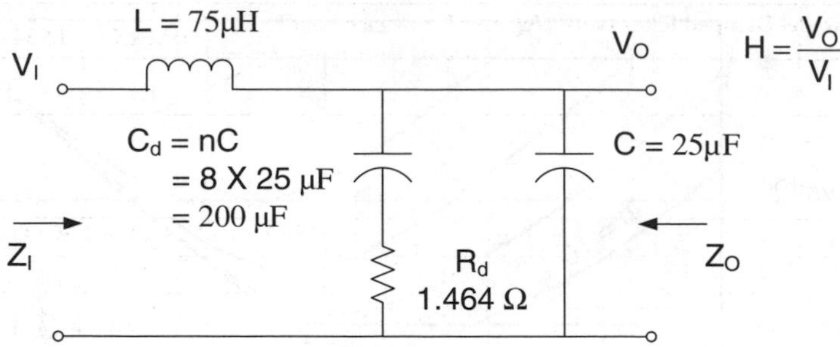

FIGURE A.10.1
Parallel damped filter example.

they indicate that requirements are still met and design is adequate. The results are again summarized:

$$H_{MAX} = +2.408 \text{ dB} \leq +3.1 \text{ dB}$$

$$Z_{I(MIN)} = 0.98\ \Omega \geq 0.59\ \Omega$$

Note that the effects of the negative resistance loading degraded the filter from the unloaded case by increasing the peaking and slightly decreasing the input impedance. As one can conclude, conservatism is generally necessary when designing an SMPS input filter.

A.4 Series Damped Single Stage Input Filter

Now take the same filter requirements of the parallel damped filter in the previous section and design a series damping circuit of the type shown in Figure A.3. Afterwards, a comparison can be made as to which might be the most desirable option for the filter damping — that is, adding *parallel* capacitance or *series*

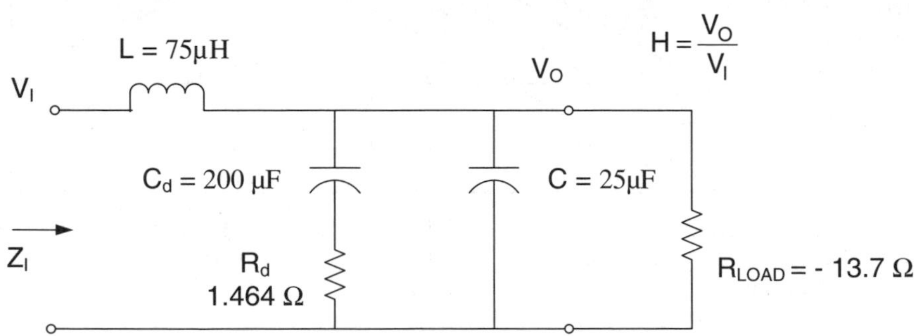

FIGURE A.10.2
Parallel damped filter example with negative resistance load.

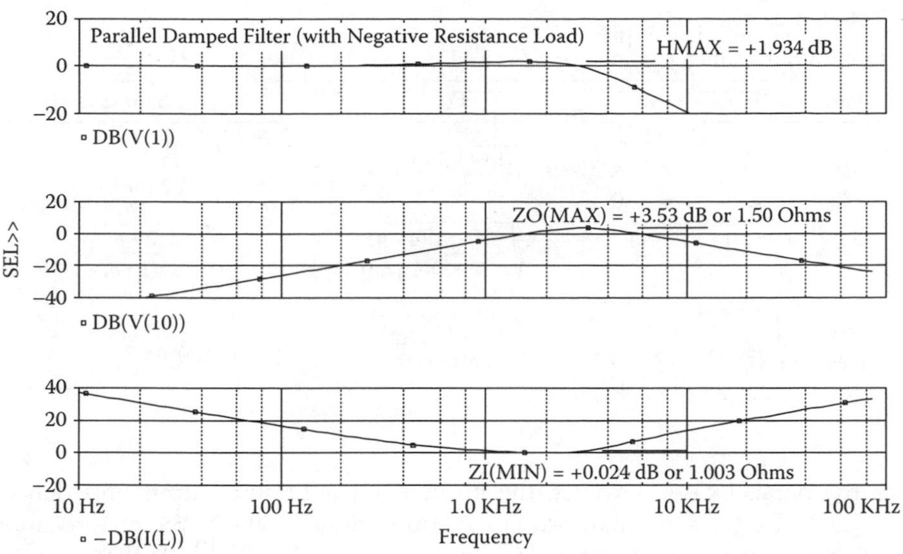

FIGURE A.11
Analysis results of parallel damped filter example (unloaded).

inductance. The series damping topology has been given the same analogous treatment as was given to the parallel damping scheme in Carsten[16] and the optimum parameter points are shown in Table A.2.

In a comparison of Table A.1 and Table A.2, the duality between the two approaches is apparent. The optimum peaking, H_{mm}, has the same relationships

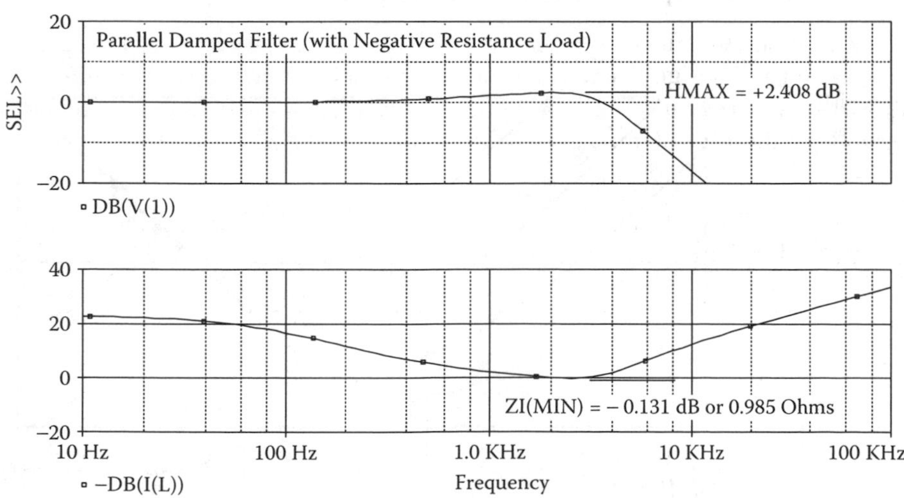

FIGURE A.12
Analysis results of parallel damped filter example (with negative resistance load).

TABLE A.2

Optimum Parameter Points for a Series Damped LC Filter

Parameter (normalized)	Q_{opt}	ω_{mm}/ω_o (normalized)
$H_{mm} = \dfrac{2+n}{n}$	$\sqrt{\dfrac{(1+n)(2+n)}{2n^2}}$	$\sqrt{\dfrac{2}{2+n}}$
$\dfrac{Z_{Omm}}{Z_C} = \sqrt{\dfrac{n^2 2(1+n)(2+n)}{n^2}}$	$\sqrt{\dfrac{2(1+n)^3(4+n)}{n^2(2+n)(4+3n)}}$	$\sqrt{\dfrac{2+n}{2(1+n)}}$
$\dfrac{Z_{Imm}}{Z_C} = \sqrt{\dfrac{n^2}{2(2+n)}}$	$\sqrt{\dfrac{(4+3n)(2+n)}{2n^2(4+n)}}$	$\sqrt{\dfrac{2}{2+n}}$

as the parallel case; however, the normalized optimum output impedance, Z_{Omm}/Z_C, for the series damped case is numerically equal to the reciprocal of the normalized parallel damped input impedance, Z_{Imm}/Z_C. Note that the normalized optimum input impedance, Z_{Imm}/Z_C, for the series damped case is numerically equal to the reciprocal of the normalized parallel damped output impedance, Z_{Omm}/Z_C. Also, note the corresponding accompanying normalized frequency, ω_{mm}/ω_O, transpositions. The curves in Figure A.13.1, Figure A.13.2, and Figure A.13.3 show the series damped normalized values of H_{PK}, $Z_{O(PK)}$, and $Z_{I(VAL)}$ as Q varies away around its optimum value. Figure A.13.4, Figure A.13.5, and Figure A.13.6 show the corresponding curves for ω/ω_O. A comparison between these curves and those of Figure A.4.1 through Figure A.4.6 shows the continued duality that exists between the two topologies.

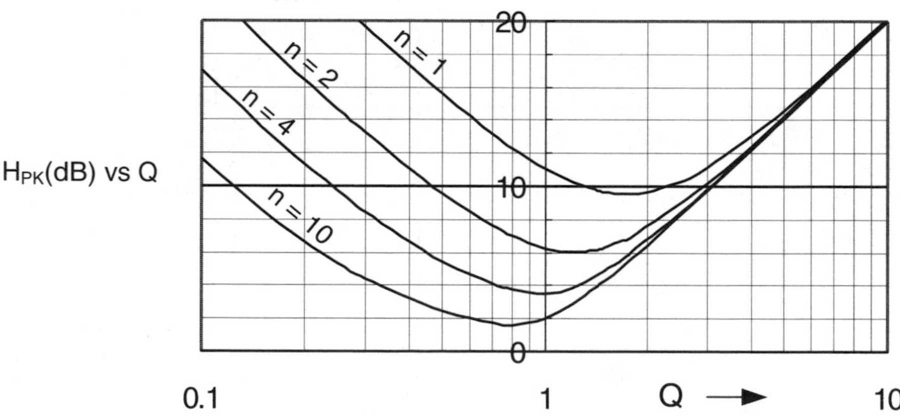

FIGURE A.13.1
Series damped. H_{PK} (dB) vs. Q.

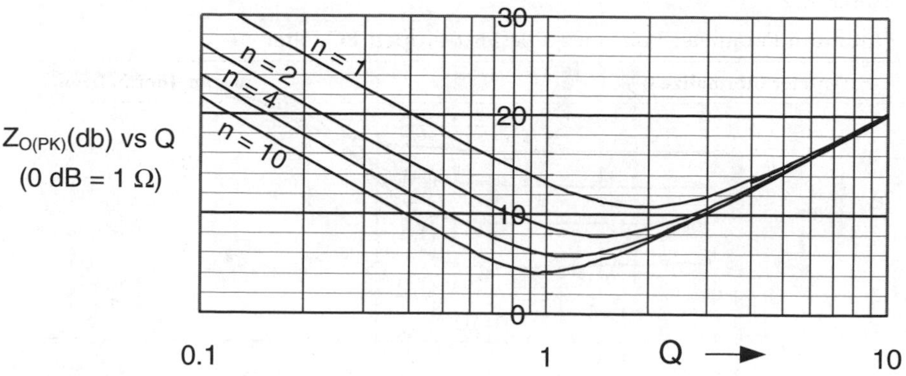

FIGURE A.13.2
Series damped. $Z_{O(PK)}$ (dB) vs. Q (0 db = 1 Ω).

Now draw these normalized design limits on the derived curves of Figure A.13.1, Figure A.13.2, and Figure A.13.3 and show the resultant intercepts in Figure A.14. With the plots for H_{PK} the same as for the parallel case, the same assessments apply; that is, $n = 4$ at the lowest, or Q_{opt}, point of 1.0 being the lowest limit. Now, examining the $Z_{O(PK)}$ case, note that a magnitude less than the limit of +8.43 dB for $n = 4$ and a Q range of 0.6 to 2.3 is possible. A considerable margin is present, but when it is compared to the parallel damped case, the margin is not as large because requirements could have been met very easily with an n of 2 for that case and a range of 0.3 to 2.6.

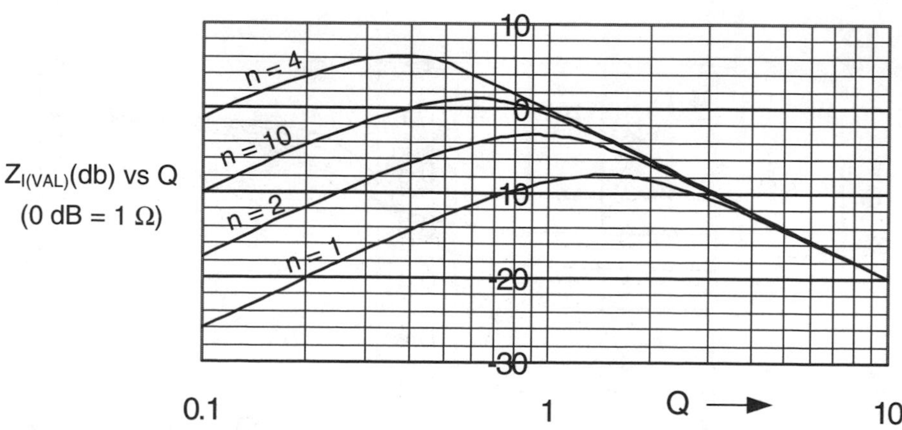

FIGURE A.13.3
Series damped. $Z_{I(VAL)}$ (dB) vs. Q (0 db = 1 Ω).

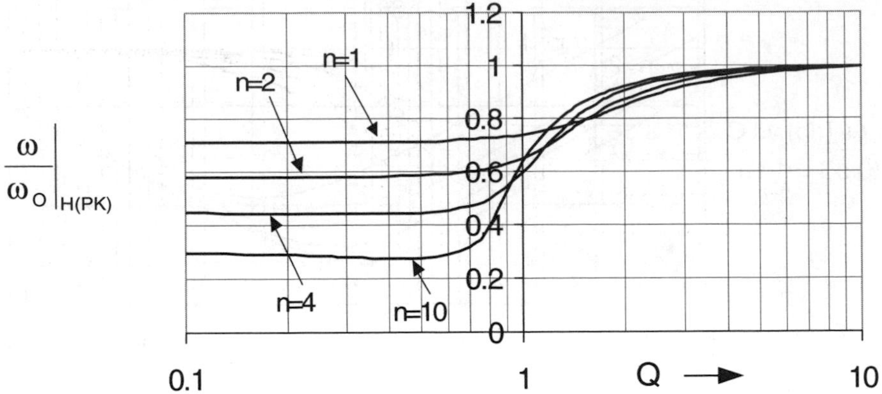

FIGURE A.13.4
Series damped.

Look at the plot for input impedance $Z_{I(VAL)}$. For values of n = 2 and Q ranging between 0.26 and 2.8, input impedance requirements of $Z_{I(VAL)}$ being greater than –9.37 dB will be satisfied. A larger degree of margin is present here than in the parallel damping case. This requirement could even be met with an n = 1 and Q range of 0.9 to 2.3. The conclusion here is that if the input impedance requirement is more restrictive, the series topology might be better than the parallel case and, conversely, if it is required to have very low filter output impedance, the parallel damped case would then be better.

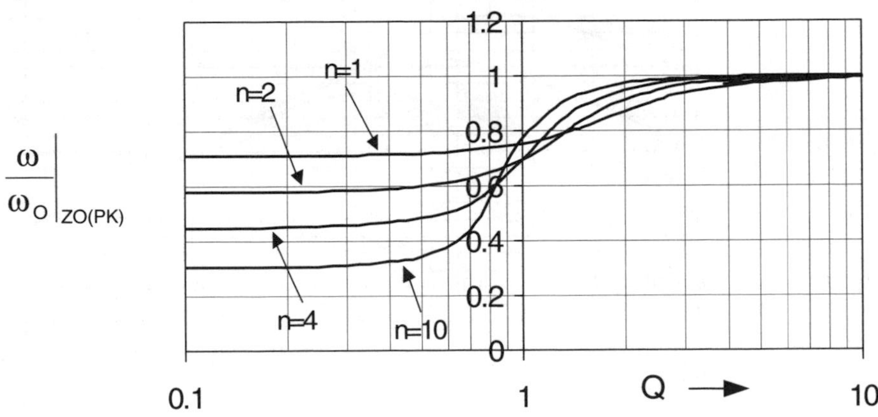

FIGURE A.13.5
Series damped.

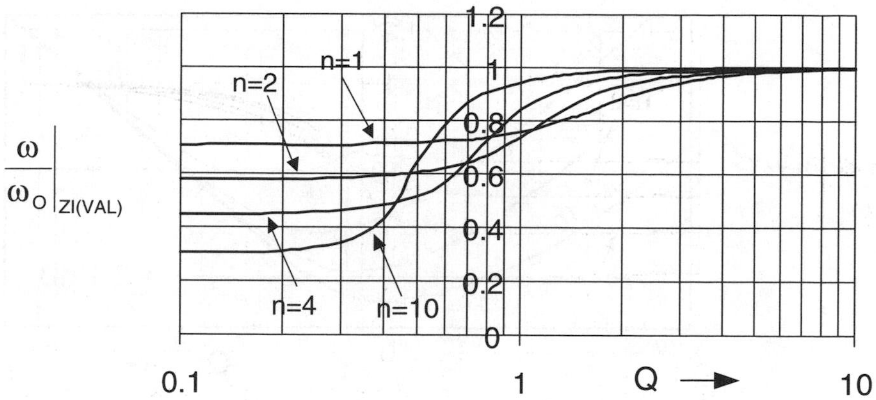

FIGURE A.13.6
Series damped.

Now try to determine the practical values n and Q. As with the parallel case, because H_{PK} is the most restrictive case, an n of 8 and a Q of $Q_{opt} = 0.845$ are selected. The value of R_d in this series damped case is:

$$R_d = \frac{Z_C}{Q} = \frac{1.732\Omega}{0.845} = 2.05\Omega \tag{A.22}$$

The circuit in Figure A.15.1 is the final series damped filter circuit. A computer analysis was run to see how this compared with requirements in the unloaded case. The results are shown in Figure A.16; thus far, unloaded filter damping design goals have been met in a practical manner without implementation of excessive design margins in any of the three parameters. The results are summarized:

$$H_{MAX} = +1.9337 \text{ dB} \leq +3.1 \text{ dB}$$
$$Z_{O(MAX)} = 3.00\ \Omega \leq 4.57\ \Omega$$
$$Z_{I(MIN)} = 2.0\ \Omega \geq 0.59\ \Omega$$

Note that, although H_{MAX} is about the same as the parallel damped case, both impedances have increased; in this example, both have approximately doubled in value.

Now consider the important worst-case tests. Load the filter with the worst-case negative resistance loading of the power converter to see whether requirements are still met. The circuit of Figure A.15.2 shows the circuit. The results of the computer analysis for H_{MAX} and $Z_{I(MIN)}$ in this loaded case are shown in Figure A.17. They indicate that the peaking requirement for H_{MAX} has been exceeded, resulting in

$$H_{MAX} = +4.014 \text{ dB} \geq +3.1 \text{ dB}$$
$$Z_{I(MIN)} = 2.84\ \Omega \geq 0.59\ \Omega$$

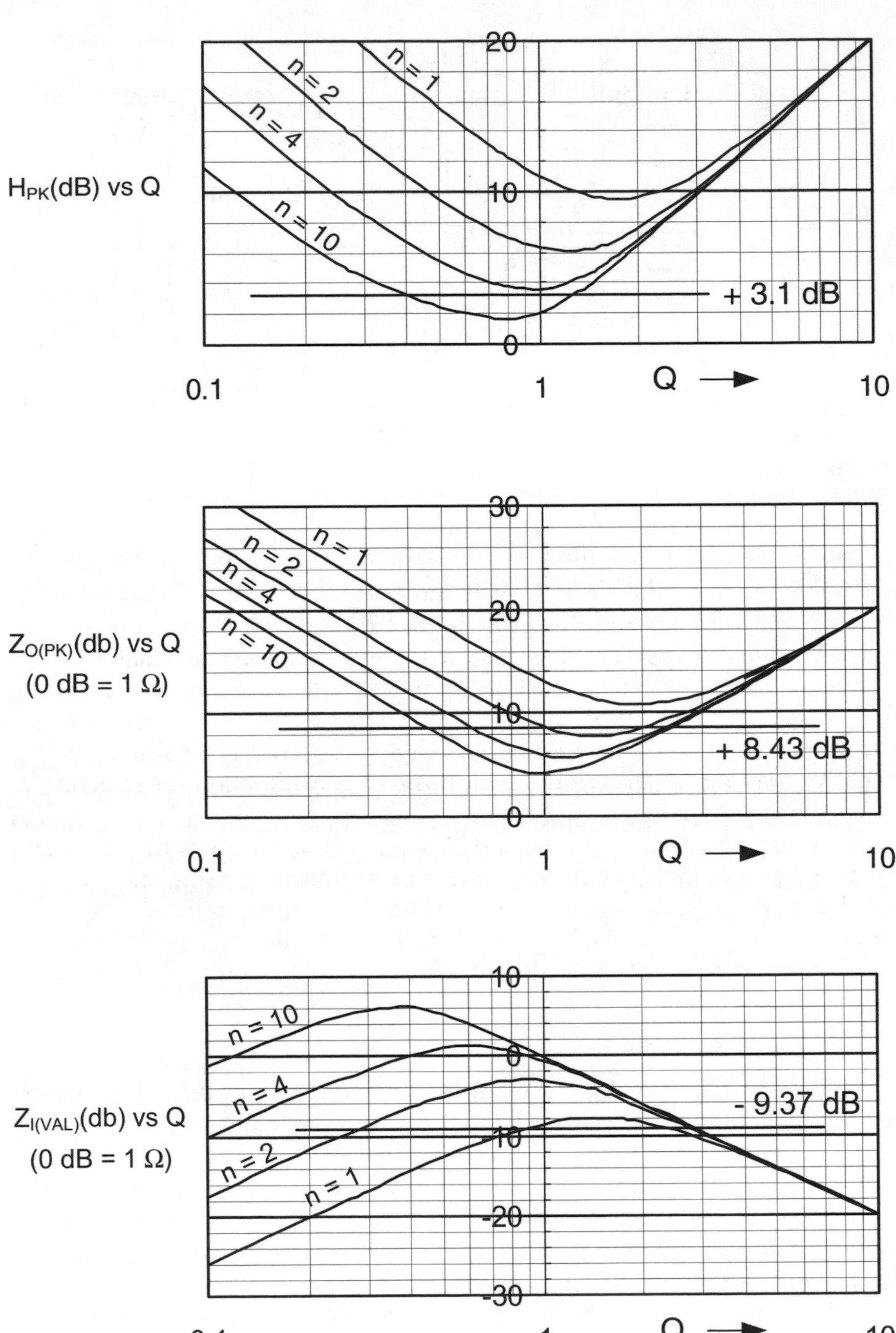

FIGURE A.14
Series damped filter design example.

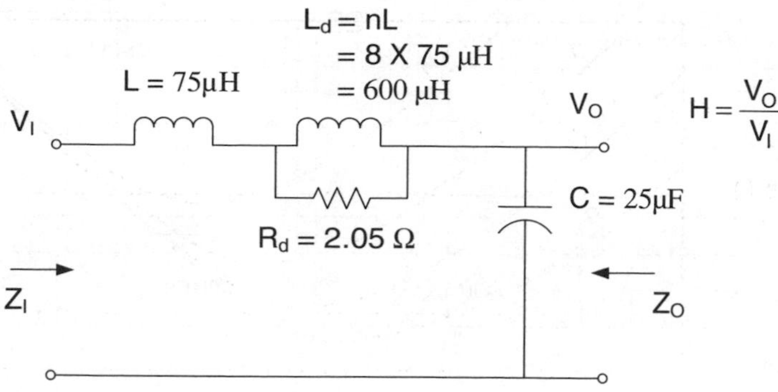

FIGURE A.15.1
Series damped filter example.

Note that the effects of the negative resistance loading again degraded the filter from the unloaded case. H_{MAX} increased to a higher level than the parallel damped case even though the unloaded parameters of H_{PK}, $Z_{O(PK)}$, and $Z_{I(VAL)}$ were similar in values. It increased to +4.014 dB and actually exceeded the +3.1 dB limit. With the higher Z_O and Z_I impedances encountered in series vs. parallel damping, it is logical to conclude that loading effects will be more pronounced and a more conservative approach might be assumed from the outset when selecting a value of n for the series damping approach. Also of note is the fact that Z_I actually increased with the negative loading, so this would be a positive consideration for selecting a series damping topology.

One more observation of note is that any loading will shift the optimum operating points slightly, but this should not be a prime consideration during the design process. Again, once a design has been attempted, an additional iteration may be necessary in an effort to hone the design for the negative resistance loaded case. This is evidenced here when noting that the peak value

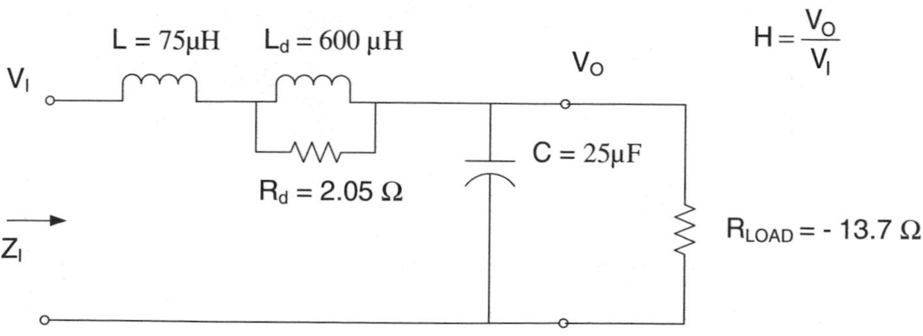

FIGURE A.15.2
Series damped filter example with negative resistance load.

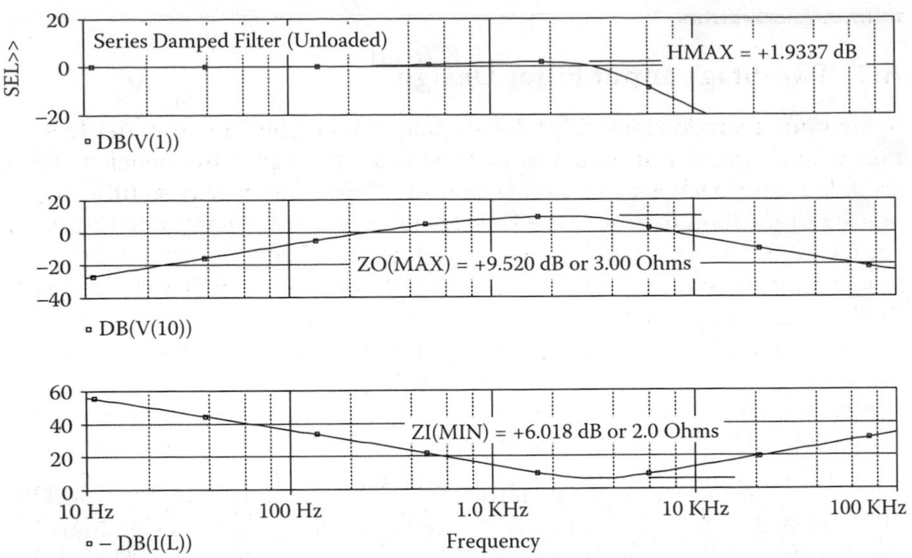

FIGURE A.16
Analysis results of series damped filter example (unloaded).

of H_{MAX} was exceeded for this loaded case. It is necessary to perform another iteration (not shown here) with a larger value of n if one wants to use the series damping approach for this example. The technique presented here makes these iterations a quick and easy process when they become necessary.

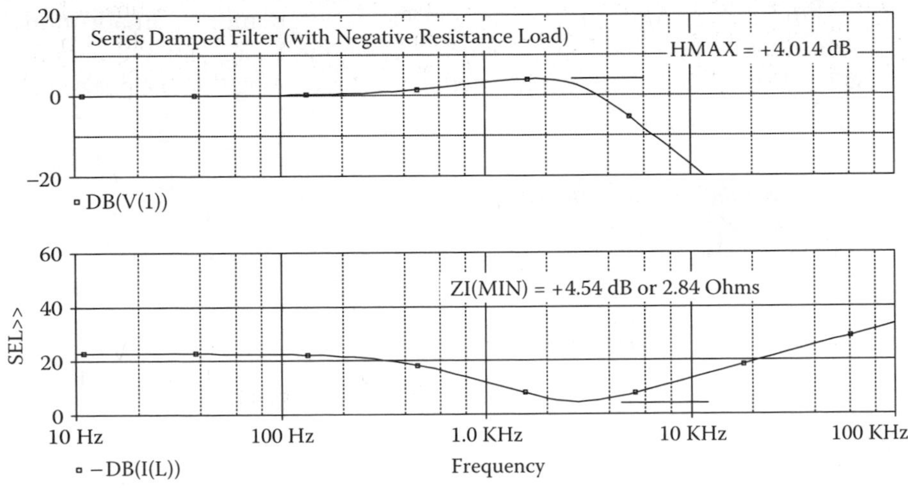

FIGURE A.17
Analysis results of series damped filter example (with negative resistance load).

A.5 Two-Stage Input Filter Design

Sometimes a single-stage filter design may not be considered a viable solution when a greater attenuation is desired at the higher frequencies. There are many approaches to implementing multiple stage low pass filter topologies, but for the purposes here an elementary two-stage approach that will suffice in many cases will be presented. Figure A.7 has shown the low- and high-frequency concerns of a single-stage filter; now that is followed with the two-stage approach.

Additional filter components may be required for any number of reasons, but consider two very common motivating factors. One is the effect of equivalent series resistance (ESR) of the filter capacitor, C. Assume that a single-stage filter with parallel damping is designed as described in the earlier example and that the damped parameters of H_{MAX}, $Z_{O(MAX)}$, and $Z_{I(MIN)}$ are acceptable at the low frequencies near resonance. Unfortunately, it is later discovered that the ESR of capacitor *C* is sufficiently large to produce a zero in the filter transfer function occurring at some frequency below the converter switching frequency of 100 kHz but considerably higher than the LC corner frequency, f_O. This reduces the attenuation at the higher frequencies and therefore the conducted emissions specification is not met at the 100 kHz switching frequency. Additional high-frequency attenuation must be provided.

The circuit of Figure A.18.1 shows a possible solution by modifying the existing parallel damped circuit as follows. Inductor *L* is split into two parts of proportion, *kL* and (1 – *k*)*L*, and an additional smaller value capacitor, C_P, is inserted to provide the necessary additional attenuation. (In most cases, the value of *k* is much smaller than (1 – k) and that stipulation is assumed here.) The attenuation plot, *H*, of Figure A.18.2 shows the effects of this addition. The ESR zero causes an attenuation loss of *A* at 100 kHz. The addition of capacitor

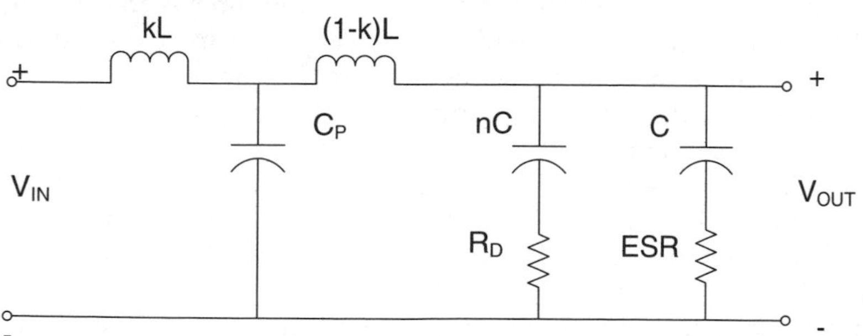

FIGURE A.18.1
Proposed two-stage extension of parallel damped single-stage filter.

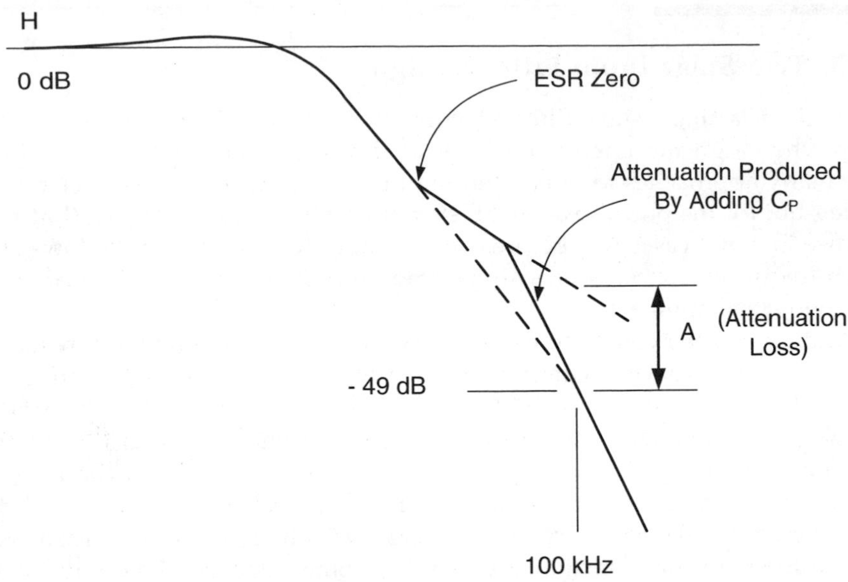

FIGURE A.18.2
Effect of adding C_P to compensate for loss of attenuation caused by ESR.

C_P, working in conjunction with input inductor, *kL*, adds an additional two-pole rolloff and restores the additionally required attenuation, *A*, at 100 kHz; now the conducted emission specification is met. It should be stated here that only the high-frequency portion of the filter has been affected; all the previously designed parameters of H_{MAX}, $Z_{O(MAX)}$, and $Z_{I(MIN)}$ at the lower frequency are negligibly affected. Another aspect of note is that the addition of C_P could produce a small amount of peaking at the corner frequency of *kL* and C_P; however, this should occur at a frequency significantly lower than the switching frequency of 100 kHz, and also considerably higher than the low-frequency corner of LC. This small effect should not affect filter performance.

A second reason for implementing a two-stage filter is that it might be possible to reduce the size of the filter physically by using more, but smaller, parts. For example, the low-frequency corner of LC can be raised by lowering these values (smaller size parts) and producing a four-pole rolloff at the higher frequencies and it will still provide the necessary attenuation at the higher switching frequency. Figure A.19 shows a comparison of a *single-stage two-pole* filter and a *two-stage four-pole* filter implemented by splitting *L* and adding C_P as previously shown. Note that the same attenuation of –49 dB is still provided at 100 kHz.

If a two-stage implementation is decided upon for whatever reason, it must be stated that when the values of *L* and *C* are selected, the procedures for designing the low-frequency damping as shown here are the same as those used for a single-stage filter.

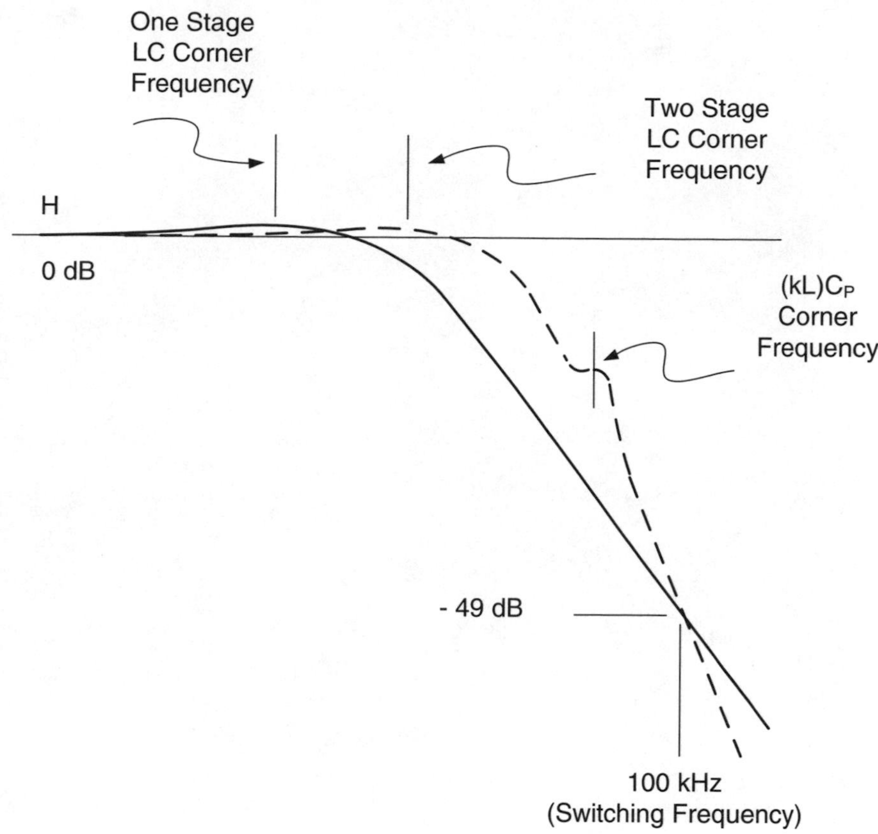

FIGURE A.19
Example of a two-stage design increasing the LC corner frequency.

Appendix B

Pulse Width Modulator Controller Macromodeling

For the efficient computer analysis of electronic circuits, a common approach used nowadays is to develop macro or behavioral models representing specific complex functions. When an SMPS that uses a commercially available integrated circuit controller is analyzed, it can be convenient and very useful to utilize a macromodel for this controller. These controllers perform the function of pulse width modulation control and are sometimes quite complex when all the ancillary functions, such as undervoltage lockout, dead band control, bias voltage supplies, soft start, and various current-limiting schemes, are included.

These controllers have two basic categories of macros. One will model the circuit functions explicitly, including the actual pulse width modulator (PWM) switching action. These are helpful when one is looking at the actual real-time SMPS switching transient performance. The second type of macro models the circuit-averaged PWM function and does not simulate the actual PWM switching action. This one is compatible with the types of analysis presented in this book and thus is considered in this appendix and in Chapter 6. If possible, a controller macromodel available from the vendor or some other source may be used for analysis. If not, it will be beneficial to develop one from the vendor's data sheet with the possible use of available test data. This section will develop two representative current mode circuit-averaged macromodel controllers for use in the SMPS analysis examples presented in Chapter 6: the UC1844 and UC1825A current mode PWM controllers.

B.1 UC 1844 Circuit-Averaging Macromodel Development

The manufacturer's specification for the commercially available UC1842/3/4/5 series of current mode PWM controllers is reproduced at the end of this appendix. This controller has been in use for several years and may be considered somewhat generic. It is of medium complexity and will serve for purposes of illustration. The block diagram shown in the manufacturer's specification notes that these controllers have the following general characteristics:

- Internal 5-V bias power supply
- Internal 2.5-V precision reference voltage
- Input undervoltage lockout circuit
- Oscillator
- General-purpose op amp for voltage feedback error amplifier
- 1.0-V peak current sense comparator
- Single-ended output

The UC1844 version of this series has been selected for this development example, but the other versions could be created by simple modifications to the macromodel being developed. The differences in these versions are the maximum allowed duty ratio and the UVLO trip levels. The internal individual macromodels will now be developed for the functions listed earlier and then assimilated into one block representing a macromodel of the entire controller. A word of caution: no macromodel exactly emulates the circuit it is modeling. Rather, the goal is to replicate the necessary functions with sufficient fidelity so as to be able to perform the desired analysis of the switching regulator.

The first function to be developed is the input undervoltage lockout. From the manufacturer's specification, the undervoltage lockout circuit has a typical input start up voltage of 16 V and a typical hysteretic drop out voltage of 10 V. (Maximum or minimum values may be used here if a worst-case analysis is desired.) Figure B.1 shows a circuit macro that will generate this function using a simple gain device configured as a voltage comparator with hysteresis. The output of this circuit is 0 V for OFF or 1 V for ON. This unit signal can be used as a multiplier for the bias and reference supplies to switch them on and off effectively. The start-up and operating supply current draw is also included as part of this macro.

The next internal macro is the 5-V bias supply. From the specification sheet, this supply will have a nominal line regulation of 6 mV for an input voltage range of 12 to 25 V or 0.46 mV/V. Nominal load regulation is 6 mV for a load change of 1.0 to 20 mA or an equivalent output impedance of 0.32 Ω. Temperature stability is nominally 0.2 mV/°C. The 2.5-V reference for the error amplifier is simply a 2:1 divider from the 5-V bias. Figure B.2 shows a macro for this circuit. An output short-circuit current limit of 100 mA is specified; however, for simplicity, this feature is not included. It may be added if desired.

The error amplifier macro is shown in Figure B.3. Although, for the sake of simplicity, this amplifier macro does not mirror the actual circuit exactly, the following characteristics are felt to be adequate for almost all cases:

- DC gain (avol): 90 dB
- Unity gain bandwidth: 1 MHz

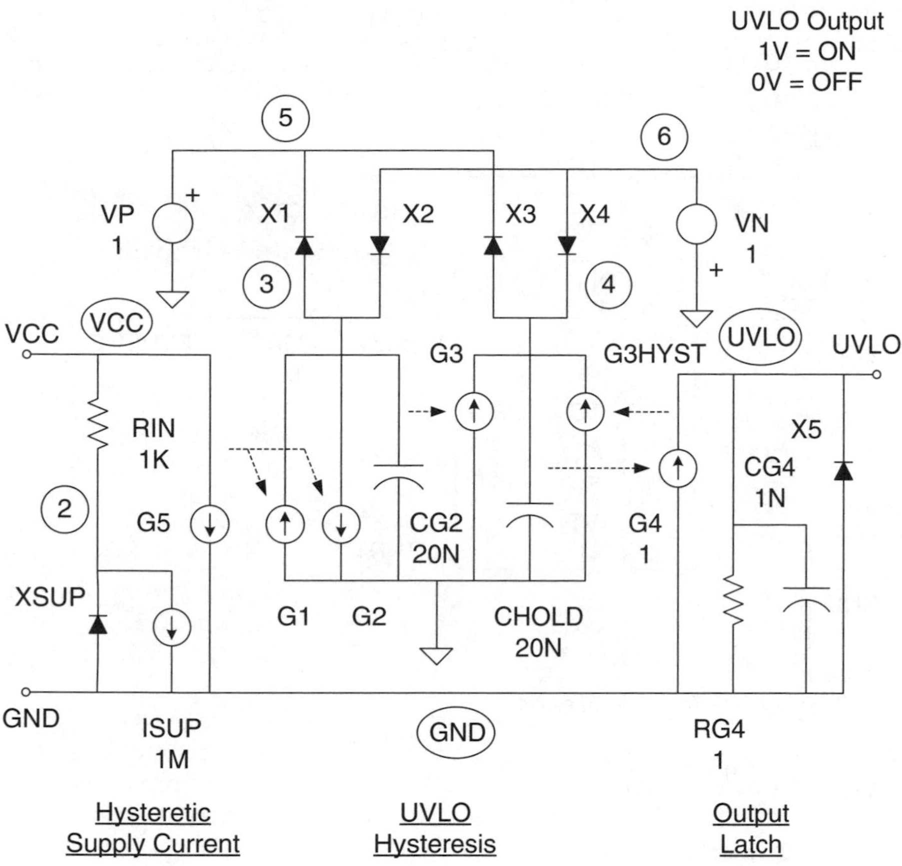

FIGURE B.1
UC1844 undervoltage lockout (UVLO) macro.

- PSSR (DC) 70 dB
- VOH 5.6 V
- VOL 0.6 V
- IOL 1.3 mA
- IOH 1.5 mA

The duty ratio generation macro, here called the duty ratio calculator (DRC), is shown in Figure B.4. The duty ratio, *d*, is determined from Equation 4.6 and Equation 4.7 in Chapter 4 for continuous and discontinuous modes, respectively:

$$d = \frac{\frac{v_C}{R_f} - i_{L2}}{\frac{T_S}{2L_2}(2m_C L_2 + v_g - v)} \quad \text{(continuous mode)} \qquad (4.6)$$

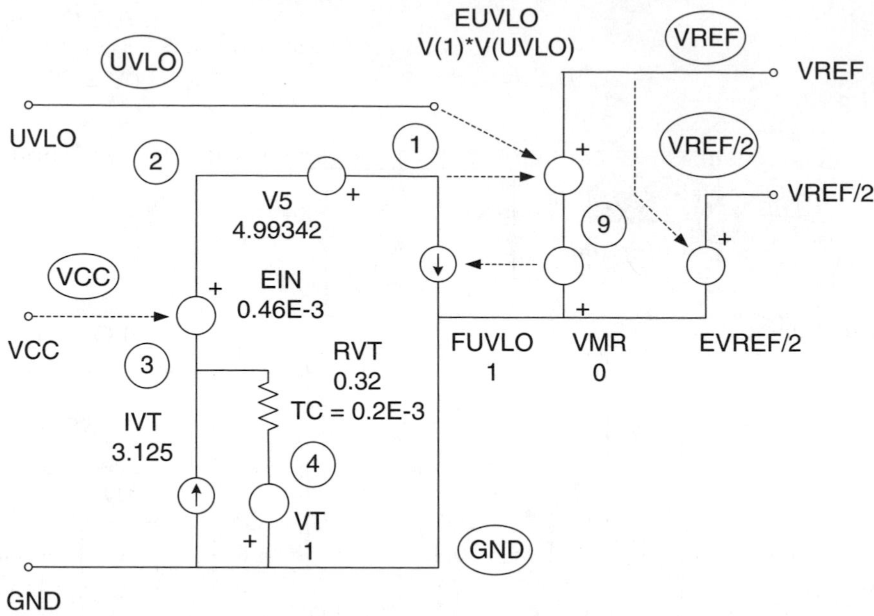

FIGURE B.2
UC1844 voltage reference (VREF) macro.

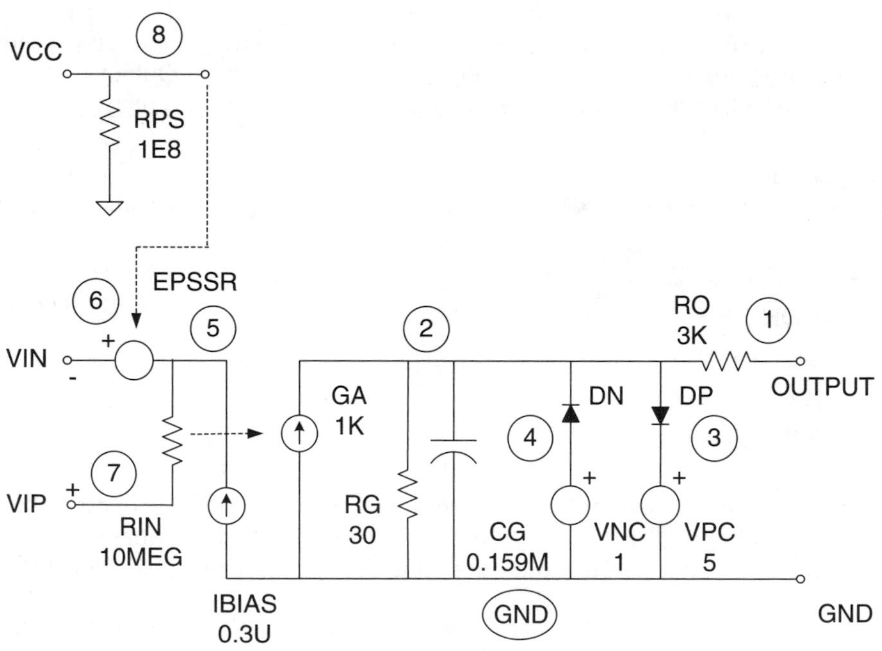

FIGURE B.3
UC1844 ERROR AMP macro.

$$d = \frac{\frac{v_C}{R_f}}{\frac{T_S}{L_2}(m_C L_2 + v_g - v)} \quad \text{(discontinuous mode)} \tag{4.7}$$

These equations are rearranged for ease of implementation in a computer simulation equation line:

$$d = \frac{v_C - i_{L2} R_f}{\frac{T_S R_f}{2L_2}(2m_C L_2 + v_g - v)} \quad \text{(continuous mode)} \tag{6.1}$$

$$d = \frac{v_C}{\frac{T_S R_f}{L_2}(m_C L_2 + v_g - v)} \quad \text{(discontinuous mode)} \tag{6.2}$$

The duty ratios calculations for the continuous and discontinuous modes are represented in Figure B.4 by GCM and GDCM, respectively. The maximum allowable duty ratio limit is imposed by IDRMAX. These three current generators are then configured in a stacked arrangement so that only the minimum of the three currents is actually monitored in the current-measuring source, VMD. The magnitude of this minimum current is numerically equal to the desired duty ratio, *d*, and is eventually converted to the *d* that provides pulse width modulation control of the power converter output voltage.

The complete PWM controller macro is shown in Figure B.5. Note that in the actual controller circuit, certain parameters, such as input voltage (v_g); output voltage (v); inductance, L; slope compensation, m_C; and scale factor, R_f, are not an *explicit* input to the actual controller block. Their *implicit* effects, however, must be added as an input to the macro controller because the sensed averaged current in the macro does not contain them. The sensed current in the actual physical circuit does contain them. These effects are necessary for producing the correct duty ratio, *d*, per Equation 4.6 and Equation 4.7.

This macro will be used to analyze some SMPS examples in Chapter 6. A circuit analysis netlist for the macro of this UC1844 is shown next. Note the *comment* entries. They define and indicate that the equations on the line just below them need to have the appropriate values of max duty ratio (*IDRMAX*), *L*, *RF*, *TS*, and *MC* inserted there for the reasons indicated in the previous paragraph. The ones shown here have representative placeholder values inserted for (*IDRMAX*), *L*, *RF*, *TS*, and *MC*. These values in the equations for GCM, GDCM, and IDRMAX must be modified to the actual ones used for a particular design.

Also, the line containing VOVRD may need to be inserted when it is desired to override the UVLO circuit. With the hysteretic latch in the UVLO

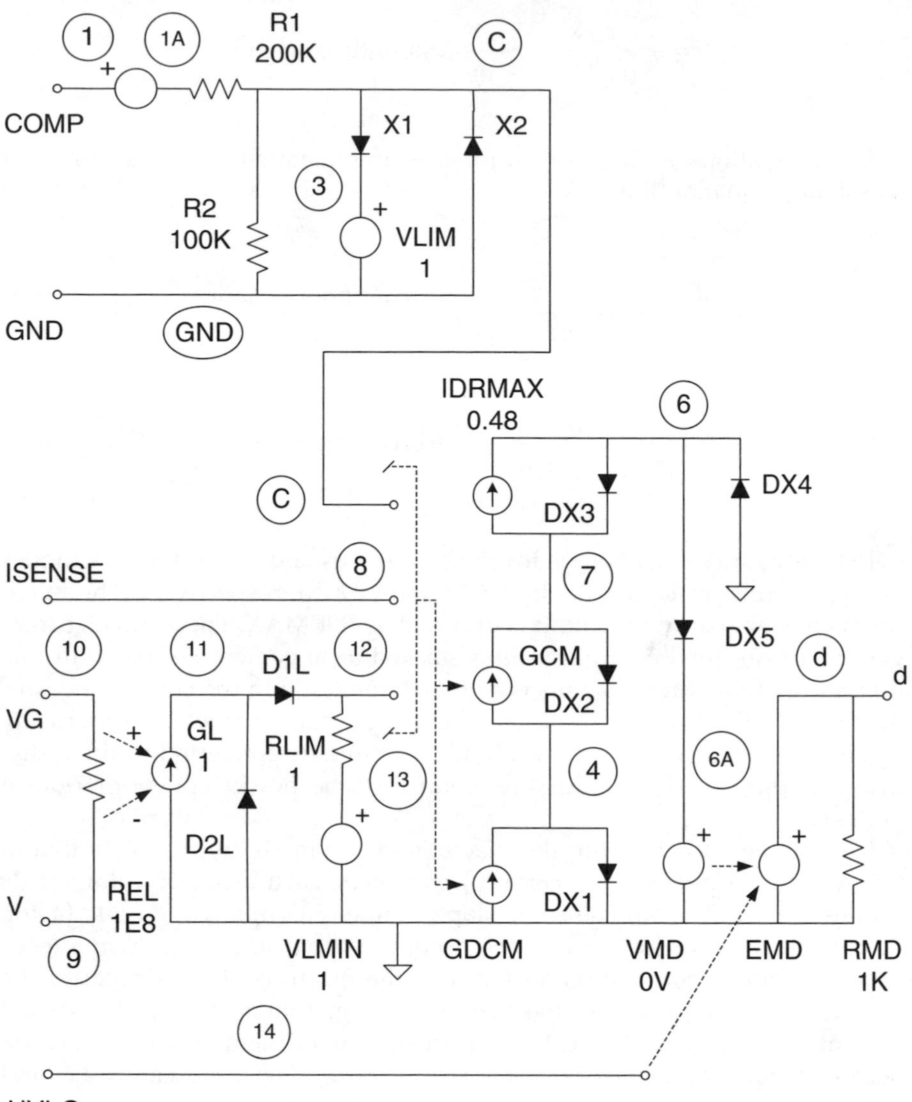

FIGURE B.4
UC1844 duty ratio calculator macro.

macro, it may sometimes be required to prevent nonconvergence when seeking a BIAS solution for an AC analysis. The start and stop UVLO trip levels are set up in lines G1 and G2, respectively. Here the arbitrary maximum start voltage of 17 V and the typical stop level of 10 V are used. G1 and G2 are in actuality high-gain devices, so the full on and off characteristics are achieved with only a slight control voltage change above the start trip level

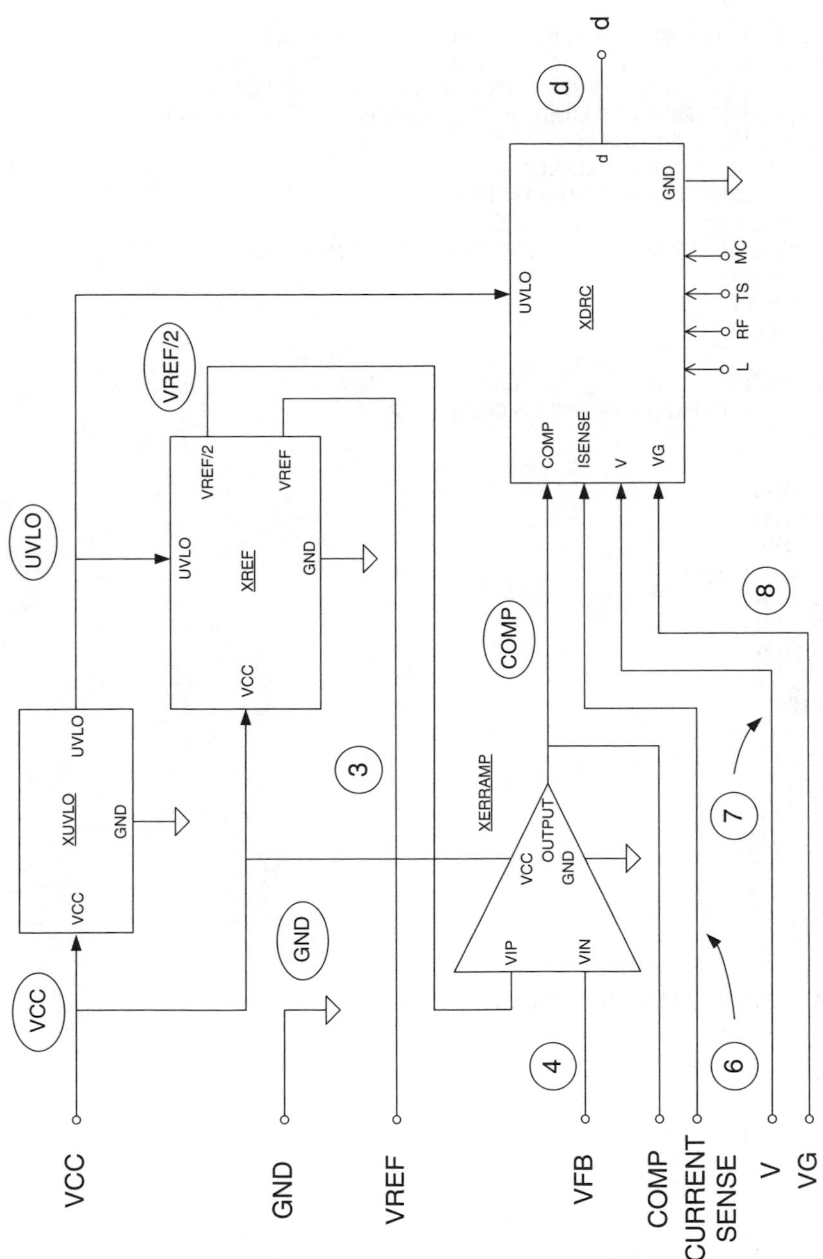

FIGURE B.5
UC1844 PWM controller macro.

or below the stop trip level. G1 and G2 values for UVLO start and stop values are indicated.

```
*
**UC 1844 PWM CONTROLLER CIRCUIT AVERAGING MACRO
.SUBCKT UC1844 VCC      3     4    COMP 6        7 8   d GND
*                  VCC    VREF VFB COMP ISENSE V VG d GND
**NOTE: ALL SIGNAL INPUTS MUST BE REFERENCED TO THE UC1844 GND
XUVLO VCC UVLO GND UVLO
XREF UVLO VCC 3 VREF/2 GND REF
XERRAMP 4 VREF/2 VCC COMP GND ERRAMP
XDRC COMP d 6 7A 8A UVLO GND DRC
**DEPENDENT GENERATORS FOR GROUND ISOLATION IF DESIRED
EVG 8A 0 8 GND 1
EV 7A 0 7 GND 1
.ENDS UC1844
*
.SUBCKT DRC 1         d 8        9  10  14      GND
*               COMP   d ISENSE V VG UVLO GND
VBK 1 1A DC 2
R1 1A C 200K
R2 C GND 100K
X1 C 3 DIDEAL
X2 GND C DIDEAL
VLIM 3 GND DC 1
RCONV1 6 0 1G
RCONV2 7 0 1G
RCONV3 4 0 1G
*
REL 10 9 1E8
GL 0 11 10 9 1
D1L 11 12 DX
D2L 0 11 DX
RLMIN 12 13 1
VLMIN 13 0 1U
*
DX1 4 0 DX
DX2 7 4 DX
DX3 6 7 DX
*
**IDRMAX VALUE = MAX LIMITED DUTY RATIO
IDRMAX 7 6 DC .48
DX4 0 6 DX
DX5 6 6A DX
VMD 6A 0
EMD d 0 VALUE = {I(VMD)*V(14)+1P}
RMD d 0 1
.MODEL DX D IS=1E-12
*
*L = 40U
*RF = 0.3
*TS = 10U
*MC = .3E6
*
```

```
**GCM 4 7 VALUE = {(V(C)-RF*V(8))/((TS*RF)/(2*L))*(2*MC*L+V(12)))}
GCM 4 7 VALUE = {LIMIT((V(C)-0.3*V(8))/((10U*0.3)/(2*40U)*(2*.3E6*40U+V(12))),0,1)}
*
**GDCM 0 4 VALUE = {V(C)/((TS*RF/L)*(MC*L+V(12))}
GDCM 0 4 VALUE = {LIMIT(V(C)/((10U*0.3/40U)*(.3E6*40U+V(12))),0,1)}
*
.ENDS DRC
*
.SUBCKT UVLO VCC UVLO   GND
*                   VCC UVLO   GND
RIN VCC 2 1K
ISUP 2 GND 1M
XSUP GND 2 DIDEAL
G5 VCC GND VALUE = {.01*V(UVLO,GND)}
**  G1 AND G2 SET UVLO START AND STOP LEVELS RESPECTIVELY
**  START = 17
**  STOP    = 10
G1 0 3 TABLE {V(VCC,GND)} = (0,0) (17,0) (18,20)
G2 3 0 TABLE {V(VCC,GND)} = (9,20) (10,0) (17,0)
CG2 3 0 20N
R1 3 0 1E3
G3 0 4 3 0 1
G3HYST 0 4 VALUE = {-100U+200U*V(UVLO,GND)}
CHOLD 4 0 20N
G4 GND UVLO 4 GND 1
**VOVRD TO BE INSERTED FOR UVLO OVERRIDE
*VOVRD UVLO GND DC 1
RG4 UVLO GND 1
CG4 UVLO GND 1N
X5 GND UVLO DIDEAL
X1 3 5 DIDEAL
X2 6 3 DIDEAL
X3 4 5 DIDEAL
X4 6 4 DIDEAL
VP 5 0 DC 1
VN 0 6 DC 1
.ENDS UVLO
*
*5.0 VOLT REFERENCE MACRO
.SUBCKT REF UVLO VCC   VREF    VREF/2       GND
*                UVLO VCC   VREF    VREF/2       GND
*
EIN 2 3 VCC GND .46E-3
RVT 3 4 .32 TC=0.2E-3
IVT GND 3 DC 3.125
VT GND 4 DC 1
V5 1 2 DC 4.99342
FUVLO 1 GND VMR 1
EUVLO VREF 9 VALUE = {V(1,GND)*V(UVLO,GND)}
VMR GND 9
EVREF/2 VREF/2 GND VREF GND .5
.ENDS REF
*
*ERROR AMP MACRO
```

```
.SUBCKT ERRAMP 6    7    8    1              GND
*                  VIN VIP VCC OUTPUT GND
RIN 5 7 10MEG
EPSRR 6 5 VALUE = {(V(8)-15)*3E-4}
IBIAS 5 GND .3U
GA GND 2 7 5 1K
RG 2 GND 30
CG 2 GND 0.159M
VNC 4 GND DC 1
DN 4 2 D1
VPC 3 GND DC 5
DP 2 3 D1
RO 1 2 3K
RPS 8 0 1E8
.ENDS ERRAMP
.MODEL D1 D IS=1E-9
*
*****************************************************
*
.SUBCKT DIDEAL 1 2
VAS 1 3 DC -1u
D1 3 2 D
D2 3 4 D
D3 4 2 D1
FAS 4 2 VAS 1
.MODEL D D IS=1E-6 EG=0 XTI=-4
.MODEL D1 D EG=0 XTI=0
CC 1 2 .1P
.ENDS DIDEAL
*
```

B.2 UC 1825A Circuit-Averaging Macromodel Development

The manufacturer's specification for the commercially available UC1823A,B/1825A,B series of current mode PWM controllers is reproduced at the end of this appendix. This controller has been in use for several years and has definitely attained the status of industry standard. It is more complex than the previously examined UC1844 and contains almost all of the general desirable features needed when an SMPS is to be designed. The block diagram shown in the manufacturer's specification notes that these controllers have the following general characteristics:

- Internal 5.1-V precision reference and bias power supply
- Input undervoltage lockout (UVLO) circuit
- Soft start circuit
- Oscillator
- General-purpose op amp for voltage feedback error amplifier

- Current limit (cycle by cycle with no soft start reset)
- Current limit (with soft start reset hiccup mode)

The UC1825A version of this device has been selected for this example; however, the other versions could be created by simple modifications of the UVLO limits and the maximum duty ratio stipulation to the macromodel being developed. Now the internal individual macromodels will now be developed for the functions listed earlier and then assimilated into one block representing a macromodel of the entire controller. Again, a word of caution: no macromodel exactly emulates the circuit it is modeling. Rather, the goal is to replicate the necessary functions with sufficient fidelity so as to be able to perform the desired analysis of the switching regulator. Because an averaging model is being developed, the oscillator and explicit pulse by pulse current limiting features are, of course, not included in this model.

The first function to be developed is the input undervoltage lockout. From the manufacturer's specification, the undervoltage lockout circuit has a typical input start up voltage of 9.2 V and a typical hysteretic dropout or stop voltage of 8.4 V. (Maximum or minimum values may be used here if a worst-case analysis is desired.) Figure B.6 shows a circuit macro that will generate this function using a simple gain device configured as a voltage comparator with hysteresis. The output of this circuit is 0 V for OFF or 1 V for ON. This unit signal can be used as a multiplier for the bias and reference supplies to switch them on and off effectively. It is also used as an on–off control signal in the soft start, overcurrent, voltage reference bias supply, and duty ratio control circuits. The start-up and operating supply current draw is also included as part of this macro.

The next internal macro is the 5.1-V reference bias supply. From the specification sheet, this supply will have a nominal line regulation of 2 mV for an input voltage range of 12 to 20 V or 0.25 mV/V. Nominal load regulation is 5 mV for a load change of 1.0 to 10 mA or an equivalent output impedance of 0.56 Ω. Temperature stability is nominally 0.2 mV/°C. Figure B.7 shows a macro for this circuit. An output short-circuit current limit of 60 mA is specified; however, for simplicity, this feature is not included. A circuit modification to provide this limit may be added if desired.

The error amplifier macro is shown in Figure B.8. Although, for the sake of simplicity, this amplifier macro does not mirror the actual circuit exactly, the following characteristics are felt to be adequate for almost all cases:

- DC gain (avol): 95 dB
- Unity gain bandwidth: 12 mHz
- PSSR (DC) 95 dB
- VOH 4.7 V
- VOL 0.6 V

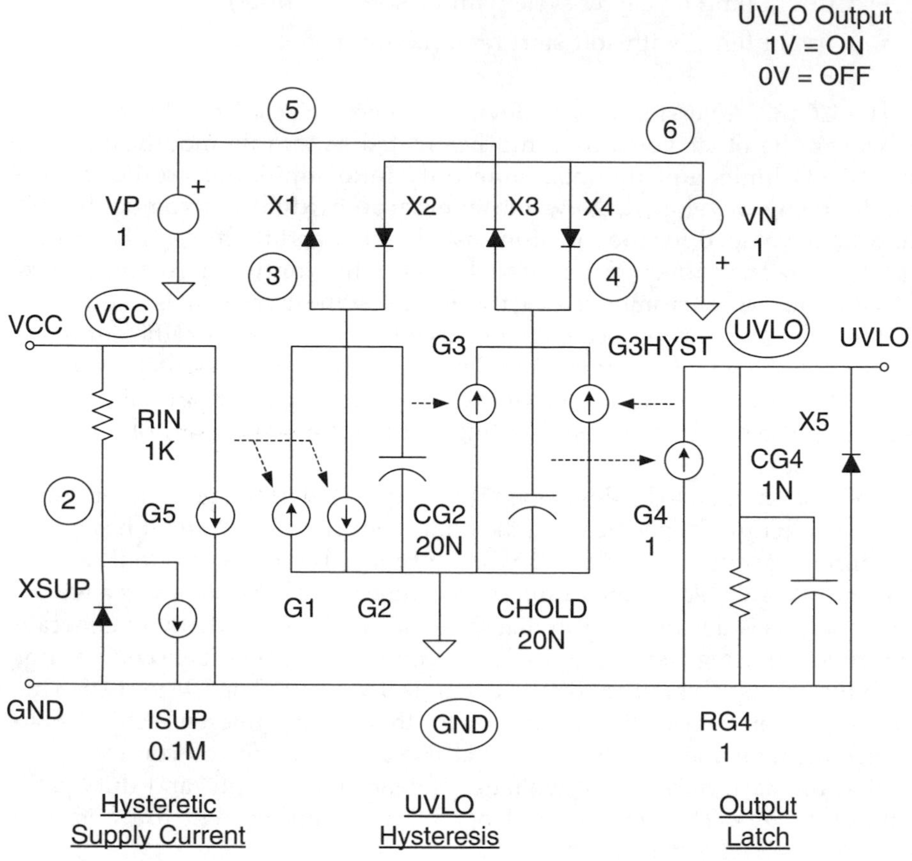

FIGURE B.6
UC1825A undervoltage lockout (UVLO) macro.

- IOL ≤ 1.0 mA
- IOH ≤ 1.0 mA

The soft start and overcurrent macro is shown in Figure B.9. At first glance, the macro may seem a little daunting, but with a good understanding of the actual soft start and overcurrent operation of the physical UC1825A from the manufacturer's data sheet, the macro's similarity is readily understood. As in the actual unit, the soft start timing capacitor is connected to the SS pin. This circuit functions by basically pulling the E/A OUT (or COMP, as it is commonly known) low whenever the UVLO indicates that a pre-start-up condition exists or if an overcurrent condition has been sensed. With this condition, a hiccup attempted recovery mode is established by rapidly discharging the soft start capacitor and then slowly recharging it during the attempted soft start-up mode. With the required sensing of peak currents in the converter for an overcurrent shutdown, a peak current calculator is

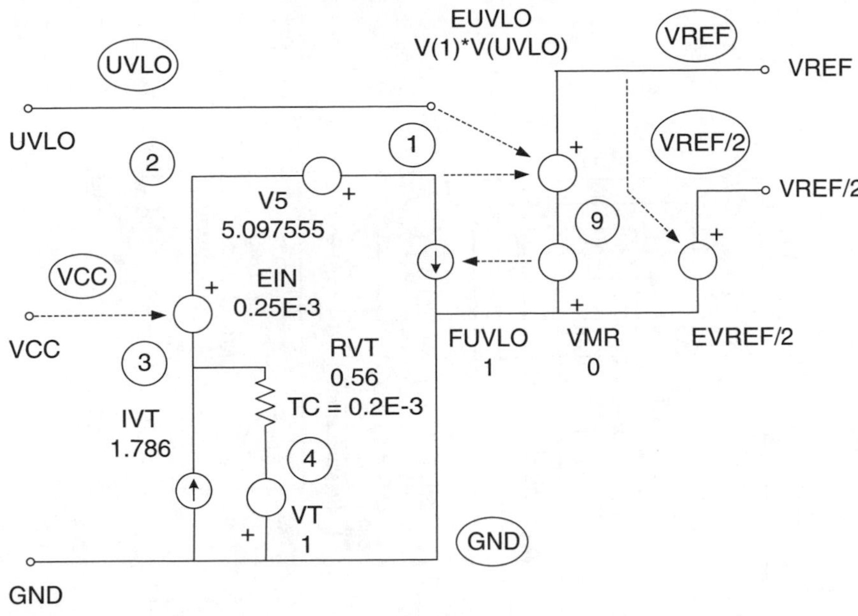

FIGURE B.7
UC1825A voltage reference (VREF) macro.

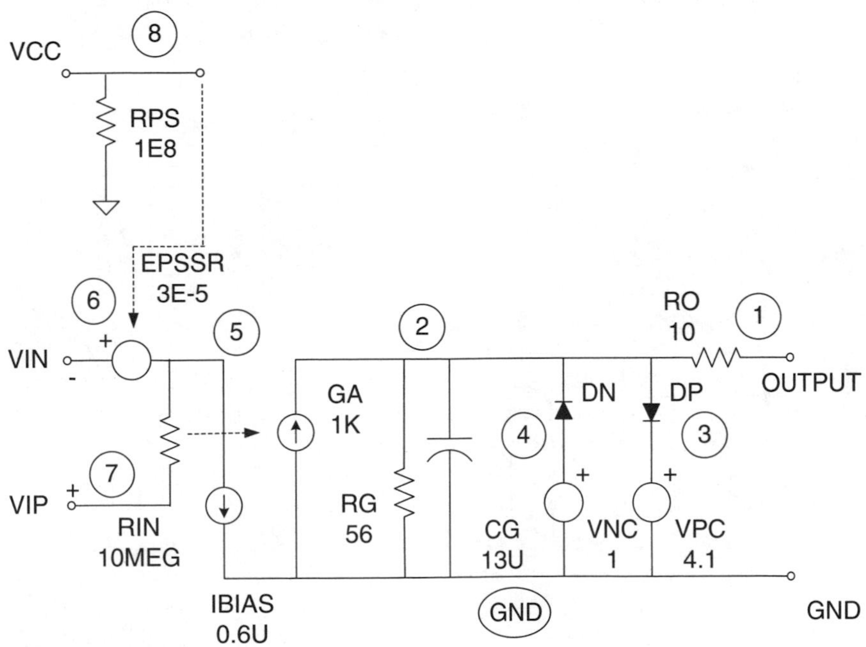

FIGURE B.8
UC1825 ERROR AMP macro.

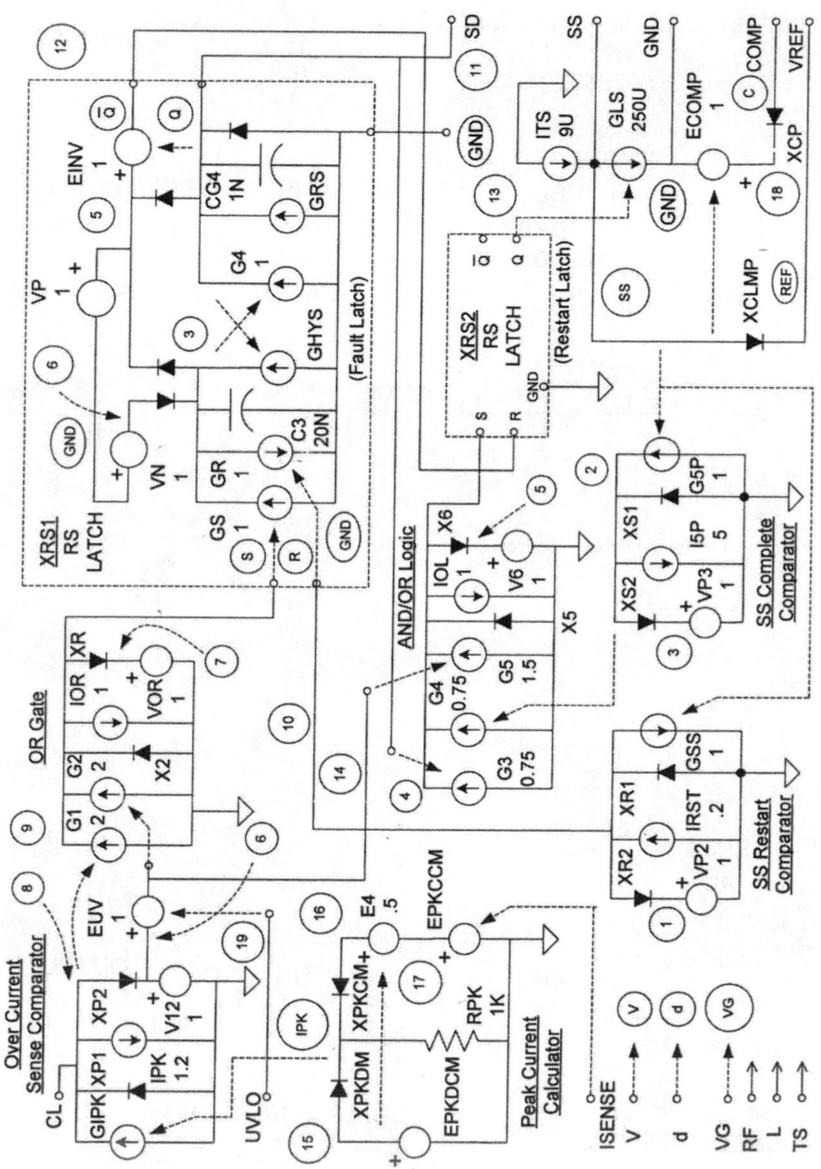

FIGURE B.9
UC1825A soft start and overcurrent macro.

necessary because the circuit-averaging macro only generates averaged currents.

The peak discontinuous mode current for the buck topology is expressed by Equation 3.2 and, for the boost and buck–boost topologies, by Equation 3.5. The continuous mode peak current is simply equal to the averaged or DC value of current plus one-half of the value expressed by the discontinuous mode peak value calculation. The peak current calculator circuit in Figure B.9 continuously calculates the continuous and discontinuous mode values of peak current; the larger of the two appears on node IPK. This is the correct value of peak current that the actual switching converter will produce and is valid for these calculations.

When the 1.2-V overcurrent threshold is sensed by the overcurrent sense comparator, the fault latch is subsequently tripped; this eventually turns on generator GLS (the 250-μA source). This pulls the SS and COMP pins low, shutting down the converter. A shutdown signal, SD, is also fed from the fault latch over to the DRC to reduce the duty ratio *d* of the converter immediately to zero. When the SS pin voltage drops below the 0.2-V threshold of the SS restart comparator, the fault latch and the restart latch are reset. This turns current sink GLS off, allowing the soft start capacitor to recharge from the 9-μA source, ITS. This allows the COMP pin to increase slowly, thus implementing a soft restart.

In the actual controller, a 1.0-V current limit comparator is used for pulse-by-pulse current limiting. This is obviously not possible to implement explicitly with the circuit-averaged simulation. For the sake of simplicity, this current limit comparator is not implemented; however, the output of the overcurrent sense comparator (threshold equal to 1.2 V) is used basically to implement the same function of rapid shutdown. Its output, CL, is fed directly to the duty ratio calculator to reduce the duty ratio *d* immediately to zero. This may be considered a practical approximation to the operation of the actual controller.

In many practical designs, noise or overshoot on the current sense circuit (ILIM input) tends to make these two trip levels of 1.0 and 1.2 dynamically indistinguishable and the shutdown and soft restart cycle occurs without actually noting any sustained pulse-by-pulse current limiting. (If it is deemed necessary to implement the 1.0 current limit rapid shutdown, instructions on how to accomplish this are shown in the macro netlist at the end of this section. This involves removing source G1 from the soft start macro, thereby disabling the overcurrent latch, and then reducing the threshold sense of source IPK from 1.2 to 1.0 amps.) In some designs, pulse-by-pulse current limiting may be provided by a saturated high value of the E/A OUT signal limiting the peak current before the 1.0 and 1.2 comparator thresholds are reached. Also, the ILIM input may be connected to ground, thus eliminating any explicit pulse-by-pulse current limiting or overcurrent hiccup mode of operation if desired.

At this point, a note about the implementation of the digital circuit functions of the soft start macro model of Figure B.9 must be made. Nowadays, most circuit simulators provide for mixed mode simulations using analog and digital components. The mixed mode is not used here, however. Because the digital logic functions in the controller are not very complex, it was decided to implement them with basic analog circuit components that all circuit simulators possess. This was done to provide a most basic and alternative approach to the model. It is a relatively easy transformation of the macro for anyone who might be interested in modifying it for use with a mixed mode simulator.

The duty ratio generation macro, duty ratio calculator (DRC), is shown in Figure B.10 and uses the same basic algorithms as the one used in the UC1844 macro derived earlier. A significant difference is that the peak current sense unit control signal, *CL*, from the soft start, overcurrent macro is fed over to reduce the duty ratio, *d*, to zero, thus shutting off the converter. The SD signal from the soft start fault latch is also fed over to reduce the converter duty ration to zero.

The complete UC1825A PWM controller macro is shown in Figure B.11. As was the case for the previous UC1844 macro, certain parameters, such as input voltage (v_g); output voltage (v); inductance, L; slope compensation, m_C; and scale factor, R_f, are not an explicit input to the actual controller block. Their implicit effects, however, must be added as an input to the macro controller because the sensed averaged current in this macro does not contain them. The sensed current in the actual physical circuit does contain them. These effects are necessary for producing the correct duty ratio, *d*, per Equation 4.6 and Equation 4.7.

This macro will also be used to analyze an SMPS example in Chapter 6. A circuit analysis netlist for the macro of this UC1825A is shown next. Note the *comment* entries. In some cases, they define and indicate that the equations on the line just below them need to have the appropriate values of max duty ratio (*IDRMAX*), *L*, *RF*, *TS*, *MC*, and *RLIM* inserted there for the reasons indicated in the previous paragraph. The ones shown here have representative placeholder values inserted for (*IDRMAX*), *L*, *RF*, *TS*, *MC*, and *RLIM*. These values in the equations for GCM, GDCM, IDRMAX, and GIPK must be modified to the actual ones used for a particular design.

When a BIAS solution, possibly for an AC analysis, is sought, it may sometimes be necessary to override the UVLO circuit. With the hysteretic latch in the UVLO macro, nonconvergence problems can arise when seeking a BIAS solution. To accomplish this, a unit voltage source, VOVRD, of 1 V is inserted at the node where the normal UVLO signal would exist. The line in the netlist containing VOVRD indicates the action required when it is desired to override the UVLO circuit.

Additionally, the soft start subcircuit, XSOFTSTART, will in most cases need to be removed from the netlist when BIAS solutions are sought. With the hysteretic nature of the R–S latches in this soft start macro, this removal will usually be required to prevent nonconvergence when a BIAS solution for an AC analysis is sought. The nonconvergent problems caused by the bistate nature of these latches may be prevented without actually removing

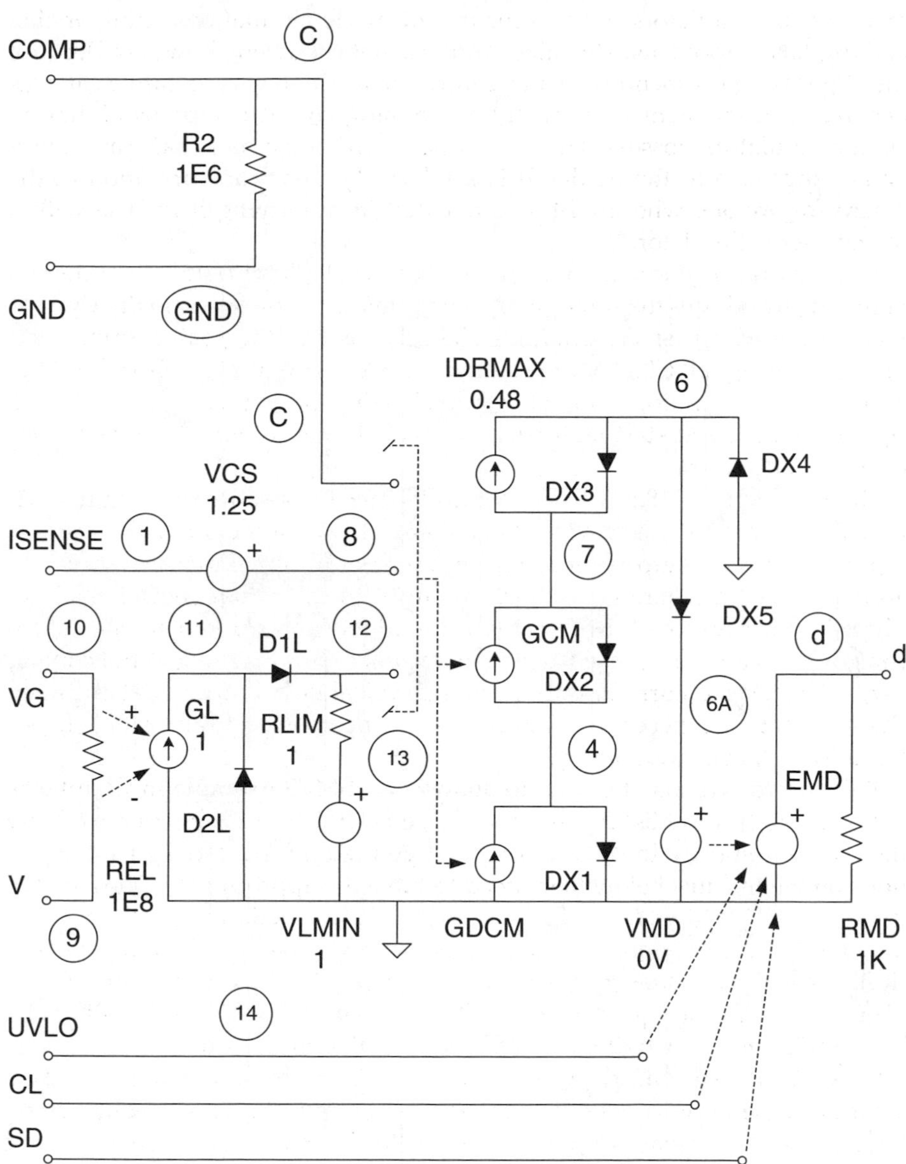

FIGURE B.10
UC1825A duty ratio calculator macro.

them from the circuit. One way might be by possibly using the NODESET command along with some limiting functions or components. The most practical way, however, is simply to remove them from the circuit as is done here. After all, soft start and UVLO are functions considered only when transient analyses are conducted and not when AC analyses are performed.

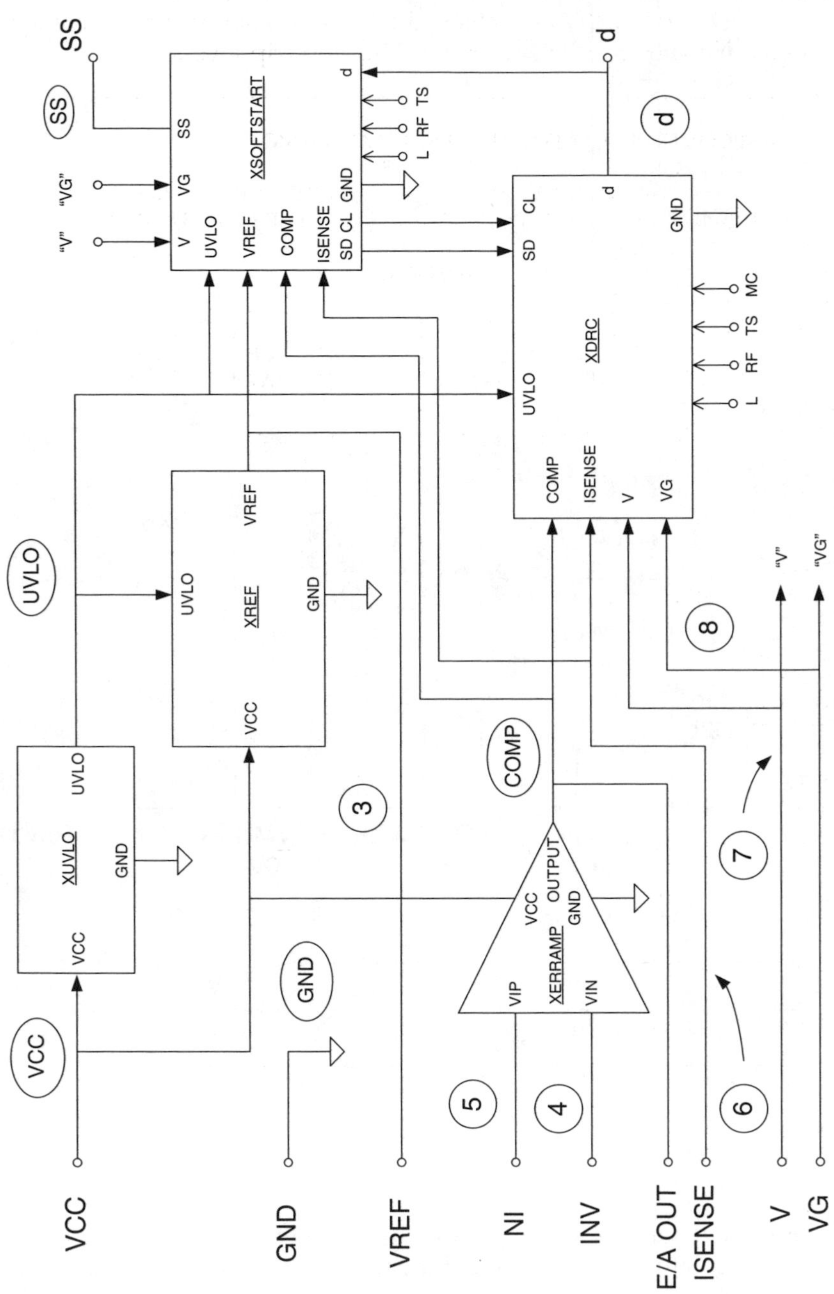

FIGURE B.11
UC1825A PWM controller macro.

The start and stop UVLO trip levels are set up in lines G1 and G2, respectively. Here an arbitrary nominal start voltage of 9.2 V and the typical stop level of 8.4 V are used. G1 and G2 are in actuality high-gain devices, so the full on and off characteristics are achieved with only a slight control voltage change above the start trip level or below the stop trip level.

```
*
**UC1825A PWM CONTROLLER CIRCUIT AVERAGING MACRO
.SUBCKT UC1825A VCC 3    4   5  COMP    6    SS 7 8  D GND
*                        VCC VREF INV NI E/A_OUT RAMP SS V VG d GND
**NOTE: ALL SIGNAL INPUTS MUST BE REFERENCED TO THE UC1825A GND
XUVLO VCC UVLO GND UVLO
XREF UVLO VCC 3 GND REF
XERRAMP 4 5 VCC COMP GND ERRAMP
XDRC COMP d 6 7A 8A UVLO SD CL GND DRC
** NOTE: SUBCKT "XSOFTSTART" SHOULD BE REMOVED FOR BIAS SOLUTIONS
XSOFTSTART 6 7A 8A 3 UVLO d COMP SS SD CL GND SOFTSTART
RSSCKT SS GND 1E9
**DEPENDENT GENERATORS FOR GROUND ISOLATION IF DESIRED
EVG 8A 0 8 GND 1
EV 7A 0 7 GND 1
.ENDS UC1825A
*
.SUBCKT DRC C      d 1      9 10 14    SD CL GND
*        COMP   d ISENSE V VG UVLO  SD CL GND
RCONV1 6 0 1G
RCONV2 7 0 1G
RCONV3 4 0 1G
R2 C GND 1E6
VCS 8 1 DC 1.25
*
REL 10 9 1E8
GL 0 11 10 9 1
D1L 11 12 DX
D2L 0 11 DX
RLMIN 12 13 1
VLMIN 13 0 1U
*
DX1 4 0 DX
DX2 7 4 DX
DX3 6 7 DX
*
**IDRMAX VALUE = MAX LIMITED DUTY RATIO
IDRMAX 7 6 DC .48
DX4 0 6 DX
DX5 6 6A DX
.MODEL DX D IS=1E-12
VMD 6A 0
EMD d 0 VALUE = {LIMIT(I(VMD)*(1-V(CL))*V(14)*(1-V(SD)),1U,.99)}
RMD d 0 1
*
**PULL DOWNS THAT ARE REQUIRED WHEN XSOFTSTART IS REMOVED FROM CIRCUIT
RCL CL 0 1E3
IRCL 0 CL DC 1P
```

```
RSD SD 0 1E3
IRSD 0 SD DC 1P
*
*L = 50U
*RF = .5
*TS = 10U
*MC = 0
*
**GCM 4 7 VALUE = {(V(C)-RF*V(8))/((TS*RF)/(2*L))*(2*MC*L+V(12)))}
GCM 4 7 VALUE = {LIMIT((V(C)-.5*V(8))/((10U*.5)/(2*50U)*(2*0*50U+V(12))),0,1)}
*
**GDCM 0 4 VALUE = {V(C)/((TS*RF/L)*(MC*L+V(12))}
GDCM 0 4 VALUE = {LIMIT(V(C)/((10U*.5/50U)*(0*50U+V(12))),0,1)}
*
.ENDS DRC
*
.SUBCKT UVLO VCC UVLO   GND
*           VCC UVLO   GND
RIN VCC 2 1K
ISUP 2 GND .1M
XSUP GND 2 DIDEAL
G5 VCC GND VALUE = {.028*V(UVLO,GND)}
**G1 AND G2 SET UVLO START AND STOP LEVELS RESPECTIVELY
**START = 9.2
**STOP  = 8.4
G1 0 3 TABLE {V(VCC,GND)} = (0,0) (9.2,0) (10.2,30)
G2 3 0 TABLE {V(VCC,GND)} = (7.4,30) (8.4,0) (9.2,0)
CG2 3 0 20N
R1 3 0 1E3
G3 0 4 3 0 1
G3HYST 0 4 VALUE = {-100U+200U*V(UVLO,GND)}
CHOLD 4 0 20N
G4 GND UVLO 4 GND 1
**VOVRD TO BE INSERTED FOR UVLO OVERRIDE
*VOVRD UVLO GND DC 1
RG4 UVLO GND 1
CG4 UVLO GND 1N
X5 GND UVLO DIDEAL
X1 3 5 DIDEAL
X2 6 3 DIDEAL
X3 4 5 DIDEAL
X4 6 4 DIDEAL
VP 5 0 DC 1
VN 0 6 DC 1
.ENDS UVLO
*
*5.0 VOLT REFERENCE MACRO
.SUBCKT REF UVLO VCC VREF GND
*          UVLO VCC VREF  GND
*
EIN 2 3 VCC GND .25E-3
RVT 3 4 .56 TC=0.2E-3
IVT GND 3 DC 1.786
VT GND 4 DC 1
V5 1 2 DC 5.097555
```

```
FUVLO 1 GND VMR 1
EUVLO VREF 9 VALUE = {V(1,GND)*V(UVLO,GND)}
VMR GND 9
.ENDS REF
*
*ERROR AMP MACRO
.SUBCKT ERRAMP  6   7   8   1      GND
*            VIN VIP VCC OUTPUT GND
RIN 6 7 10MEG
EPSRR 6 5 VALUE = {(V(8)-12)*3E-5}
IBIAS 5 GND .6U
GA GND 2 7 6 1K
RG 2 GND 56
CG 2 GND 13U
VNC 4 GND DC 1.1
DN 4 2 D1
VPC 3 GND DC 4.1
DP 2 3 D1
RO 1 2 10
RPS 8 0 1E8
.ENDS ERRAMP
.MODEL D1 D IS=1E-9
*
.SUBCKT SOFTSTART ISENSE V VG VREF UVLO d C SS 11 8  GND
*              ISENSE V VG VREF UVLO d C SS SD CL GND
*
** L = 50U
** RF = .5
** TS = 10U
*
**PEAK CURRENT CALCULATOR
**EPKDCM 15 0 VALUE = {V(d)*(V(VG)-V(V))*(TS/L)*RF}
**NOTE: V(V) IS ZERO FOR ALL TOPOLOGIES EXCEPT THE BUCK
**     THE BUCK EQUATION IS SHOWN HERE FOR THE GENERAL CASE
EPKDCM 15 0 VALUE = {V(d)*(V(VG)-V(V))*(10U/50U)*.5}
XPKDCM 15 IPK DIDEAL
XPKCCM 16 IPK DIDEAL
E4 16 17 15 0 .5
EPKCCM 17 0 ISENSE 0 1
RPK IPK 0 1E3
*
**OVER CURRENT SENSE COMPARATOR 1.2
** RLIM = 0.5
**GIPK 0 8 VALUE = {V(IPK)*RLIM}
GIPK 0 8 VALUE = {V(IPK)*.5}
XP1 0 8 DIDEAL
IPK 8 0 DC 1.2
XP2 8 6 DIDEAL
V12 6 0 DC 1
EUV 6 19 UVLO GND 1
*
**OR GATE
** NOTE: G1 MAY BE REMOVED HERE TO DISABLE "HICCUP" CURRENT LIMIT MODE.
**          AN EQUIVALENT OF PULSE-BY-PULSE CURRENT LIMITING WILL RESULT.
**          THE VALUE OF IPK ABOVE MAY BE CHANGED FROM 1.2 TO 1.0 TO THE
```

```
**          ACTUAL CURRENT LIMIT COMPARATOR THRESHOLD IF DESIRED.
G1 0 9 8 0 2
G2 0 9 19 0 2
X2 0 9 DIDEAL
IOR 9 0 DC 1
XR 9 7 DIDEAL
VOR 7 0 DC 1
*
**AND/OR LOGIC
G3 0 4 11 0 .75
G4 0 4 2 0 .75
G5 0 4 19 GND 1.5
X5 0 4 DIDEAL
IOL 4 0 DC 1
X6 4 5 DIDEAL
V6 5 0 DC 1
*
**SS RESTART COMPARATOR
GSS 10 0 SS GND 1
XR1 0 10 DIDEAL
IRS2 0 10 DC .2
XR2 10 1 DIDEAL
VP2 1 0 DC 1
*
*SS COMPLETE COMPARATOR
G5P 0 2 SS 0 1
XS1 0 2 DIDEAL
I5P 2 0 DC 5
XS2 2 3 DIDEAL
VP3 3 0 DC 1
*
*SS/COMP
XCP C 18 DIDEAL
ECOMP 18 GND SS GND 1
ITS 0 SS DC 9U
GLS SS GND VALUE = {LIMIT(V(13)*250U,1U,260U)}
XCLPH SS VREF DIDEAL
XCLPL GND SS DIDEAL
.MODEL D D
*
XRS1 10 9 11 12 GND RS-LATCH
XRS2 12 4 13 14 0 RS-LATCH
*
.ENDS SOFTSTART
*
.SUBCKT RS-LATCH R S Q QBAR GND
GR 3 GND R GND 1
GS GND 3 S GND 1
C3 3 GND 20N
G4 GND Q 3 GND 1
CG4 Q GND 20N
GHYS GND 3 VALUE = {–100U+200U*V(Q,GND)}
GRS GND Q VALUE = {.75*(V(R,GND)+V(S,GND))}
X1 GND Q DIDEAL
X2 Q 5 DIDEAL
```

```
X3 6 3 DIDEAL
X4 3 5 DIDEAL
EINV 5 QBAR VALUE = {V(Q)*(1-V(R)*V(S))}
VN GND 6 DC 1
VP 5 GND DC 1
.ENDS RS-LATCH
*
.SUBCKT DIDEAL 1 2
VAS 1 3 DC -1u
D1 3 2 D
D2 3 4 D
D3 4 2 D1
FAS 4 2 VAS 1
.MODEL D D IS=1E-6 EG=0 XTI=-4
.MODEL D1 D EG=0 XTI=0
CC 1 2 .1P
.ENDS DIDEAL
*
```

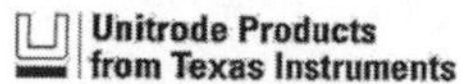

Current Mode PWM Controller

FEATURES

- Optimized for Off-line and DC to DC Converters
- Low Start Up Current (<0.5mA)
- Trimmed Oscillator Discharge Current
- Automatic Feed Forward Compensation
- Pulse-by-Pulse Current Limiting
- Enhanced Load Response Characteristics
- Under-Voltage Lockout With Hysteresis
- Double Pulse Suppression
- High Current Totem Pole Output
- Internally Trimmed Bandgap Reference
- 500kHz Operation
- Low Ro Error Amp

DESCRIPTION

The UC1842A/3A/4A/5A family of control ICs is a pin for pin compatible improved version of the UC3842/3/4/5 family. Providing the necessary features to control current mode switched mode power supplies, this family has the following improved features. Start up current is guaranteed to be less than 0.5mA. Oscillator discharge is trimmed to 8.3mA. During under voltage lockout, the output stage can sink at least 10mA at less than 1.2V for Vcc over 5V.

The difference between members of this family are shown in the table below.

Part #	UVLO On	UVLO Off	Maximum Duty Cycle
UC1842A	16.0V	10.0V	<100%
UC1843A	8.5V	7.9V	<100%
UC1844A	16.0V	10.0V	<50%
UC1845A	8.5V	7.9V	<50%

BLOCK DIAGRAM

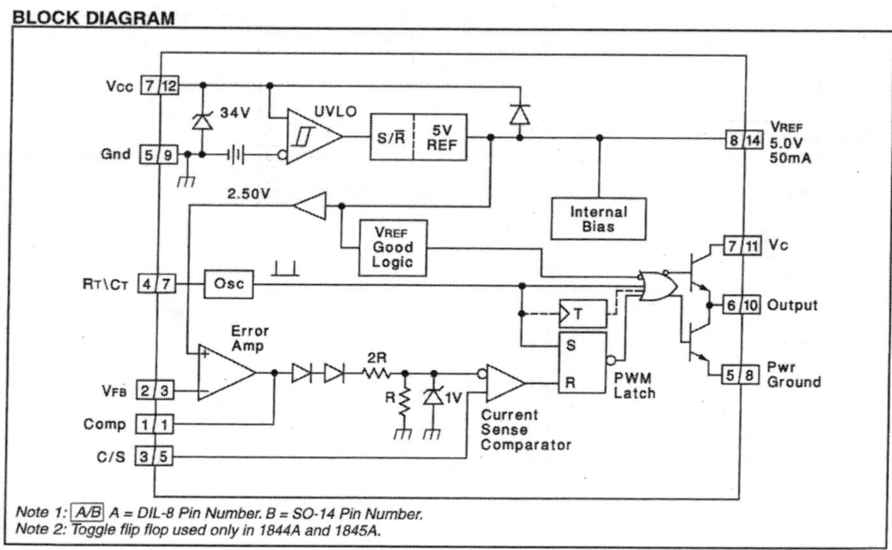

Note 1: A/B A = DIL-8 Pin Number. B = SO-14 Pin Number.
Note 2: Toggle flip flop used only in 1844A and 1845A.

SLUS224A - SEPTEMBER 1994 - REVISED APRIL 2002

CONNECTION DIAGRAMS

ABSOLUTE MAXIMUM RATINGS (Note 1)

Supply Voltage (Low Impedance Source) 30V
Supply Voltage (Icc mA) . . . . Self Limiting
Output Current. . . . . ±1A
Output Energy (Capacitive Load). 5μJ
Analog Inputs (Pins 2, 3). . . . . -0.3V to +6.3V
Error Amp Output Sink Current 10mA
Power Dissipation at $T_A \leq 25°C$ (DIL-8) 1W
Storage Temperature Range. . . . . -65°C to +150°C
Lead Temperature (Soldering, 10 Seconds) . . . . 300°C

Note 1. All voltages are with respect to Ground, Pin 5. Currents are positive into, negative out of the specified terminal. Consult Packaging Section of Databook for thermal limitations and considerations of packages. Pin numbers refer to DIL package only.

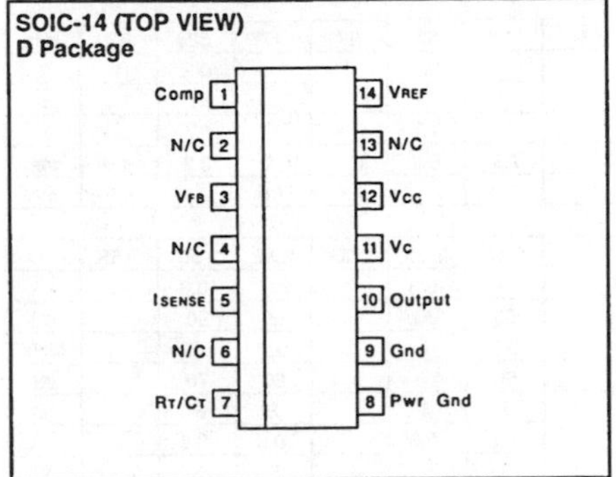

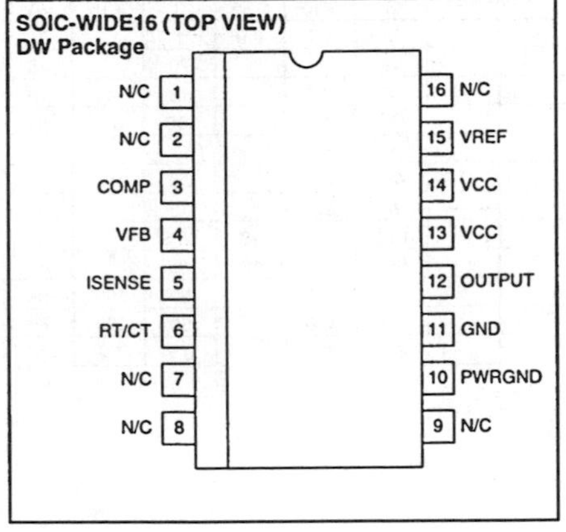

PLCC-20, LCC-20
(TOP VIEW)
Q, L Packages

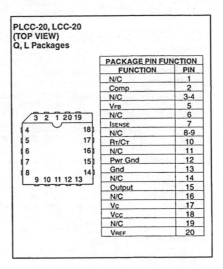

3 2 1 20 19
4 18
5 17
6 16
7 15
8 14
9 10 11 12 13

PACKAGE PIN FUNCTION	
FUNCTION	PIN
N/C	1
Comp	2
N/C	3-4
V_{FB}	5
N/C	6
I_{SENSE}	7
N/C	8-9
R_T/C_T	10
N/C	11
Pwr Gnd	12
Gnd	13
N/C	14
Output	15
N/C	16
V_C	17
V_{CC}	18
N/C	19
V_{REF}	20

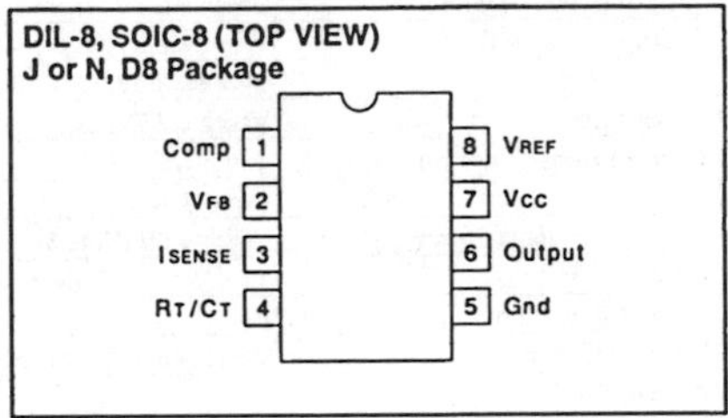

ELECTRICAL CHARACTERISTICS Unless otherwise stated, these specifications apply for $-55°C \le T_A \le 125°C$ for the UC184xA; $-40°C \le T_A \le 125°C$ for the UC284xAQ; $-40°C \le T_A \le 85°C$ for the UC284xA; $0 \le T_A \le 70°C$ for the UC384xA; V_{CC} = 15V (Note 5); R_T = 10k; C_T = 3.3nF; $T_A = T_J$; Pin numbers refer to DIL-8.

PARAMETER	TEST CONDITIONS	UC184xA\UC284xA			UC384xA			UNITS
		MIN.	TYP.	MAX.	MIN.	TYP.	MAX.	
Reference Section								
Output Voltage	T_J = 25°C, I_O = 1mA	4.95	5.00	5.05	4.90	5.00	5.10	V
Line Regulation	$12 \le V_{IN} 25V$		6	20		6	20	mV
Load Regulation	$1 \le I_O \le 20mA$		6	25		6	25	mV
Temp. Stability	(Note 2, Note 7)		0.2	0.4		0.2	0.4	mV/°C
Total Output Variation	Line, Load, Temp.	4.9		5.1	4.82		5.18	V
Output Noise Voltage	$10Hz \le f \le 10kHz$ T_J = 25°C (Note 2)		50			50		µV
Long Term Stability	T_A = 125°C, 1000Hrs. (Note 2)		5	25		5	25	mV
Output Short Circuit		-30	-100	-180	-30	-100	-180	mA
Oscillator Section								
Initial Accuracy	T_J = 25°C (Note 6)	47	52	57	47	52	57	kHz
Voltage Stability	$12 \le V_{CC} \le 25V$		0.2	1		0.2	1	%
Temp. Stability	$T_{MIN} \le T_A \le T_{MAX}$ (Note 2)		5			5		%
Amplitude	$V_{PIN 4}$ peak to peak (Note 2)		1.7			1.7		V
Discharge Current	T_J = 25°C, $V_{PIN 4}$ = 2V (Note 8)	7.8	8.3	8.8	7.8	8.3	8.8	mA
	$V_{PIN 4}$ = 2V (Note 8)	7.5		8.8	7.6		8.8	mA
Error Amp Section								
Input Voltage	$V_{PIN 1}$ = 2.5V	2.45	2.50	2.55	2.42	2.50	2.58	V
Input Bias Current			-0.3	-1		-0.3	-2	µA
A_{VOL}	$2 \le V_O \le 4V$	65	90		65	90		dB
Unity Gain Bandwidth	T_J = 25°C (Note 2)	0.7	1		0.7	1		MHz
PSRR	$12 \le V_{CC} \le 25V$	60	70		60	70		dB
Output Sink Current	$V_{PIN 2}$ = 2.7V, $V_{PIN 1}$ = 1.1V	2	6		2	6		mA
Output Source Current	$V_{PIN 2}$ = 2.3V, $V_{PIN 1}$ = 5V	-0.5	-0.8		-0.5	-0.8		mA
V_{OUT} High	$V_{PIN 2}$ = 2.3V, R_L = 15k to ground	5	6		5	6		V
V_{OUT} Low	$V_{PIN 2}$ = 2.7V, R_L = 15k to Pin 8		0.7	1.1		0.7	1.1	V
Current Sense Section								
Gain	(Note 3, Note 4)	2.85	3	3.15	2.85	3	3.15	V/V
Maximum Input Signal	$V_{PIN 1}$ = 5V (Note 3)	0.9	1	1.1	0.9	1	1.1	V
PSRR	$12 \le V_{CC} \le 25V$ (Note 3)		70			70		dB
Input Bias Current			-2	-10		-2	-10	µA
Delay to Output	$V_{PIN 3}$ = 0 to 2V (Note 2)		150	300		150	300	ns
Output Section								
Output Low Level	I_{SINK} = 20mA		0.1	0.4		0.1	0.4	V
	I_{SINK} = 200mA		15	2.2		15	2.2	V
Output High Level	I_{SOURCE} = 20mA	13	13.5		13	13.5		V
	I_{SOURCE} = 200mA	12	13.5		12	13.5		V
Rise Time	T_J = 25°C, C_L = 1nF (Note 2)		50	150		50	150	ns
Fall Time	T_J = 25°C, C_L = 1nF (Note 2)		50	150		50	150	ns
UVLO Saturation	V_{CC} = 5V, I_{SINK} = 10mA		0.7	1.2		0.7	1.2	V

ELECTRICAL CHARACTERISTICS Unless otherwise stated, these specifications apply for $-55°C \le T_A \le 125°C$ for the UC184xA; $-40°C \le T_A \le 125°C$ for the UC284xAQ; $-40°C \le T_A \le 85°C$ for the UC284xA; $0 \le T_A \le 70°C$ for the UC384xA; V_{CC} = 15V (Note 5); R_T = 10k; C_T = 3.3nF; $T_A = T_J$; Pin numbers refer to DIL-8.

PARAMETER	TEST CONDITIONS	UC184xA\UC284xA			UC384xA			UNITS
		MIN.	TYP.	MAX.	MIN.	TYP.	MAX.	
Under-Voltage Lockout Section								
Start Threshold	x842A/4A	15	16	17	14.5	16	17.5	V
	x843A/5A	7.8	8.4	9.0	7.8	8.4	9.0	V
Min. Operation Voltage After	x842A/4A	9	10	11	8.5	10	11.5	V
Turn On	x843A/5A	7.0	7.6	8.2	7.0	7.6	8.2	V
PWM Section								
Maximum Duty Cycle	x842A/3A	94	96	100	94	96	100	%
	x844A/5A	47	48	50	47	48	50	%
Minimum Duty Cycle				0			0	%
Total Standby Current								
Start-Up Current			0.3	0.5		0.3	0.5	mA
Operating Supply Current	$V_{PIN\,2} = V_{PIN\,3}$ = 0V		11	17		11	17	mA
Vcc Zener Voltage	I_{CC} = 25mA	30	34		30	34		V

Note 2: *Ensured by design, but not 100% production tested.*
Note 3: *Parameter measured at trip point of latch with $V_{PIN2} = 0$.*
Note 4: *Gain defined as:* $A = \dfrac{\Delta VPIN\ 1}{\Delta VPIN\ 3}$; $0 \le V_{PIN\,3} \le 0.8V$.
Note 5: *Adjust Vcc above the start threshold before setting at 15V.*
Note 6: *Output frequency equals oscillator frequency for the UC1842A and UC1843A. Output frequency is one half oscillator frequency for the UC1844A and UC1845A.*
Note 7: *"Temperature stability, sometimes referred to as average temperature coefficient, is described by the equation:*

$$Temp\ Stability = \frac{VREF\ (max) - VREF\ (min)}{TJ\ (max) - TJ\ (min)}.$$

V_{REF} (max) and V_{REF} (min) are the maximum & minimum reference voltage measured over the appropriate temperature range. Note that the extremes in voltage do not necessarily occur at the extremes in temperature."
Note 8: *This parameter is measured with R_T = 10kΩ to V_{REF}. This contributes approximately 300µA of current to the measurement. The total current flowing into the R_T/C pin will be approximately 300µA higher than the measured value.*

Error Amp Configuration

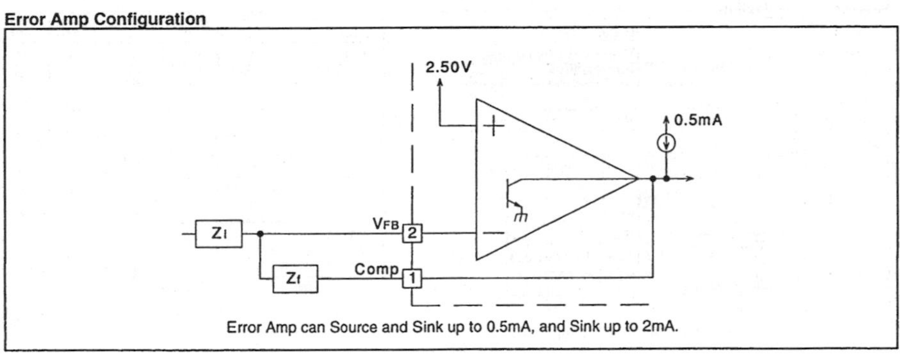

Error Amp can Source and Sink up to 0.5mA, and Sink up to 2mA.

Under-Voltage Lockout

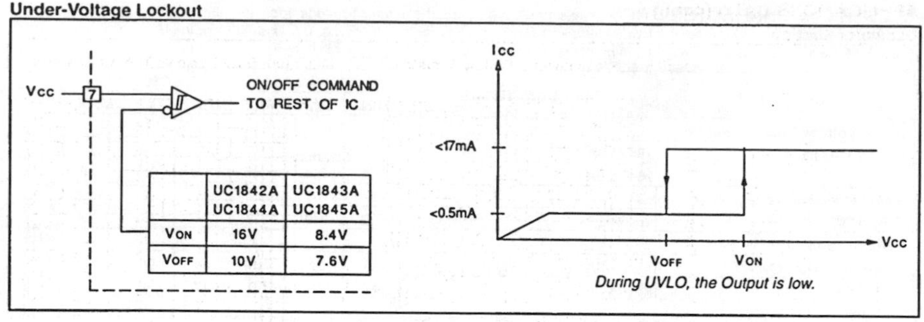

Current Sense Circuit

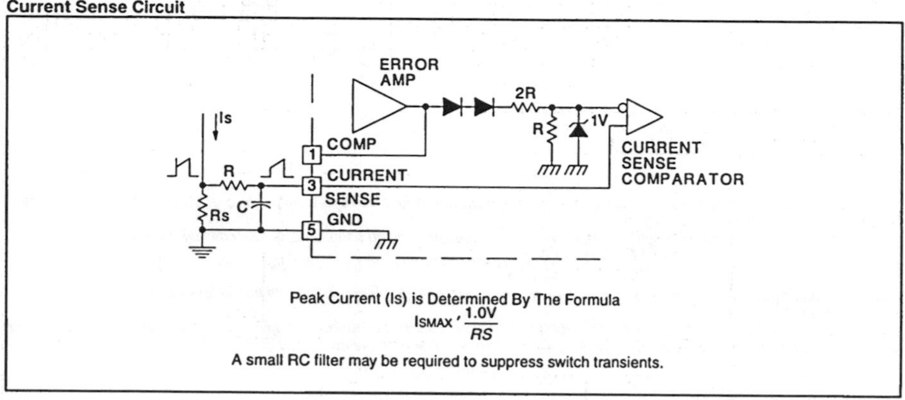

Output Saturation Characteristics

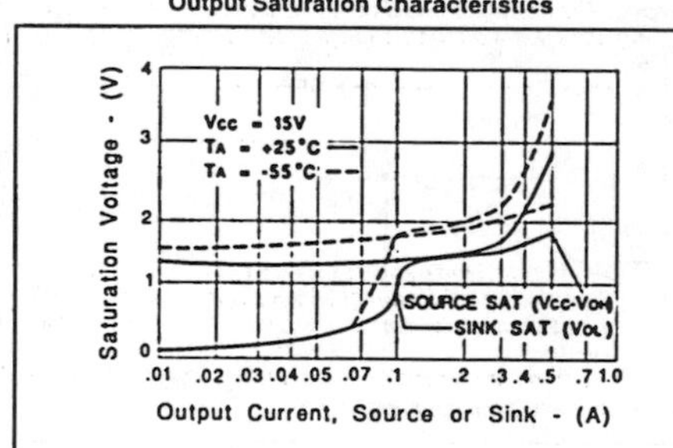

Error Amplifier Open-Loop Frequency Response

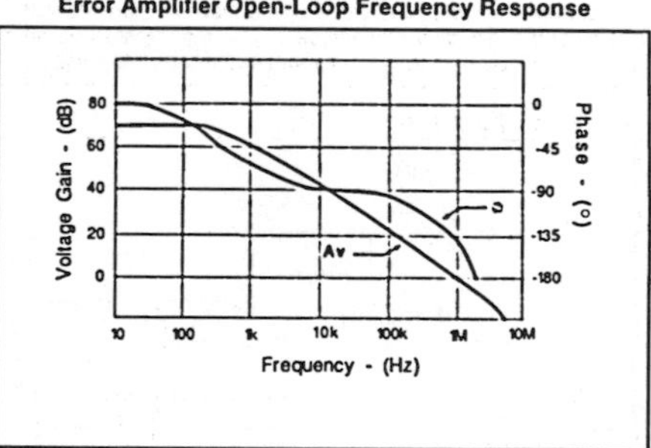

APPLICATIONS DATA (cont.)

Oscillator Section

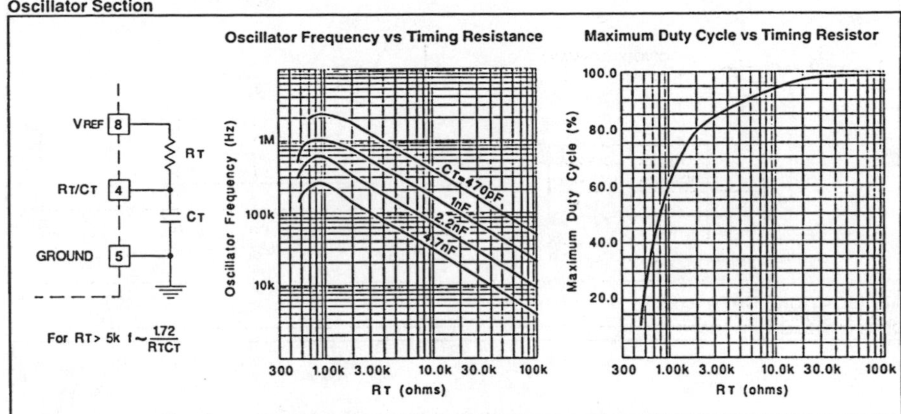

Open-Loop Laboratory Test Fixture

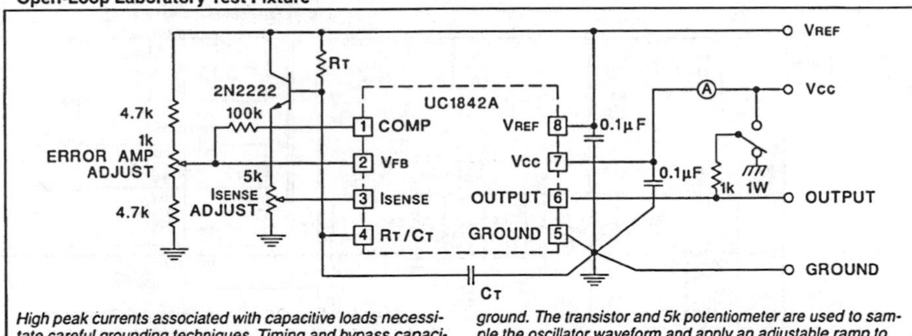

High peak currents associated with capacitive loads necessitate careful grounding techniques. Timing and bypass capacitors should be connected close to pin 5 in a single point ground. The transistor and 5k potentiometer are used to sample the oscillator waveform and apply an adjustable ramp to pin 3.

Slope Compensation

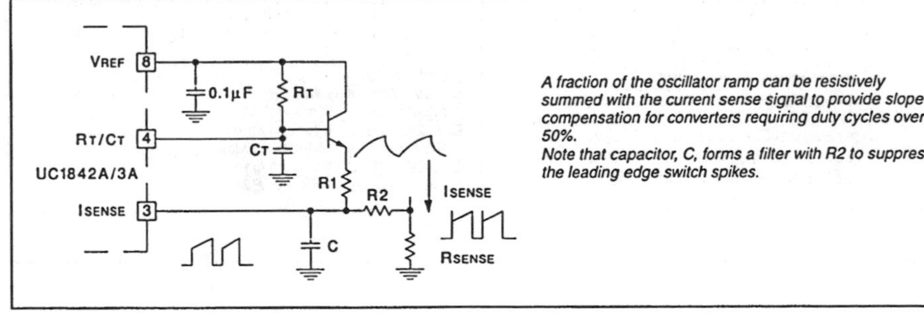

A fraction of the oscillator ramp can be resistively summed with the current sense signal to provide slope compensation for converters requiring duty cycles over 50%.

Note that capacitor, C, forms a filter with R2 to suppress the leading edge switch spikes.

APPLICATIONS DATA (cont.)

Off-line Flyback Regulator

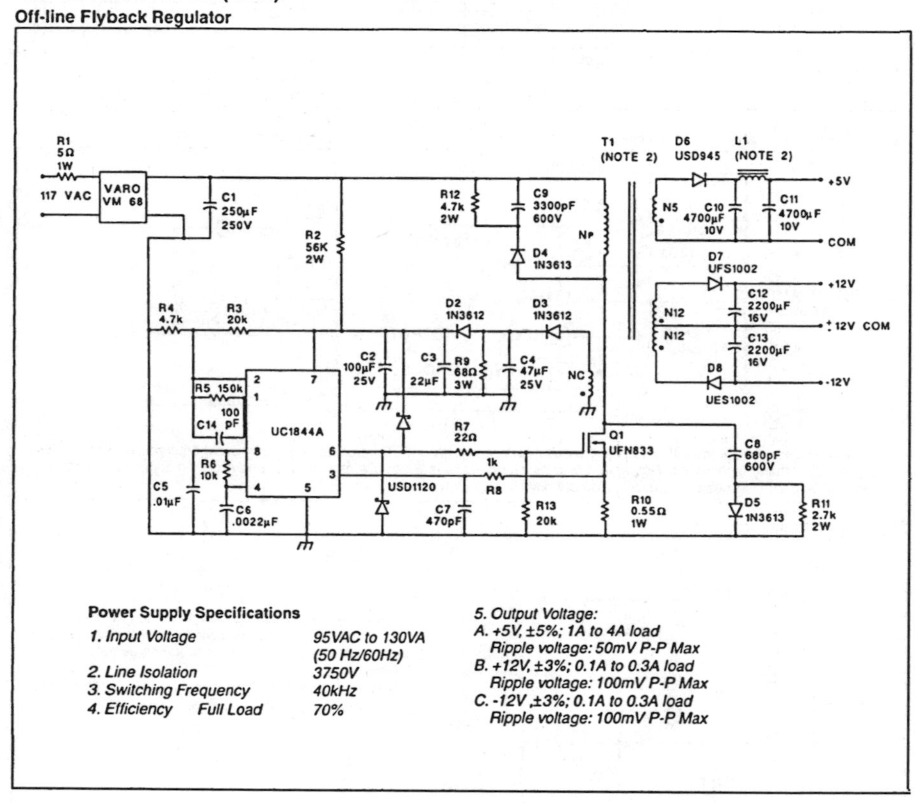

Power Supply Specifications

1. Input Voltage — *95VAC to 130VA (50 Hz/60Hz)*

2. Line Isolation — *3750V*

3. Switching Frequency — *40kHz*

4. Efficiency Full Load — *70%*

5. Output Voltage:

A. +5V, ±5%; 1A to 4A load
Ripple voltage: 50mV P-P Max

B. +12V, ±3%; 0.1A to 0.3A load
Ripple voltage: 100mV P-P Max

C. -12V ,±3%; 0.1A to 0.3A load
Ripple voltage: 100mV P-P Max

High Speed PWM Controller

FEATURES

- Improved versions of the UC3823/UC3825 PWMs
- Compatible with Voltage or Current-Mode Topologies
- Practical Operation at Switching Frequencies to 1MHz
- 50ns Propagation Delay to Output
- High Current Dual Totem Pole Outputs (2A Peak)
- Trimmed Oscillator Discharge Current
- Low 100μA Startup Current
- Pulse-by-Pulse Current Limiting Comparator
- Latched Overcurrent Comparator With Full Cycle Restart

DESCRIPTION

The UC3823A & B and the UC3825A & B family of PWM control ICs are improved versions of the standard UC3823 & UC3825 family. Performance enhancements have been made to several of the circuit blocks. Error amplifier gain bandwidth product is 12MHz while input offset voltage is 2mV. Current limit threshold is guaranteed to a tolerance of 5%. Oscillator discharge current is specified at 10mA for accurate dead time control. Frequency accuracy is improved to 6%. Startup supply current, typically 100μA, is ideal for off-line applications. The output drivers are redesigned to actively sink current during UVLO at no expense to the startup current specification. In addition each output is capable of 2A peak currents during transitions.

Functional improvements have also been implemented in this family. The UC3825 shutdown comparator is now a high-speed overcurrent comparator with a threshold of 1.2V. The overcurrent comparator sets a latch that ensures full discharge of the soft start capacitor before allowing a restart. While the fault latch is set, the outputs are in the low state. In the event of continuous faults, the soft start capacitor is fully charged before discharge to insure that the fault frequency does not exceed the designed soft start period. The UC3825 Clock pin has become CLK/LEB. This pin combines the functions of clock output and leading edge blanking adjustment and has been buffered for easier interfacing.

(continued)

BLOCK DIAGRAM

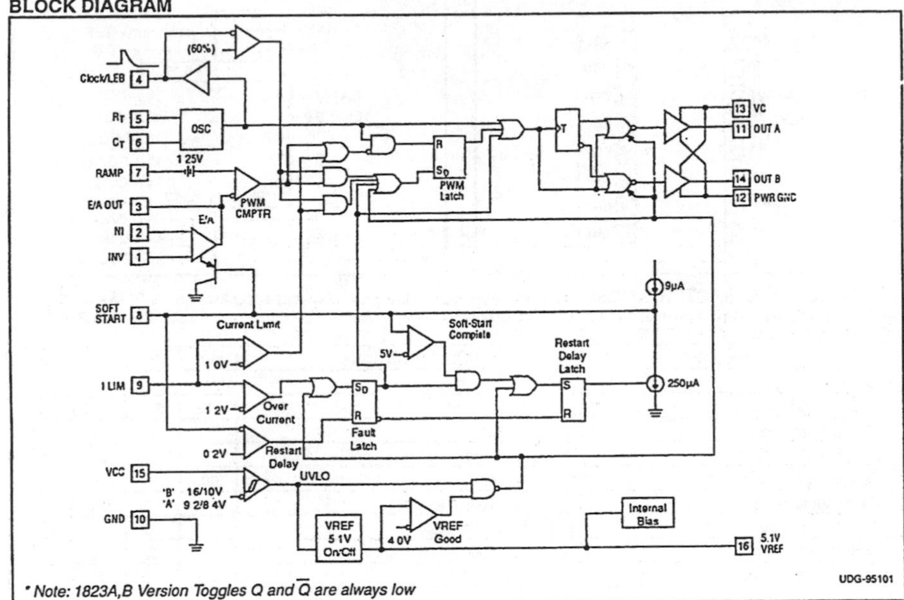

* Note: 1823A,B Version Toggles Q and $\overline{Q}$ are always low

SLUS334A - AUGUST 1995 - REVISED NOVEMBER 2000

DESCRIPTION (cont.)

The UC3825A,B has dual alternating outputs and the same pin configuration of the UC3825. The UC3823A,B outputs operate in phase with duty cycles from zero to less than 100%. The pin configuration of the UC3823A,B is the same as the UC3823 except pin 11 is now an output pin instead of the reference pin to the current limit comparator. "A" version parts have UVLO thresholds identical to the original UC3823/25. The "B" versions have UVLO thresholds of 16 and 10V, intended for ease of use in off-line applications.

Consult Application Note U-128 for detailed technical and applications information. Contact the factory for further packaging and availability information.

Device	UVLO	Dmax
UC3823A	9.2V/8.4V	< 100%
UC3823B	16V/10V	< 100%
UC3825A	9.2V/8.4V	< 50%
UC3825B	16V/10V	< 50%

ABSOLUTE MAXIMUM RATINGS

Supply Voltage (VC, VCC) 22V
Output Current, Source or Sink (Pins OUTA, OUTB)
DC 0.5A
Pulse (0.5μs) 2.2A
Power Ground (PGND).......................... ±0.2V
Analog Inputs
(INV, NI, RAMP).......................... –0.3V to 7V
(ILIM, SS).......................... –0.3V to 6V
Clock Output Current (CLK/LEB) .... –5mA
Error Amplifier Output Current (EAOUT) 5mA
Soft Start Sink Current (SS) 20mA
Oscillator Charging Current (RT) .... –5mA
Power Dissipation at T_A = 60°C 1W
Storage Temperature Range .... –65°C to +150°C
Junction Temperature.................... –55°C to +150°C
Lead Temperature (Soldering, 10 sec.).............. 300°C

All currents are positive into, negative out of the specified terminal. Consult Packaging Section of Databook for thermal limitations and considerations of packages.

CONNECTION DIAGRAMS

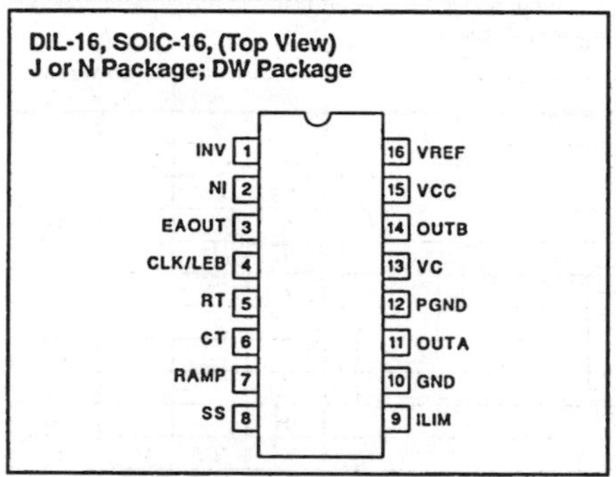

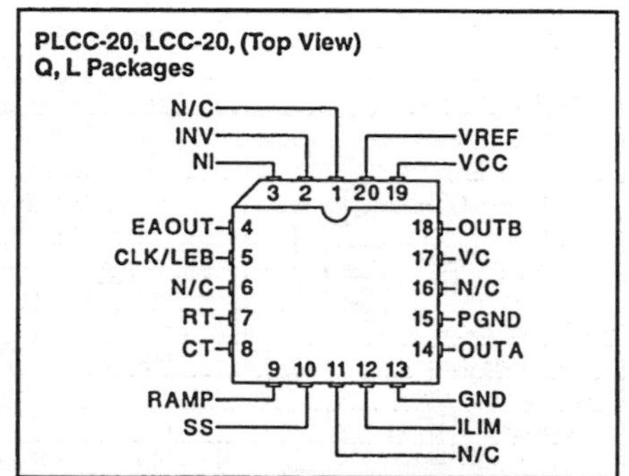

ELECTRICAL CHARACTERISTICS: Unless otherwise stated, these specifications apply for T_A = –55°C to +125°C for the UC1823A,B and UC1825A,B; –40°C to +85°C for the UC2823A,B and UC2825A,B; 0°C to +70°C for the UC3823A,B and UC3825A,B; RT = 3.65k, CT = 1nF, VCC = 12V, $T_A = T_J$.

PARAMETER	TEST CONDITIONS	MIN	TYP	MAX	UNITS
Reference Section					
Output Voltage	T_J = 25°C, Io = 1mA	5.05	5.1	5.15	V
Line Regulation	12 < VCC < 20V		2	15	mV
Load Regulation	1mA < I_O < 10mA		5	20	mV
Total Output Variation	Line, Load, Temp	5.03		5.17	V
Temperature Stability	$T_{MIN} < T_A < T_{MAX}$ (Note 1)		0.2	0.4	mV/°C
Output Noise Voltage	10Hz < f < 10kHz (Note 1)		50		μVRMS
Long Term Stability	T_J = 125°C, 1000 hours (Note 1)		5	25	mV
Short Circuit Current	VREF = 0V	30	60	90	mA

ELECTRICAL CHARACTERISTICS: Unless otherwise stated, these specifications apply for T_A = –55°C to +125°C for the UC1823A,B and UC1825A,B; –40°C to +85°C for the UC2823A,B and UC2825A,B; 0°C to +70°C for the UC3823A,B and UC3825A,B; RT = 3.65k, CT = 1nF, VCC = 12V, $T_A = T_J$.

PARAMETER	TEST CONDITIONS	MIN	TYP	MAX	UNITS
Oscillator Section					
Initial Accuracy	T_J = 25°C (Note 1)	375	400	425	kHz
Total Variation	Line, Temperature (Note 1)	350		450	kHz
Voltage Stability	12V < VCC < 20V			1	%
Temperature Stability	$T_{MIN} < T_A < T_{MAX}$ (Note 1)		5		%
Initial Accuracy	RT = 6.6k, CT = 220pF, T_A = 25°C (Note 1)	0.9	1	1.1	MHz
Total Variation	RT = 6.6k, CT = 220pF (Note 1)	0.85		1.15	MHz
Clock Out High		3.7	4		V
Clock Out Low			0	0.2	V
Ramp Peak		2.6	2.8	3	V
Ramp Valley		0.7	1	1.25	V
Ramp Valley to Peak		1.6	1.8	2	V
Oscillator Discharge Current	RT = Open, V_{CT} = 2V	9	10	11	mA
Error Amplifier Section					
Input Offset Voltage			2	10	mV
Input Bias Current			0.6	3	μA
Input Offset Current			0.1	1	μA
Open Loop Gain	1V < V_O < 4V	60	95		dB
CMRR	1.5V < V_{CM} < 5.5V	75	95		dB
PSRR	12V < VCC < 20V	85	110		dB
Output Sink Current	V_{EAOUT} = 1V	1	2.5		mA
Output Source Current	V_{EAOUT} = 4V	–0.5	–1.3		mA
Output High Voltage	I_{EAOUT} = –0.5mA	4.5	4.7	5	V
Output Low Voltage	I_{EAOUT} = 1mA	0	0.5	1	V
Gain Bandwidth Product	F = 200kHz	6	12		MHz
Slew Rate	(Note 1)	6	9		V/μs
PWM Comparator					
RAMP Bias Current	V_{RAMP} = 0V		–1	–8	μA
Minimum Duty Cycle				0	%
Maximum Duty Cycle		85			%
Leading Edge Blanking	R = 2k, C = 470pF	300	375	450	ns
LEB Resistor	$V_{CLK/LEB}$ = 3V	8.5	10	11.5	kohm
EAOUT Zero D.C. Threshold	V_{RAMP} = 0V	1.1	1.25	1.4	V
Delay to Output	V_{EAOUT} = 2.1V, V_{RAMP} = 0 to 2V Step (Note 1)		50	80	ns
Current Limit/Start Sequence/Fault Section					
Soft Start Charge Current	V_{SS} = 2.5V	8	14	20	μA
Full Soft Start Threshold		4.3	5		V
Restart Discharge Current	V_{SS} = 2.5V	100	250	350	μA
Restart Threshold			0.3	0.5	V
ILIM Bias Current	0 < V_{ILIM} < 2V			15	μA
Current Limit Threshold		0.95	1	1.05	V

ELECTRICAL CHARACTERISTICS: Unless otherwise stated, these specifications apply for $T_A = -55°C$ to $+125°C$ for the UC1823A,B and UC1825A,B; $-40°C$ to $+85°C$ for the UC2823A,B and UC2825A,B; 0°C to +70°C for the UC3823A,B and UC3825A,B; RT = 3.65k, CT = 1nF, VCC = 12V, $T_A = T_J$.

PARAMETER	TEST CONDITIONS	MIN	TYP	MAX	UNITS
Current Limit/Start Sequence/Fault Section (cont.)					
Over Current Threshold		1.14	1.2	1.26	V
ILIM Delay to Output	V_{ILIM} = 0 to 2V Step (Note 1)		50	80	ns
Output Section					
Output Low Saturation	I_{OUT} = 20mA		0.25	0.4	V
	I_{OUT} = 200mA		1.2	2.2	V
Output High Saturation	I_{OUT} = 20mA		1.9	2.9	V
	I_{OUT} = 200mA		2	3	V
UVLO Output Low Saturation	I_O = 20mA		0.8	1.2	V
Rise/Fall Time	C_L = 1nF (Note 1)		20	45	ns
UnderVoltage Lockout					
Start Threshold	UCX823B and X825B only		16	17	V
Stop Threshold	UCX823B and X825B only	9	10		V
UVLO Hysteresis	UCX823B and X825B only	5	6	7	V
Start Threshold	UCX823A and X825A only	8.4	9.2	9.6	V
UVLO Hysteresis	UCX823A and X825A only	0.4	0.8	1.2	V
Supply Current					
Startup Current	VC = VCC = V_{TH}(start) −0.5V		100	300	μA
Icc			28	36	mA

Note 1:Guaranteed by design. Not 100% tested in production.

APPLICATIONS INFORMATION

OSCILLATOR

The UC3823A,B/3825A,B oscillator is a saw tooth. The rising edge is governed by a current controlled by the RT pin and value of capacitance at the CT pin. The falling edge of the sawtooth sets dead time for the outputs. Selection of RT should be done first, based on desired maximum duty cycle. CT can then be chosen based on desired frequency, RT, and D_{MAX}. The design equations are:

$$RT = \frac{3V}{(10mA)(1 - D_{MAX})}$$

$$CT = \frac{(1.6 \bullet D_{MAX})}{(RT \bullet F)}$$

Recommended values for RT range from 1k to 100k. Control of D_{MAX} less than 70% is not recommended.

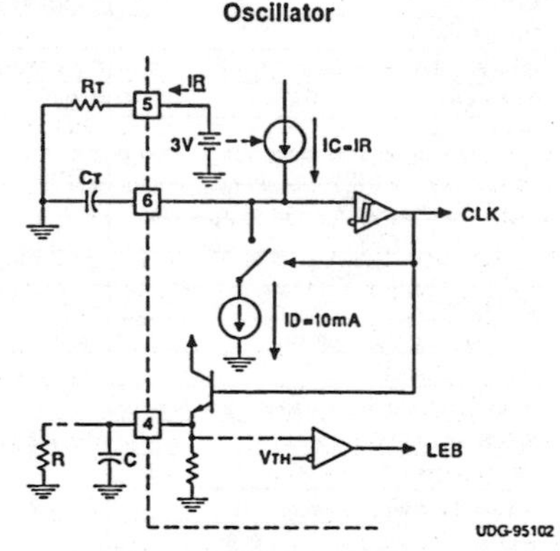

APPLICATIONS INFORMATION (cont.)

OSCILLATOR (cont.)

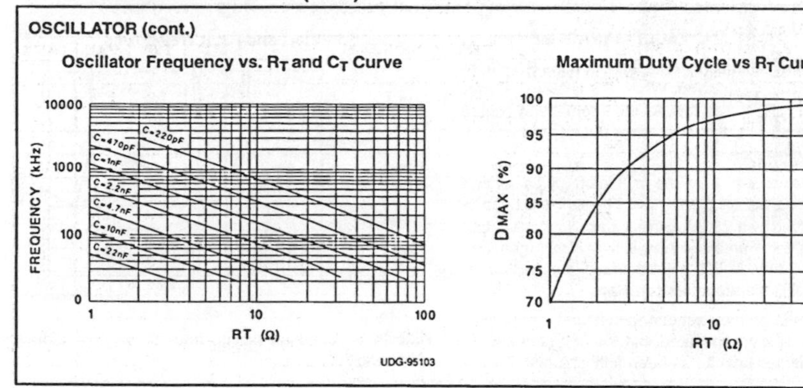

LEADING EDGE BLANKING

The UC3823A,B/3825A,B performs fixed frequency pulse width modulation control. The UC3823A,B outputs operate together at the switching frequency and can vary from 0 to some value less than 100%. The UC3825A,B outputs are alternately controlled. During every other cycle, one output will be off. Each output then, switches at one-half the oscillator frequency, varying in duty cycle from 0 to less than 50%.

To limit maximum duty cycle, the internal clock pulse blanks both outputs low during the discharge time of the oscillator. On the falling edge of the clock, the appropriate output(s) is driven high. The end of the pulse is controlled by the PWM comparator, current limit comparator, or the overcurrent comparator.

Normally the PWM comparator will sense a ramp crossing a control voltage (error amp output) and terminate the pulse. Leading edge blanking (LEB) causes the PWM comparator to be ignored for a fixed amount of time after the start of the pulse. This allows noise inherent with switched mode power conversion to be rejected. The PWM ramp input may not require any filtering as result of leading edge blanking.

To program a Leading Edge Blanking period, connect a capacitor, C, to CLK/LEB. The discharge time set by C and the internal 10k resistor will determine the blanked interval. The 10k resistor has a 10% tolerance. For more accuracy, an external 2k 1% resistor, R, can be added, resulting in an equivalent resistance of 1.66k with a tolerance of 2.4%. The design equation is:

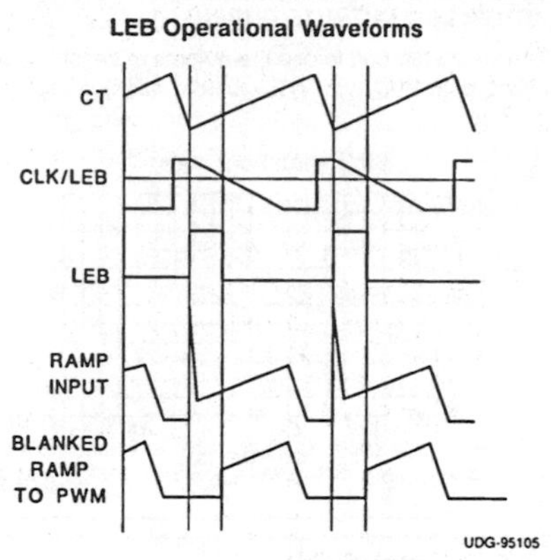

$$t_{LEB} = 0.5 \cdot (R \,||\, 10k) \cdot C.$$

Values of R less than 2k should not be used

Leading edge blanking is also applied to the current limit comparator. After LEB, if the ILIM pin exceeds the one volt threshold, the pulse is terminated. The over current comparator, however, is not blanked. It will catch catastrophic over current faults without a blanking delay. Any time the ILIM pin exceeds 1.2V, the fault latch will be set and the outputs driven low. For this reason, some noise filtering may be required on the ILIM pin.

APPLICATIONS INFORMATION (cont.)

UVLO, SOFT START AND FAULT MANAGEMENT

Soft start is programmed by a capacitor on the SS pin. At power up, SS is discharged. When SS is low, the error amp output is also forced low. As the internal 9μA source charges the SS pin, the error amp output follows until closed loop regulation takes over.

Anytime ILIM exceeds 1.2V, the fault latch will be set and the output pins will be driven low. The soft start cap is then discharged by a 250μA current sink. No more output pulses are allowed until soft start is fully discharged, and ILIM is below 1.2V. At this point the fault latch will be reset and the chip will execute a soft start.

Should the fault latch be set during soft start, the outputs will be immediately terminated, but the soft start cap will not be discharged until it has been fully charged. This results in a controlled hiccup interval for continuous fault conditions.

Soft Start and Fault Waveforms

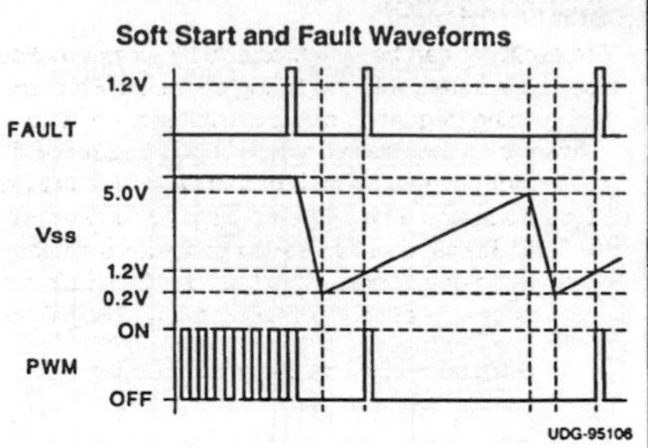

ACTIVE LOW OUTPUTS DURING UVLO

The UVLO function forces the outputs to be low and considers both VCC and VREF before allowing the chip to operate.

Simplified Schematic

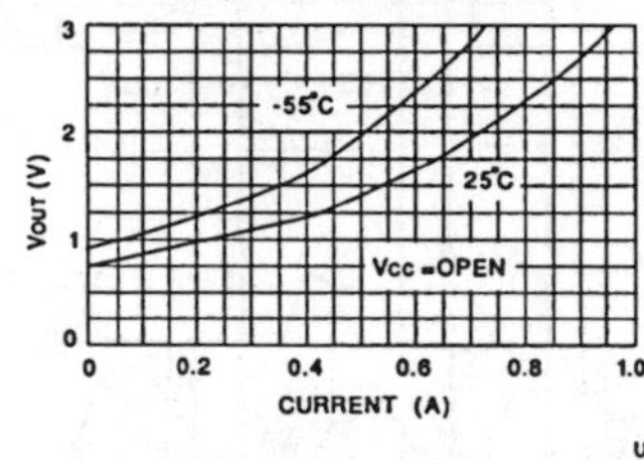

Output V and I During UVLO

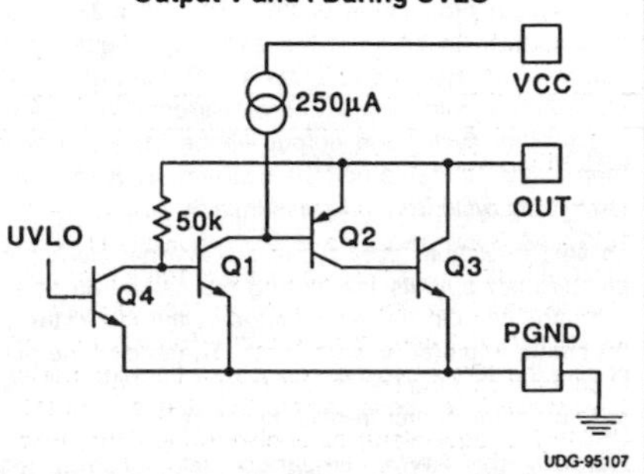

PWM APPLICATIONS

Current Mode

ISWITCH
CT
6
OSCILLATOR
CT
1.25V
7
RAMP
RSENSE
FROM E/A
UDG-95109

Voltage Mode

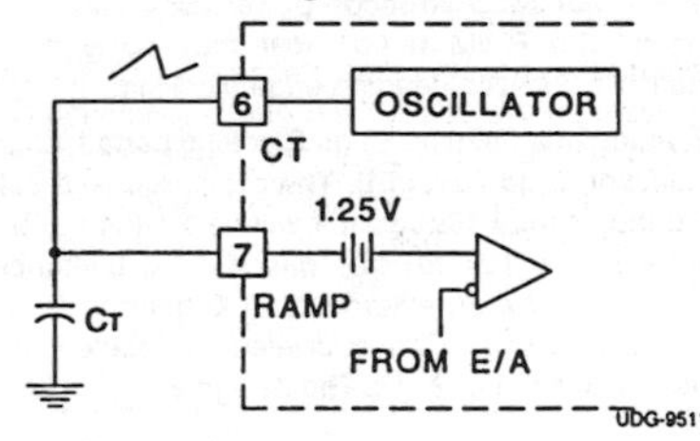

APPLICATIONS INFORMATION (cont.)

SYNCHRONIZATION

The oscillator can be synchronized by an external pulse inserted in series with the timing capacitor. Program the free running frequency of the oscillator to be 10 to 15% slower than the desired synchronous frequency. The pulse width should be greater than 10ns and less than half the discharge time of the oscillator. The rising edge of the CLK/LEB pin can be used to generate a synchronizing pulse for other chips. Note that, the CLK/LEB pin will no longer accept an incoming synchronizing signal.

General Oscillator Synchronization

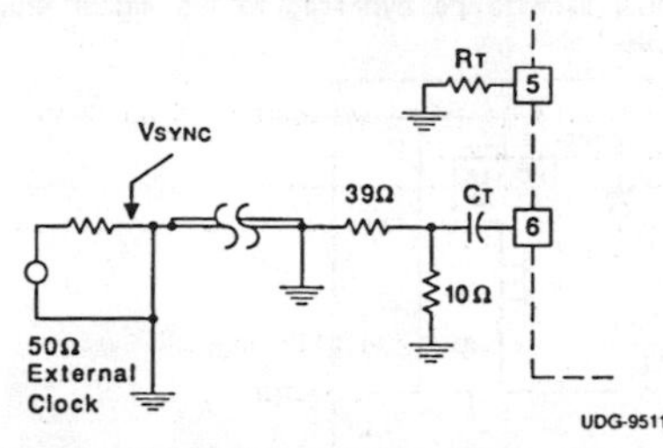

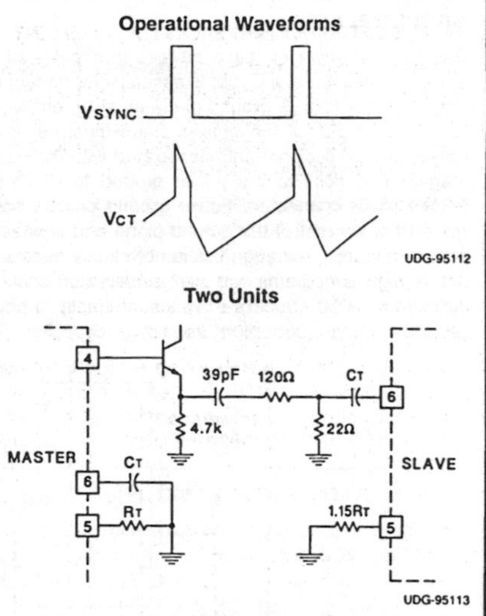

HIGH CURRENT OUTPUTS

Each totem pole output of the UC3823A,B and UC3825A,B can deliver a 2 amp peak current into a capacitive load. The output can slew a 1000pF capacitor 15 volts in approximately 20 nanoseconds. Separate collector supply (VC) and power ground (PGND) pins help decouple the IC's analog circuitry from the high power gate drive noise. The use of 3 Amp Schottky diodes (1N5120, USD245 or equivalent) as shown in the figure from each output to both VC and PGND are recommended. The diodes clamp the output swing to the supply rails, necessary with any type of inductive/capacitive load, typical of a MOSFET gate. Schottky diodes must be used because a low forward voltage drop is required. DO NOT USE standard silicon diodes.

Although a "single ended" device, two output drivers are available on the UC3823A,B devices. These can be "paralleled" by the use of a one-half ohm (noninductive) resistor connected in series with each output for a combined peak current of 4 amps.

Power MOSFET Drive Circuit

VC, 1nF, 10μF, D1, OUT, 6.8Ω, D2, PGND, GND

D1, D2 = 1N5820

UDG-95114

APPLICATIONS INFORMATION (cont.)

GROUND PLANES

Each output driver of these devices is capable of 2A peak currents. Careful layout is essential for correct operation of the chip. A ground plane must be employed. A unique section of the ground plane must be designated for high di/dt currents associated with the output stages. This point is the power ground to which the PGND pin is connected. Power ground can be separated from the rest of the ground plane and connected at a single point, although this is not strictly necessary if the high di/dt paths are well understood and accounted for. VCC should be bypassed directly to power ground with a good high frequency capacitor. The sources of the power MOSFET should connect to power ground as should the return connection for input power to the system and the bulk input capacitor. The output should be clamped with a high current Schottky diode to both VCC and PGND. Nothing else should be connected to power ground.

VREF should be bypassed directly to the signal portion of the ground plane with a good high frequency capacitor. Low ESR/ESL ceramic 1μF capacitors are recommended for both VCC and VREF. All analog circuitry should likewise be bypassed to the signal ground plane.

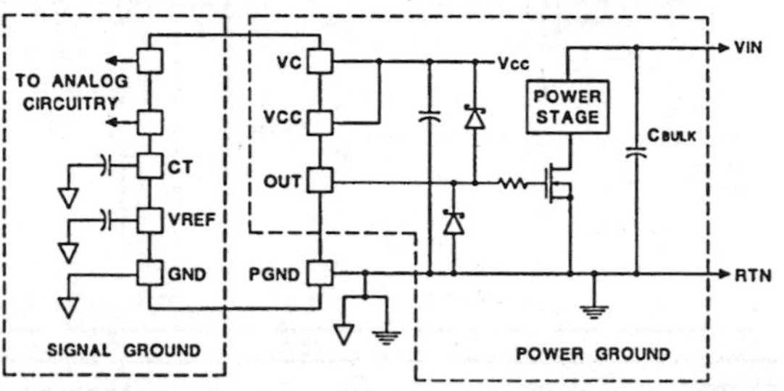

Open Loop Test Circuit

This test fixture is useful for exercising many of the UC3823A,B, UC3825A,B functions and measuring their specifications. As with any wideband circuit, careful grounding and bypass procedures should be followed. The use of a ground plane is highly recommended.

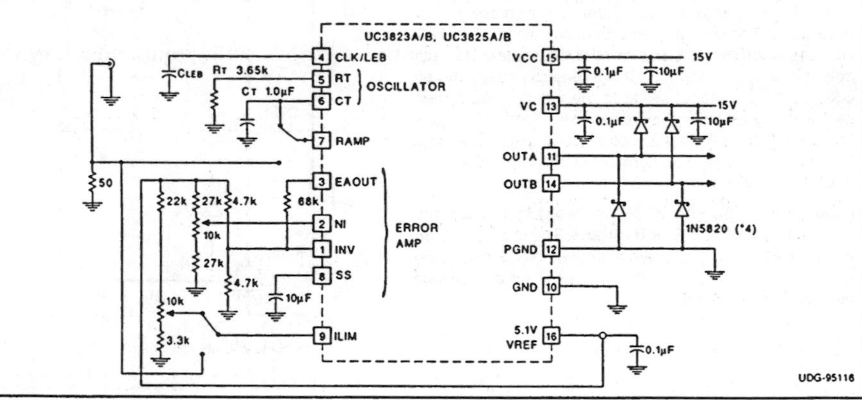

UNITRODE CORPORATION
7 CONTINENTAL BLVD. • MERRIMACK, NH 03054
TEL. (603) 424-2410 • FAX (603) 424-3460

References

1. R.D. Middlebrook and S. Cuk, A general unified approach to modelling switching-converter power stages, IEEE Power Electronics Specialists Conference, NASA Lewis Research Center, Cleveland, OH, June 8–10, 1976.
2. R.D. Middlebrook and S. Cuk, Input filter considerations in design and application of switching regulators, *IEEE Ind. Appli. Soc. Annu. Meeting,* 1976 Record, 366–382 (IEEE Publication 76 CH 1122-1-IA).
3. S. Cuk and R.D. Middlebrook, A general unified approach to modeling switching Dc-to-Dc converters in discontinuous conduction mode, IEEE Power Electronics Specialists Conference, Palo Alto, CA, June 1977.
4. S. Cuk and R.D. Middlebrook, A new optimum topology switching Dc-to-Dc converter, IEEE Power Electronics Specialists Conference, Palo Alto, CA, June 14–16, 1977.
5. S-P. Hsu, A. Brown, L. Rensink, and R.D. Middlebrook, Modelling and analysis of switching Dc-to-Dc converters in constant-frequency current-programmed mode, IEEE Power Electronics Specialists Conference, San Diego, CA, June 1979.
6. S. Cuk, Discontinuous inductor current mode in the optimum topology switching converter, IEEE Power Electronics Specialists Conference, Syracuse, NY, June 13–15, 1978.
7. L. Dixon, Pulse width modulator control methods with complementary optimization, PCI 81, Munich, Germany, September 1981.
8. D. Maksimovic, R. Erickson, and G. Griesbach, Modeling of cross-regulation in converters containing coupled inductors, *IEEE Trans. Power Electron.,* 15(4), 607–615, July 2000.
9. V. Bello, Computer program adds SPICE to switching-regulator analysis, *Electron. Design,* March 5, 1981.
10. B. Kuo, *Automatic Control Systems,* Prentice Hall, Englewood Cliffs, NJ.
11. J.J. D'Azzo and C.H. Houpis, *Linear Control System Analysis and Design,* McGraw–Hill, New York.
12. R.D. Middlebrook, Topics in multiple loop regulators and current mode programming, IEEE Power Electronics Specialists Conference, June 1985.
13. R.D. Middlebrook, Design techniques for preventing input-filter oscillations in switched-mode regulators, *Proc. Powecon* 5, San Francisco, May 4–6, 1978.
14. R.P. Severns and G.E. Bloom, *Modern DC to DC Switchmode Power Converter Circuits,* Van Nostrand Reinhold, New York, 1985.
15. H.W. Ott, *Noise Reduction Techniques in Electronic Systems,* John Wiley & Sons, New York.
16. B. Carsten, High-frequency conductor losses in switchmode magnetics. High-Frequency Power Conversion Conference, CA, May 1986.
17. K. O'Meara, Proximity losses in AC magnetic devices, *PCIM,* December 1996.

18. S.D. Schmit, Curve fit equations for ferrite materials allow computer-aided design, *PCIM*, July, 1997.
19. W. Shockley, The theory of p–n junctions in semiconductors and p–n junction transistors, *Bell Syst. Tech. J.*, 28, 435–489, July 1949.
20. W.M. Polivka, P.R.K. Chetty, and R.D. Middlebrook, State space average modelling of converters with parasitics and storage-time modulation, IEEE Power Electronics Specialists Conference, Atlanta, GA, June 16–20, 1980.
21. L. Dixon, High-power factor preregulator using the SEPIC converter, Unitrode Seminar SEM-1000, 1994.
22. P.C. Todd, UC3854 controlled power factor correction circuit design, Unitrode Application Note U-134.
23. B. Andreycak, Power correction using the UC3852 controlled on-time zero current switching technique, Unitrode Application Note U-132.
24. J.G. Kassakian, M.F. Schlect, and G.C. Verghese, *Principles of Power Electronics*, Addison–Wesley, Reading, MA, 1991.
25. R.A. Mammano and C.E. Mullett, Using an integrated controller in the design of mag-amp output regulators, Unitrode Application Note U-109.
26. R. Erickson and D. Maksimovic, *Fundamentals of Power Electronics*, 2nd ed., Kluwer Academic Publishers, Dordrecht, 2000.
27. L. Dixon, Average current mode control of switching power supplies. Unitrode Application Note U-140.
28. B. Arbetter and D. Maksimovic, Feed-forward pulse-width modulators for switching power converters, *IEEE Trans. Power Electron.*, 12(2), 361–368, March 1997.
29. R.D. Middlebrook, Measurement of loop gain in feedback systems, *Int. J. Electron.*, 38(4), 485–512, 1975.

Index

C

D

E

F

H

I

L

M

N

O

P

Q

R

S

T

U

V